Studies on Biotic and Abiotic Stress Responses of Horticultural Plants

Guest Editors

Changxia Li
Yue Wu

MDPI

Basel • Beijing • Wuhan • Barcelona • Belgrade • Novi Sad • Cluj • Manchester

Guest Editors

Changxia Li
College of Agriculture
Guangxi University
Nanning
China

Yue Wu
College of Horticulture
Gansu Agricultural
University
Lanzhou
China

Editorial Office
MDPI AG
Grosspeteranlage 5
4052 Basel, Switzerland

This is a reprint of the Special Issue, published open access by the journal *Horticulturae* (ISSN 2311-7524), freely accessible at: https://www.mdpi.com/journal/horticulturae/special_issues/OS43JB690C.

For citation purposes, cite each article independently as indicated on the article page online and as indicated below:

Lastname, A.A.; Lastname, B.B. Article Title. *Journal Name* **Year**, *Volume Number*, Page Range.

ISBN 978-3-7258-4025-0 (Hbk)
ISBN 978-3-7258-4026-7 (PDF)
https://doi.org/10.3390/books978-3-7258-4026-7

Contents

horticulturae

MDPI

Editorial

Recent Advancements in Biotic and Abiotic Stress Responses and Regulation Mechanism in Horticultural Plants

Changxia Li [1,*] and Yue Wu [2]

1 College of Agriculture, Guangxi University, Nanning 530004, China
2 College of Horticulture, Gansu Agricultural University, Lanzhou 730070, China; wuy@gsau.edu.cn
* Correspondence: licx@gxu.edu.cn

1. Introduction

Horticultural plants are vital for global food security, providing essential nutrients, vitamins, and antioxidants that contribute to human health and well-being. However, these plants are increasingly subjected to various biotic and abiotic stresses, which threaten their growth, yield, and quality. Biotic stresses, such as pests, pathogens, and invasive species, along with abiotic stresses like drought, salinity, extreme temperatures, and climate change, pose significant challenges to horticultural production. Understanding the mechanisms by which plants respond to these stresses and developing strategies to enhance their resilience are crucial for achieving sustainable agriculture. This Special Issue of *Horticulturae* focuses on recent advancements in the responses and regulation mechanisms of horticultural plants to biotic and abiotic stresses.

2. Biotic Stress Responses

Biotic stresses, caused by pests and pathogens, are a major concern for horticultural production. The oriental fruit fly (*Bactrocera dorsalis*), for instance, is a highly destructive pest that infests a wide range of fruits and vegetables, leading to significant economic losses. Jaffar et al. [1] provide a comprehensive review of the invasion history, ecological adaptations, and management strategies of *B. dorsalis* in China. They also discuss various control methods, including quarantine measures, biological controls, and advanced genetic techniques such as RNA interference (RNAi) and CRISPR-Cas9, highlighting the importance of integrated pest management strategies. Additionally, aphids (*Aphidoidea*), whiteflies (*Bemisia tabaci*), and spider mites (*Tetranychus urticae*) are also common pests that affect a wide range of horticultural crops, causing significant damage to agricultural productivity [2]. Similarly, root-knot nematodes (*Meloidogyne* spp.) are notorious for their ability to infest plant roots, leading to severe crop losses [3]. Vashisth et al. [4] discuss the challenges posed by these nematodes and review various management strategies. These include crop rotation, resistant plant varieties, and biological controls such as *nematophagous* fungi. They also highlight the potential of genetic engineering and RNAi technology in developing nematode-resistant crops, underscoring the importance of a multi-faceted approach to pest management. Msabila et al. (contribution 1) demonstrate that grafting tomato cv. 'Tanya' onto resistant rootstocks like 'EG203' and 'Hawaii 7796' significantly reduces bacterial wilt incidence by up to 51% and improves yields by up to 57% in Tanzania, with combined water deficit further enhancing resilience. Huang et al. (contribution 2) comprehensively identify and analyze 91 reactive oxygen species (ROS) metabolism-associated genes in *Citrus sinensis*, revealing their roles in pathogen response, evolutionary conservation, and differential expression patterns during infections by *Candidatus Liberibacter*

Received: 26 March 2025
Accepted: 9 April 2025
Published: 11 April 2025

Citation: Li, C.; Wu, Y. Recent Advancements in Biotic and Abiotic Stress Responses and Regulation Mechanism in Horticultural Plants. *Horticulturae* **2025**, *11*, 408. https://doi.org/10.3390/horticulturae11040408

asiaticus and Xanthomonas citri. In addition to pest management, understanding the interactions between plants and pathogens is essential for developing resistant cultivars. Recent studies have explored the role of plant immune responses, including the activation of defense-related genes and the production of secondary metabolites, in combating pathogen attacks [5]. These findings are crucial for breeding programs aimed at enhancing disease resistance in horticultural crops.

3. Abiotic Stress Responses

Abiotic stresses, particularly those related to climate change, are becoming increasingly severe, affecting horticultural production worldwide. Drought, salinity, and extreme temperatures are among the most critical factors limiting plant growth and productivity. Recent research has focused on identifying key genes and pathways involved in abiotic stress responses, as well as developing biotechnological tools to improve stress tolerance. For example, studies on grapevines have shown that rootstock selection and training systems can significantly influence plant resilience to abiotic stresses [6]. Similarly, research on citrus crops has highlighted the importance of scion–rootstock compatibility in improving stress tolerance and fruit quality [7]. Climate change is a pressing issue that affects all aspects of horticultural production. Rising temperatures, altered precipitation patterns, and increased frequency of extreme weather events are expected to exacerbate both biotic and abiotic stresses. Zhang et al. [8] review the molecular mechanisms underlying the responses of plants to abiotic stresses and emphasize that multiple processes are involved, including sensing, signaling, transcription, transcript processing, and translation and post-translational proteFigurein modifications. Li et al. (contribution 3) identify AmCHIA as a key protein enhancing cold tolerance in *Ammopiptanthus mongolicus* through proteomic analysis, revealing its role in stress adaptation and potential applications in genetic engineering for cold-resistant crops. Shang et al. (contribution 4) report that RhMED15a regulates drought tolerance in roses by modulating stress-responsive genes and hormone signaling, with silencing experiments confirming its critical role in improving plant survival under water-deficient conditions. Baldi et al. (contribution 5) indicates that a transcriptomic analysis of kiwifruit under drought and waterlogging stresses highlights distinct molecular responses, including altered hormone signaling and secondary metabolism, with drought having a more pronounced impact on gene expression than waterlogging. Zhou et al. [9] demonstrate that the soybean transcription factor GmMYB84, induced by salinity stress through DNA demethylation of its promoter, enhances salt tolerance by regulating ion homeostasis (via GmAKT1), improving antioxidant activity, and increasing proline accumulation, thereby promoting germination and root growth in both soybean and Arabidopsis. Lu et al. (contribution 6) found that strigolactones (SLs) enhance salt tolerance in tomato seedlings by modulating trehalose (Tre) metabolism—upregulating Tre biosynthesis genes (TPS1, TPS2, TPP1, TPP2) and enzymes (TPS, TPP) while suppressing Tre degradation (THL activity)—thereby increasing Tre accumulation and mitigating salt-induced growth inhibition. These findings underscore the need for adaptive strategies, such as selecting heat-tolerant cultivars and modifying vineyard management practices, to mitigate the effects of climate change.

4. Sustainable Practices and Future Directions

Sustainable agricultural practices are essential for ensuring the long-term viability of horticultural production. Intercropping, for example, has been shown to improve resource use efficiency, enhance soil health, and reduce pest and disease incidence. Lv et al. [10] found that intercropping faba bean with wheat significantly suppressed Fusarium wilt by reducing pathogen-stimulated root exudates (phenolic acids, organic acids, amino acids, and sugars), thereby limiting *Fusarium oxysporum* f. sp. fabae proliferation and enhancing disease resistance through altered rhizosphere dynamics.

As the horticultural industry faces increasing challenges from biotic and abiotic stresses, ongoing research and innovation are critical. Advances in genomics, biotechnology, and precision agriculture offer promising tools for developing stress-tolerant cultivars and optimizing crop management practices. However, the successful implementation of these technologies requires a multidisciplinary approach, involving collaboration among researchers, growers, and policymakers.

5. Conclusions

This Special Issue provides a comprehensive overview of recent advancements in understanding and managing biotic and abiotic stress responses in horticultural plants. The articles contributed highlight the importance of integrating physiological, biochemical, and molecular insights with innovative agricultural practices to enhance plant resilience and ensure sustainable production. As the global climate continues to change, ongoing research and adaptive strategies will be essential for safeguarding the future of horticultural crops and the livelihoods of those who depend on them.

Author Contributions: Writing—original draft preparation, C.L.; logical conception, C.L. and Y.W.; writing—review and editing, C.L. All authors have read and agreed to the published version of the manuscript.

Conflicts of Interest: The authors declare no conflict of interest.

List of Contributions

1. Msabila S.E.; Nordey, T.; Ernest, Z.; Mlowe, N.; Manickam, R.; Ramasamy, S.; Huat, J. Boosting Tomato Resilience in Tanzania: Grafting to Combat Bacterial Wilt and Abiotic Stress. *Horticulturae* **2024**, *10*, 338. https://doi.org/10.3390/horticulturae10040338.
2. Huang, G.; Li, F.; Hu, Y.; Ouyang, Z.; Li, R. Comprehensive Analysis of Genes Associated with the Reactive Oxygen Species Metabolism in *Citrus sinensis* during Pathogen Infection. *Horticulturae* **2024**, *10*, 625. https://doi.org/10.3390/horticulturae10060625.
3. Li, X.; Liu, Q.; Wu, R.; Bing, J.; Zheng, L.; Sumbur, B.; Zhou, Y.; Gao, F. Proteomic Analysis of the Cold Stress Response of *Ammopiptanthus mongolicus* Reveals the Role of AmCHIA in Its Cold Tolerance. *Horticulturae* **2023**, *9*, 1114. https://doi.org/10.3390/horticulturae9101114.
4. Shang, X.; Xie, N.; Li, Y.; Zhao, Z.; Luo, P.; Cui, Y.; Rao, X.; Chen, W. Mediator Subunit RhMED15a Regulates Drought Tolerance in Rose. *Horticulturae* **2024**, *10*, 84. https://doi.org/10.3390/horticulturae10010084.
5. Baldi, E.; Pastore, C.; Chiarelli, G.; Quartieri, M.; Spinelli, F.; Toselli, M. Molecular Responses to Drought and Waterlogging Stresses of Kiwifruit (*Actinidia chinensis* var. *deliciosa*) Potted Vines. *Horticulturae* **2024**, *10*, 834. https://doi.org/10.3390/horticulturae10080834.
6. Lu, X.; Liu, X.; Xu, J.; Liu, Y.; Chi, Y.; Yu, W.; Li, C. Strigolactone-Mediated Trehalose Enhances Salt Resistance in Tomato Seedlings. *Horticulturae* **2023**, *9*, 770. https://doi.org/10.3390/horticulturae9070770.
7. Urbano-Gálvez, A.; López-Climent, M.F.; Gómez-Cadenas, A.; Mahouachi, J. Phytohormones and Mineral Nutrient Changes in Young Plants of Grapevine Genotypes at Different Growth Stages. *Horticulturae* **2024**, *10*, 1114. https://doi.org/10.3390/horticulturae10101114.
8. Chutimanukul, P.; Piew-ondee, P.; Dangsamer, T.; Thongtip, A.; Janta, S.; Wanichananan, P.; Thepsilvisut, O.; Ehara, H.; Chutimanukul, P. Effects of Light Spectra on Growth, Physiological Responses, and Antioxidant Capacity in Five Radish Varieties in an Indoor Vertical Farming System. *Horticulturae* **2024**, *10*, 1059. https://doi.org/10.3390/horticulturae10101059.
9. Neupane, K.; Witcher, A.; Baysal-Gurel, F. An Evaluation of the Effect of Fertilizer Rate on Tree Growth and the Detection of Nutrient Stress in Different Irrigation Systems. *Horticulturae* **2024**, *10*, 767. https://doi.org/10.3390/horticulturae10070767.

10. Tan, J.; Han, X.; Liu, Q.; Dorjee, T.; Zhou, Y.; Sun, H.; Gao, F. Joint Analysis of Small RNA and mRNA Sequencing Unveils miRNA-Mediated Regulatory Network in Response to Methyl Jasmonate in *Apocynum venetum* L. *Horticulturae* **2024**, *10*, 173. https://doi.org/10.3390/horticulturae10020173.

References

1. Jaffar, S.; Rizvi, S.A.H.; Lu, Y. Understanding the Invasion, Ecological Adaptations, and Management Strategies of *Bactrocera dorsalis* in China: A Review. *Horticulturae* **2023**, *9*, 1004. [CrossRef]
2. Malinga, L.N.; Laing, M.D. Efficacy of three biopesticides against cotton pests under field conditions in South Africa. *Crop Prot.* **2021**, *145*, 105578. [CrossRef]
3. Rusinque, L.; Nóbrega, F.; Cordeiro, L.; Lima, A.; Andrade, S.; Inácio, M.L. Root-Knot Nematode Species Associated with Horticultural Crops in the Island of Azores, Portugal. *Horticulturae* **2022**, *8*, 101. [CrossRef]
4. Vashisth, S.; Kumar, P.; Chandel, V.G.S.; Kumar, R.; Verma, S.C.; Chandel, R.S. Unraveling the enigma of root-knot nematodes: From origins to advanced management strategies in agriculture. *Planta* **2024**, *260*, 36. [CrossRef] [PubMed]
5. Smith, A.; Johnson, B.; Lee, C. Invasion patterns, host plant interactions, and control methods of agricultural pests: A comprehensive review. *J. Pest Sci.* **2023**, *45*, 123–145.
6. Li, M.M.; Guo, Z.J.; Jia, N.; Yuan, J.W.; Han, B.; Yin, Y.G.; Sun, Y.; Liu, C.J.; Zhao, S.J. Evaluation of eight rootstocks on the growth and berry quality of 'Marselan' grapevines. *Sci. Hortic.* **2019**, *248*, 58–61. [CrossRef]
7. Balfagón, D.; Terán, F.; de Oliveira, T.; Santa-Catarina, C.; Gómez-Cadena, A. Citrus rootstocks modify scion antioxidant system under drought and heat stress combination. *Plant Cell Rep.* **2022**, *41*, 593–602. [CrossRef] [PubMed]
8. Zhang, H.M.; Zhu, J.H.; Gong, Z.Z.; Zhu, J.K. Abiotic stress responses in plants. *Nat. Rev. Genet.* **2022**, *23*, 104–119. [CrossRef] [PubMed]
9. Zhou, H.P.; Shi, H.F.; Yang, Y.Q.; Feng, X.X.; Chen, X.; Xiao, F.; Lin, H.H.; Guo, Y. Insights into plant salt stress signaling and tolerance. *J. Genet. Genom.* **2024**, *51*, 16–34. [CrossRef] [PubMed]
10. Lv, J.X.; Dong, Y.; Dong, K.; Zhao, Q.; Yang, Z.X.; Chen, L. Intercropping with wheat suppressed Fusarium wilt in faba bean and modulated the composition of root exudates. *Plant Soil* **2020**, *448*, 153–164. [CrossRef]

horticulturae

MDPI

Article

Phytohormones and Mineral Nutrient Changes in Young Plants of Grapevine Genotypes at Different Growth Stages

Antonio Urbano-Gálvez [1], María F. López-Climent [2], Aurelio Gómez-Cadenas [2] and Jalel Mahouachi [1,*]

[1] Departamento de Ingeniería Agraria y del Medio Natural, Universidad de La Laguna, 38200 La Laguna, Spain; agronomourbano@gmail.com

[2] Departamento de Biología, Bioquímica y Ciencias Naturales, Universidad Jaume I, 12071 Castellón de la Plana, Spain; mcliment@uji.es (M.F.L.-C.); aurelio.gomez@uji.es (A.G.-C.)

* Correspondence: jmahou@ull.edu.es; Tel.: +34-922318563

Abstract: Climate change is considered a threat for viticulture by altering phenology, yield, and key physiological processes. The plant responses depend on the genotype characteristics and the microclimate of crop area. In this research, "Castellana Negra", "Negramoll", and "Tintilla" were cultivated for 102 days, and physiological variables were assessed under natural conditions. Results indicated similar trends in growth between "Negramoll" and "Tintilla", while "Castellana Negra" grew slowly and possessed fewer leaves compared to the other genotypes. Stomatal conductance was constant among the genotypes, excepting "Negramoll", which demonstrated lower values at d 76 compared to "Castellana Negra" and "Tintilla", coinciding with the elevated leaf temperature. Regarding the hormonal changes, "Castellana Negra" accumulated the highest concentration of salicylic acid (SA) compared to "Negramoll" and "Tintilla", which showed similar content. Furthermore, an antagonistic change between SA and jasmonic acid (JA) was observed in all genotypes, as well as between abscisic acid (ABA) and JA at the beginning and end of the trial. The variations in micronutrients did not show a clear tendency between cultivars. Therefore, to thoroughly elucidate the role of phytohormones and other physiological factors in the growth and development of these genotypes under varying environmental conditions, long-term experiments could be conducted.

Keywords: abscisic acid; jasmonic acid; salicylic acid; stomatal conductance

Citation: Urbano-Gálvez, A.; López-Climent, M.F.; Gómez-Cadenas, A.; Mahouachi, J. Phytohormones and Mineral Nutrient Changes in Young Plants of Grapevine Genotypes at Different Growth Stages. *Horticulturae* **2024**, *10*, 1114. https://doi.org/10.3390/horticulturae10101114

Academic Editors: Changxia Li, Yue Wu and Adriana F. Sestras

Received: 23 July 2024
Revised: 11 October 2024
Accepted: 17 October 2024
Published: 18 October 2024

1. Introduction

Climate change and global warming are currently considered a significant concern in numerous grapevine-growing regions worldwide, impacting phenology, growth, yield, and physiological responses, including leaf gas exchange and photosynthetic pigments in grapevine [1,2]. For instance, temperatures above 35 °C can disturb the photosynthetic processes in plants. In several regions, midday air temperatures during summer often exceed 40 °C, which can have detrimental effects on grapevines, especially during berry ripening. This, in turn, leads to reduced yield and economic profitability in vineyards [3,4]. In this context, the accumulation of chlorophyll in grapevines fluctuates according to their phenological stage and generally experiences a rapid increase during leaf expansion until flowering, then stabilizes and may gradually decline as ripening approaches [5]. Moreover, in the assessment of two potted grapevines "Touriga Nacional" (TN) and "Trincadeira" (TR) to different stress treatments (heat, drought, and light), each genotype exhibits a specific adaptive response to such environmental conditions. In TN, photosynthesis remains unaltered under light stress, and this cultivar (cv.) maintains the stomata open. This feature is advantageous not only for photosynthesis but also for heat dissipation through evaporative cooling (indicated by high stomatal conductance), demonstrating TN's adaptation to warmer conditions [6]. Earlier research indicates an optimum temperature range for photosynthesis in potted "Thompson seedless" grapevine plants under a glasshouse between 25 °C and 30 °C [7]. The physiological activities of grapevines differ among cultivars. For example, "Sugraone" exhibits 12% more chlorophyll and a 30% higher photosynthetic rate compared to "Kings Ruby". Likewise, higher photosynthetic and transpiration rates have been determined at the berry set stage, while water-use efficiency (WUE) peaks around blooming [1].

Regarding the patterns of change in plant hormones, it is well known that ABA plays a crucial role in regulating diverse plant development and physiological processes, including vegetative growth, leaf senescence and abscission, stomatal opening, photosynthetic activity, and as a signaling mediator for plants' adaptive responses to environmental stress [8–10]. In different grapevine genotypes, throughout the progressive drought conditions, negative correlations between stomatal conductance and foliar ABA concentrations have been reported [11–13]. In addition, the role of ABA is emphasized as a connection between the berry ripening process and the grapevine's response to stress. Since the stress response is predominantly mediated by ABA, applying exogenous ABA to grapevine organs is reviewed as a strategy to enhance grape quality and manage abiotic stress [14].

Salicylic acid (SA) is a fundamental endogenous hormone and signaling molecule identified in most plant species. It influences various physiological processes, including seed germination, photosynthesis, ion uptake, and transport, thereby regulating plant growth and development [15,16]. SA can promote the accumulation of phenylalanine (PAL) mRNA, the synthesis of new PAL proteins, and enhance PAL activity under high-temperature stress. Additionally, SA treatment increases the levels of phenolic compounds in grape berries [17].

Jasmonic acid (JA) plays a central role in regulating various plant defense mechanisms, including the detoxification of reactive oxygen species and the rise of osmoprotectants against abiotic stress. Recent evidence suggests that plant tolerance is linked to the modulation of JA responses to stress. This modulation promptly influences JA metabolism and responses by upregulating the expression of genes involved in JA biosynthesis and signaling [18–21]. The interaction between JA and ABA to modulate the response to abiotic stress has been suggested [22,23]. Specifically, the concentrations and gene expression of ABA increase in response to JA, and in contrast, ABA induces the expression of JA biosynthetic genes and contents [24,25].

Regarding the involvement of auxin in grapevine growth and physiological processes, it has been shown that exogenous naphthalene acetic acid (NAA) increases growth, proline production, total soluble sugars contents, maintains photosynthetic activity, and enhances water-use efficiency, as well as IAA and ABA levels. In addition, its application induces a high antioxidant enzyme activity, and it decreases hydrogen peroxide and malondialdehyde content [10]. Exogenous application of IAA at preharvest induces changes in sugar content, acidity, and color in grape berries [26]. Other treatments, such as ethephon, promote IAA biosynthesis [27], and kaolin increases IAA and ABA concentrations in grapevines [28–30].

On the other hand, the uptake of mineral nutrients can differ between the genotypes of *Vitis vinifera*. Mainly, the accumulation of N, P, and Zn^{2+} displays only slight variations. Nevertheless, the absorption of K^+, Ca^{2+}, Mg^{2+}, and Cl^- shows major changes. Particularly, by using several cvs. ("Airén", "Cencibel", "Garnacha", "Cabernet Sauvignon", and "Chardonnay"), the results indicate that the uptake of Ca^{2+} is high in "Cencibel", P is low in "Airén", and K^+ significantly differs in almost all of the cvs. [31]. Under a hydroponic culture experiment, it has been indicated that the uptake of K^+ by "Dattier de Beiruth" grafted on SO4 rootstock is high, and the accumulation of Mg^{2+} is inhibited by the rise of the K^+:Mg^{2+} ratio due to the antagonism between both nutrients [32]). Potassium within the plant is mostly found in a free form that can quickly be redistributed to growing organs like berries, or it can be stored in branches and roots [33]. Potassium is crucial for plants, activating enzymes in photosynthesis and respiration and maintaining cellular osmotic potential [34]. The variations in temperature and evaporation affect soil water availability for grapevines and impact internal water movement, influencing nutrient absorption and mobilization [35].

This research aimed to investigate the physiological responses of three grapevine genotypes, focusing on changes in phytohormones, mineral nutrients, growth, leaf temperature, and stomatal conductance. By revealing these traits under natural conditions, the study may identify the common and specific attributes of these genotypes, as well as their potential for cultivation in determined crop regions.

2. Materials and Methods

2.1. Plant Material and Experimental Conditions

V. vinifera "Castellana Negra" ("Cas"), "Negramoll" ("Neg"), and "Tintilla" ("Tin") are traditionally cultivated in several viticulture areas, including subtropical Canary Island conditions with great agronomical potential and productivity. The plants were propagated from cuttings taken from shoots pruned in field-grown vineyards. Each cutting, with four buds, was placed in 1 L plastic bags filled with substrate, leaving two buds exposed above the surface, while the other two were buried to promote rooting. After 6 months of growth, the plants were transferred to 5 L plastic pots. Once the plants reached 1 year of age, they were pruned, maintaining two-to-three buds on each of the two shoots per plant. Pruning was performed on February 15, and at this stage, the length of the stems containing the buds was around 15 cm. Plants were grown in 5 L plastic pots filled with a peat substrate (Leader potting soil, Düren, Germany) under a greenhouse. Such substrate contained the following amounts of N (200 mg L^{-1}), P_2O_5 (200 mg L^{-1}), and K_2O (300 mg L^{-1}), and 2 weeks before pruning was added to each pot 60 g of granular fertilizer (NPK fertilizer containing Mg with trace elements: 18-9-10 + 2MgO + Te (Osmocote Pro 5-6M, ICL Specialty Fertilizers, Murcia, Spain)). The experimental system was conducted for 102 days, and the climatic variables recorded, such as temperature that ranged an average value between 16 and 33 °C, relative humidity varied in an interval of 50–90%, and maximum photosynthetically active radiation (PAR) registered around 1200 μmol m^{-2} s^{-1}.

2.2. Plant Growth and Sampling

Plant growth expressed as functional leaf number per plant was determined along three dates (22, 76, and 102 days) after the beginning of bud sprouting. At each date, the total of functional leaves per plant and three plants per treatment were harvested, frozen in liquid nitrogen, freeze-dried, and used for the analyses of mineral nutrients and phytohormones.

2.3. Stomatal Conductance

Stomatal conductance was determined by means of a steady-state porometer (SC-1 Leaf Porometer, Meter Group Inc., Pullman, WA, USA) on fully expanded leaves, typically between the third and fifth leaves from the plant apex. Measurements were usually taken between 08:30 and 11:00 a.m., with leaf chamber temperatures averaging around 25.7 ± 0.8 °C and humidity at approximately 59.4 ± 4.0%.

2.4. Analyses of Phytohormones

The levels of ABA, JA, SA, and IAA were analyzed in leaf tissues following the procedure described previously [36]. Thus, 50 mg of dry and powdered leaves were homogenized in 2 mL of ultrapure water with a mill ball equipment (MillMix 20, Domel Železniki, Slovenija) together with the addition as internal standards of 25 ng of [$^2H_6$]-ABA, [$^{13}C_6$]-SA, dehydrojasmonic acid (DHJA), and 2.5 ng of [$^2H_5$]-IAA. After centrifuging at 4000× *g* and 4 °C for 10 min, the supernatants were collected, and the pH was adjusted to 2.8–3.2 using 80% acetic acid. The extract was partitioned twice against 2 mL of diethyl ether (Fisher Scientific, Hampton, NH, USA), and the organic layer was recovered and dried under vacuum using a centrifuge concentrator (Speed Vac, Jouan, Saint Herblain Cedex, France). When dried, the pellet was resuspended in 0.5 mL of a 10:90 methanol:H_2O solution by sonication. The mixture was filtered through 0.22 μm of polytetrafluoroethylene membrane syringe filters (Albert S.A., Barcelona, Spain) and injected into an ultraperformance LC system (Acquity SDS; Waters Corp., Milford, MA, USA). Chromatographic separations were conducted on a reversed-phase C18 column (Gravity, 50 × 2.1 mm, 1.8 μm particle size; Macherey–Nagel GmbH, Düren, Germany) by means of methanol:H_2O gradient (both supplemented with 0.1% acetic acid) at a flow rate of 300 μL min−1. The studied phytohormones were quantified with a TQS triple quadrupole mass spectrometer (Xevo TQ-S, Waters Corp., Milford, MA, USA) through an orthogonal Z-spray electrospray ion source. Standard curves were performed by injecting commercial standards into the UPLC–MS system and used to quantify hormone concentrations in the samples. Results were processed with MassLynx v4.1 software. Data represent the average of three biological replicates for each experimental group.

2.5. Determination of Foliar Mineral Elements

Ground leaf tissues (0.2 g) collected at d 22, 76, and 102 were used to determine the ion contents. Thus, the powder was cremated in a muffle oven at 450 °C and the ashes were dissolved in an aqueous HCl solution for analyzing the mineral elements. Nitrogen was determined using Kjeldahl digestion in sulphuric medium with copper and selenium as catalysts, gathering in boric acid and titration with a 0.05 N aqueous sulphuric acid solution (Sigma-Aldrich, Madrid, Spain) [37,38]. Phosphorus was analyzed by using the vanadium phosphomolybdate method [39]. The cations, such as K^+, Ca^{2+}, Mg^{2+}, Na^+, Fe^{2+}, Mn^{2+}, Cu^{2+}, and Zn^{2+}) by inductively coupled plasma atomic emission spectroscopy (ICP-AES). Lastly, boron was measured via the azomethine H colorimetric [40].

2.6. Statistical Analyses and Experimental Design

A total of 81 plants (27 per cultivar) were distributed in three blocks with 9 plants per cultivar in each one. Three plants per cultivar and block were used for measuring leaf number and stomatal conductance, as well as three plants per treatment and block were sampled to analyze their concentrations in minerals and phytohormones in leaf tissues at each stage of growth. Data were statistically analyzed by using the software IBM® SSPSS® Statistics version 25 for Windows (IMB Corporation, Armonk, NY, USA). All the parameters were examined through one-way ANOVA, and post hoc Duncan test was used for mean separation. The significance level was indicated at $p < 0.05$.

3. Results and Discussion

3.1. Plant Growth

To assess the growth of *Vitis vinifera* plants, the number of leaves was counted at three stages following initial bud sprouting: on days 22, 76, and 102 (Figure 1). As expected, the number of functional leaves increased progressively in proportion to plant age, except for "Cas", which maintained the same number of leaves on both d 76 and d 102. Therefore, "Cas" showed an average of 9 leaves at the beginning of the experiment and 22 by the end, while "Neg" and "Tin" exhibited similar leaf numbers at each stage, ranging from 10–12 leaves on d 22 to 35–32 leaves on d 102. Compared to "Cas", both "Neg" and "Tin" had a higher number of leaves per plant (25% and 20% more on d 76, and 38% and 33% more on d 102), respectively. These cultivars are widely cultivated in wide regions of vineyards and are particularly well-suited to subtropical conditions in many regions of the Canary Islands; however, they differ in their growth and development characteristics [41]. It is also important to note that, under our experimental conditions, "Cas" appears to exhibit slower growth in terms of leaf emergence and shoot development compared to "Tin" and "Neg", at least during the early stages. However, Rodríguez-Torres [41] suggests that adult "Cas" plants in the field may be significantly more vigorous than the other genotypes mentioned. Therefore, since plants grew in non-stressful conditions the differences in leaf number observed between these cultivars, specifically the lower number in "Cas" and the higher in "Neg" and "Tin" may be related to the regional microclimate and the specific behavioral traits of the genotypes. In previous research, it has been reported that the temperature variations affect shoot growth in "Semillon", leading to the largest canopies at 35 °C and the smallest in those at 40 °C, showing a two-fold dissimilarity [42]. In general, global warming can negatively affect grapevine development through the disruption of phenology, growth, yield, and physiological processes [1,2]. In several viticulture locations worldwide midday air temperatures often exceed 40 °C in summer, impacting grapevines along berry ripening, and leading to yield decrease and consequently vineyards productivity [3,4].

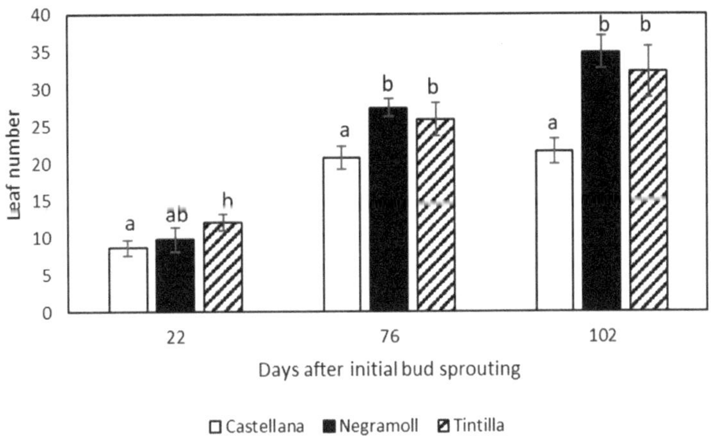

Figure 1. Functional leaf number per plant in grapevines cultivars: Castellana, Negramoll and Tintilla at three dates during a period of 102 days after the onset of sprouting. Data are means ± standard errors obtained from the total of leaves per plants (average of 9 plants per genotype). For each date, different letters designate significant differences at $p < 0.05$, analyzed by ANOVA and Duncan test.

3.2. Stomatal Conductance and Leaf Temperature

Stomatal conductance (gs) remained consistent throughout the different growth stages, ranging from 220 to 314 mmol m^{-2} s^{-1} in the leaves of all genotypes; however, the only exception was on d 76, when "Neg" showed the lowest value (162 mmol m^{-2} s^{-1}) (Figure 2A). At this date, compared to "Neg" this variable was 35% and 26% higher in "Cas" and "Tin", respectively, although, by d 76, leaf temperature was similar in the different cvs. as well as throughout the experimental period (Figure 2B). The decrease in gs at d 76 coincided with the highest leaf temperature (around 30 °C), suggesting that 'Ne' may be slightly sensitive to temperature equal or higher to 30 °C. Nevertheless, the determined gs value in "Neg" seems adequate for supporting the photosynthesis activity. Overall, the data suggest that a leaf temperature range (20–30 °C) is suitable for maintaining stomatal function in these grapevine genotypes. Similar findings show that the optimal temperature for photosynthesis in potted "Thompson" grapevine plants under glasshouse generally ranged between 25 °C and 30 °C [7]. On the other hand, it has been reported in adult plants that the increasing temperatures lead to a decrease in photosynthesis, with an optimal rate at 30 °C under saturated light conditions. Stomatal closure contributed to just 15–30% of this decline [43,44]. Particularly, in "Semillon" leaves, the decline in photosynthesis caused by high temperatures is primarily non-stomatal, contrasting with the response perceived during short periods of high-temperature exposure [45]. In photosynthesis in "Chardonnay" leaves after fruit removal, optimal photosynthesis occurs at 30 °C, with increased rates at higher temperatures, aligning with current climate conditions. Additionally, stomatal limitation of photosynthesis varies from 20% at low to 45% at high temperatures [46]. For "Semillon", photosynthesis and carbon allocation exhibit a high temperature threshold at 33 °C, with extended exposure to higher temperatures proving detrimental to overall performance [46]. Under our experimental conditions, temperature did not seem to significantly impact stomatal conductance, leaf number, or plant growth of the different genotypes.

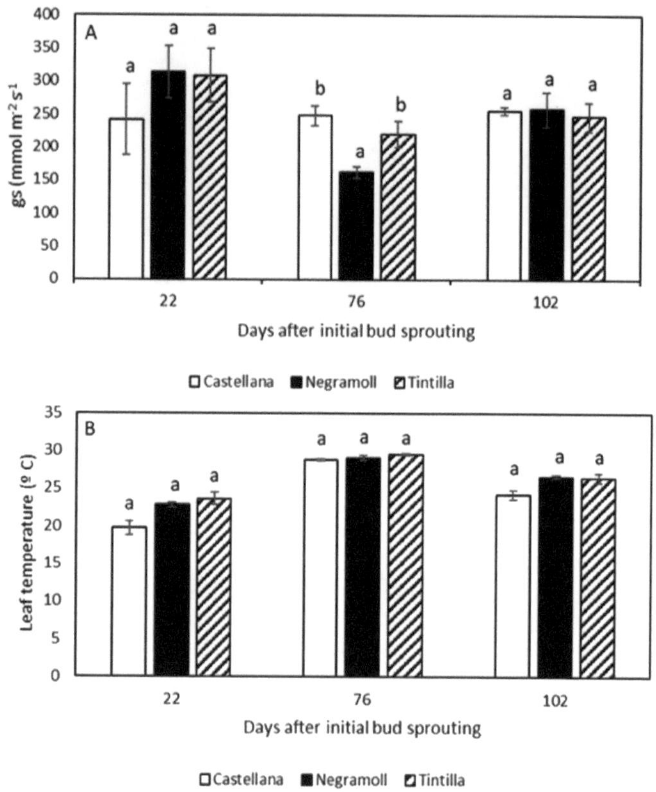

Figure 2. Stomatal conductance (gs) (**A**) and leaf temperature (**B**) in grapevines cultivars: Castellana, Negramoll and Tintilla at three dates during a period of 102 days after the onset of sprouting. Data are means ± standard errors obtained from the total of leaves per plants (average of 9 plants per genotype). For each date, different letters designate significant differences at $p < 0.05$, analyzed by ANOVA and Duncan test.

3.3. Plant Hormones Changes

The foliar ABA concentrations showed a similar pattern of change in the studied cultivars throughout the experimental period (Figure 3A). A slight increase in ABA content was observed from the initial to the final sampling dates This increase was particularly noticeable between d 22 (93 and 68 ng.ng^{-1} DW) and 102 (137 and 115 ng.g^{-1} DW) in "Cas" and "Tin", respectively. At each stages of growth, "Cas" and "Neg" displayed similar levels of ABA; however, "Tin" accumulated reduced contents of this phytohormone than "Cas" at d 22 and d 102. These small differences between genotypes seem to occur temporary, but generally under non-stressful environments similar amounts of ABA would be accumulated in these cvs. The current scientific literature extensively highlights the role of ABA in regulating several physiological processes, including plant growth, leaf senescence, abscission, stomatal closure, photosynthetic activity, among others. ABA also plays a crucial role as a signaling mediator in plant responses to abiotic stress [8–10]. In grapevines, stress-induced ABA movement within the leaf mediates the stomatal response, regulating both transpiration and CO_2 assimilation [47].

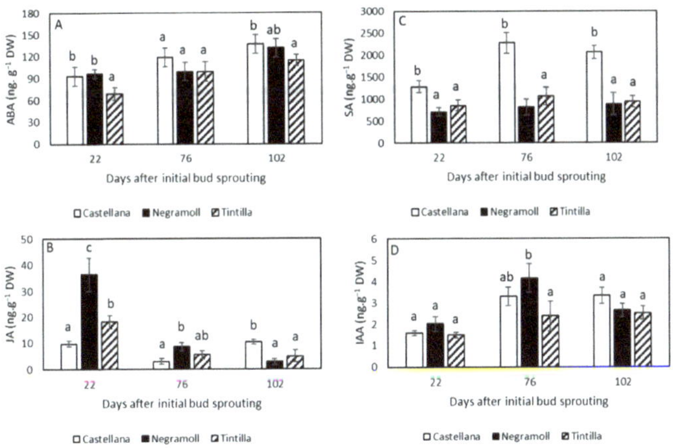

Figure 3. Levels of foliar ABA (**A**), JA (**B**), SA (**C**) and IAA (**D**) in grapevines cultivars: Castellana, Negramoll and Tintilla at three dates during a period of 102 days after the onset of sprouting. Data are means ± standard errors obtained from the total of leaves per plants (average of 9 plants per genotype). For each date, different letters designate significant differences at $p < 0.05$, analyzed by ANOVA and Duncan test.

Nevertheless, JA concentration peaked during the early growth stage (d 22), with the highest levels observed in "Neg" followed by "Tin" and then "Cas" (Figure 3B). Afterwards, JA levels in "Neg" progressively decreased by around 76% and 92% at d 76 and d 102, respectively compared to d 22. "Tin" leaves also accumulated their maximum concentrations at the first stage of growth and displayed lower levels later. In "Cas" leaves, JA exhibited a slightly different pattern compared to the other genotypes, keeping contents steady at 22 as 102 days of growth, with a small decrease observed at d 76. By comparing d 22 and d 102, the levels of JA and ABA in all the genotypes showed an antagonistic change probably linked to plant age. While JA was higher and lower at d 22 and d 102, respectively, ABA displayed the opposite trend. These results are compatible with previous scenarios where the interaction between JA and ABA modulates the plant responses to environmental stress in various experimental systems [22,23]. JA also regulates plant tolerance mechanisms through the detoxification of reactive oxygen species and synthesis of osmoprotectants under abiotic stress conditions. In addition, this modulation rapidly affects JA responses by upregulating the expression of genes involved in JA biosynthesis and signaling [18–21].

On the other hand, SA was the plant hormone that exhibited higher leaf content with respect to the other analyzed hormones (Figure 3C). At d 22, "Cas" leaf tissues accumulated around 1280 ng.g^{-1} DW and subsequently its concentration increased over 44% and 38% at d 76 and d 102, respectively. In this genotype, the accumulation of SA seems to increase with plant age, in contrast to JA. In addition, the concentrations in "Cas" leaf tissues were greater at d 76 (2.8–2.2-fold) and d 102 (2.3–2.2-fold) compared to those in "Neg" and "Tin", respectively. Finally, the average of SA content in "Neg" and "Tin" leaves were similar throughout the experimental period. It is now well established that SA regulates plant growth and physiological processes, through enhancing photosynthesis, ion uptake and transport, proline and soluble sugar levels, the antioxidant system, and reducing oxidative stress [15,16]. In addition, SA can enhance phenylalanine activity and promote the accumulation of phenolics substances in grape berries under high-temperature field conditions [17]. Leaf IAA content showed a similar trend in all genotypes during the three growth stages, with a slight accumulation on d 22, followed by a moderate increase thereafter (Figure 3D). The IAA levels were consistently low during the growth period, ranging from 1.5 to 4.2 ng.g^{-1} DW. These foliar levels seem to be suitable for plant growth and physiology in these genotypes under natural conditions, as no stress was applied in this experimental system. The involvement of IAA in grape quality was reported by Deytieux-Belleau et al. [26], showing that the supply of IAA at preharvest induces changes in sugar content, acidity and color in grape berries. Other experiments indicate that kaolin increases IAA and ABA contents in grapevines [28–30]. In addition, another auxin (NAA) increases IAA and ABA concentrations, water use efficiency, proline, total soluble sugars, plant growth, and maintains photosynthetic activity [10].

3.4. *Leaf Mineral Nutrients Changes*

Foliar N levels remained similar in all cvs. at each date, ranging from 3–3.2% on d 22 to 2.4–2.5% of DW on d 102 (Table 1). This indicates a slightly higher levels at the beginning of the trial compared to the end. In contrast, leaf P content followed an opposite trend, although in this nutrient the levels were low, initiating at 0.2–0.3% on d 22 and increasing to 0.4–0.5% of DW by d 102. Leaf K^+ concentrations were initially consistent around 0.8% of DW on d 22 (Table 1), with a non-significant decrease in "Tin" compared to "Cas" and "Neg" by d 76. Later, on d 102, "Cas" leaves had slightly lower K^+ levels compared to "Neg". Throughout the entire growth period, K^+ concentrations in all genotypes ranged from 0.7% to 0.9% of DW (Table 1). Overall, the concentrations of the measured macronutrients (N, P and K^+) appeared optimal for all genotypes during these growth stages, as no deficiency symptoms were observed under the experimental conditions. Tagliavini and Scandellari [33] indicate that accurately determining K^+ contents in leaves is often challenging, as most K^+ exists in a free form that can quickly be redistributed to growing organs, like berries or be stored in reserve organs, including branches and roots. It has also been reported that the fluctuation of temperature and evaporation can influence the internal water movement processes within the grapevine, impacting nutrient uptake and mobilization [35]. As well, potassium can activate enzymes involved in photosynthesis and respiration, and maintaining cellular osmotic potential [34].

Table 1. Contents of foliar macronutrients in grapevines cultivars: Castellana, Negramoll and Tintilla at three dates during a period of 102 days after the onset of sprouting. Data are means ± standard errors obtained from the total of leaves per plants (average of 9 plants per genotype). For each date, different letters designate significant differences at $p < 0.05$, analyzed by ANOVA and Duncan test.

	Days After Initial Bud Sprouting		
Macronutrients	**22**	**76**	**102**
"Cas" N (%DW)	3.02 ± 0.01 a	2.66 ± 0.14 a	2.53 ± 0.17 a
"Neg" N (%DW)	3.17 ± 0.05 a	2.80 ± 0.15 a	2.51 ± 0.07 a
"Tin" N (%DW)	3.11 ± 0.19 a	2.66 ± 0.08 a	2.36 ± 0.08 a
"Cas" P (%DW)	0.20 ± 0.01 a	0.38 ± 0.09 a	0.51 ± 0.07 a
"Neg" P (%DW)	0.24 ± 0.02 ab	0.27 ± 0.04 a	0.38 ± 0.04 a
"Tin" P (%DW)	0.29 ± 0.02 b	0.31 ± 0.06 a	0.43 ± 0.04 a
"Cas" K^+ (%DW)	0.85 ± 0.02 a	0.87 ± 0.06 a	0.75 ± 0.05 a
"Neg" K^+ (%DW)	0.84 ± 0.03 a	0.91 ± 0.05 a	0.94 ± 0.05 a
"Tin" K^+ (%DW)	0.78 ± 0.05 a	0.72 ± 0.06 a	0.83 ± 0.07 a
"Cas" Ca^{2+} (%DW)	0.90 ± 0.08 a	1.57 ± 0.13 a	1.84 ± 0.09 b
"Neg" Ca^{2+} (%DW)	0.96 ± 0.07 a	1.19 ± 0.14 a	1.37 ± 0.02 a
"Tin" Ca^{2+} (%DW)	1.20 ± 0.04 b	1.51 ± 0.18 a	1.81 ± 0.11 b
"Cas" Mg^{2+} (%DW)	0.12 ± 0.01 a	0.15 ± 0.01 a	0.16 ± 0.02 a
"Neg" Mg^{2+} (%DW)	0.14 ± 0.01 a	0.15 ± 0.01 a	0.15 ± 0.01 a
"Tin" Mg^{2+} (%DW)	0.17 ± 0.01 b	0.18 ± 0.01 a	0.20 ± 0.00 b

Foliar Ca^{2+} levels progressively increased during the experimental period in all cvs., ranging from 0.9 to 1.8% of DW, regardless of the date (Table 1). The data pointed out that Ca^{2+} content in "Tin" was approximately 25% and 21% higher at d 22 compared to its concentration in "Cas" and "Neg" leaves, respectively. In addition, the diagnosed levels of

Ca^{2+} at d 102 were about 25% lower in "Neg" in comparison with "Cas" and "Tin". Notably, Ca^{2+} exhibited the second-highest major higher level in leaf tissues in all cvs., following N, while K^+ ranked third. The average of these concentrations was relatively higher compared to those reported by Amiri et al. [48], which were around 0.88% for "Qarah Shani" and 0.86% for "Thompson Seedless".

The foliar levels of Mg^{2+} were statistically similar in "Cas" and "Neg" at each date throughout the experimental period (Table 1). However, "Tin" accumulated higher concentrations of this element compared to the other genotypes at d 22 (29% and 18%) and d 102 (25%), respectively. The foliar concentrations of this nutrient were markedly lower than those determined in "Qarah Shani" and "Thompson Seedless", which were about 1.95% and 2.01%, respectively [48]. In an experimental system using *Vitis vinifera* "Dattier de Beiruth" grafted on SO4 rootstock under a hydroponic culture, the results show that the uptake of K^+ is high, and the buildup of Mg^{2+} is inhibited by the increase of K^+:Mg^{2+} ratio indicating an antagonism between both nutrients [32].

The foliar concentrations of Na^+ in the three cvs. varied between 0.02% and 0.06% DW throughout the different stages of growth, and such levels appear to be appropriate for these genotypes since no ion toxic symptoms were detected in the leaves (Table 2). Throughout the trial, no statistical changes in the contents of this micronutrient were observed in any of the genotypes. In this sense, Amiri et al. [48] show similar leaf Na contents in "Qarah Shani" and "Thompson Seedless" under non-stressful conditions.

Table 2. Contents of foliar micronutrients in grapevines cultivars: Castellana, Negramoll and Tintilla at three dates during a period of 102 days after the onset of sprouting. Data are means ± standard errors obtained from the total of leaves per plants (average of 9 plants per genotype). For each date, different letters designate significant differences at $p < 0.05$, analyzed by ANOVA and Duncan test.

	Days After Initial Bud Sprouting		
Micronutrients	**22**	**76**	**102**
"Cas" Na^+ (%DW)	0.04 ± 0.01 a	0.03 ± 0.00 a	0.03 ± 0.01 a
"Neg" Na^+ (%DW)	0.06 ± 0.01 a	0.03 ± 0.00 a	0.04 ± 0.01 a
"Tin" Na^+ (%DW)	0.04 ± 0.00 a	0.02 ± 0.00 a	0.04 ± 0.00 a
"Cas" Fe^{2+} (ppm)	44.47 ± 0.78 a	53.41 ± 2.64 a	0.03 ± 0.01 a
"Neg" Fe^{2+} (ppm)	47.89 ± 3.13 a	63.43 ± 2.48 bc	0.04 ± 0.01 a
"Tin" Fe^{2+} (ppm)	48.97 ± 3.37 a	57.31 ± 1.21 ab	0.04 ± 0.00 a
"Cas" Mn^{2+} (ppm)	36.33 ± 4.63 a	88.70 ± 1.67 b	87.41 ± 10.13 bc
"Neg" Mn^{2+} (ppm)	40.02 ± 2.66 a	39.85 ± 4.78 a	48.43 ± 11.41 a
"Tin" Mn^{2+} (ppm)	33.16 ± 1.98 a	39.01 ± 3.00 a	64.41 ± 9.98 ab
"Cas" Cu^{2+} (ppm)	1.86 ± 0.24 ab	1.14 ± 0.31 a	4.11 ± 1.94 ab
"Neg" Cu^{2+} (ppm)	2.48 ± 0.19 b	1.25 ± 0.27 a	8.47 ± 1.11 b
"Tin" Cu^{2+} (ppm)	1.27 ± 0.30 a	1.59 ± 0.34 a	3.53 ± 0.18 a
"Cas" Zn^{2+} (ppm)	6.08 ± 0.46 b	8.43 ± 1.37 a	7.71 ± 0.82 a
"Neg" Zn^{2+} (ppm)	7.14 ± 0.40 b	7.38 ± 0.79 a	10.20 ± 1.83 a
"Tin" Zn^{2+} (ppm)	4.78 ± 0.29 a	8.06 ± 1.69 a	7.25 ± 0.36 a
"Cas" B (ppm)	17.70 ± 1.07 b	26.56 ± 6.34 a	27.40 ± 3.93 a
"Neg" B (ppm)	13.60 ± 0.77 a	18.30 ± 1.19 a	21.99 ± 3.96 a
"Tin" B (ppm)	16.08 ± 1.21 ab	19.58 ± 3.52 a	28.65 ± 8.77 a

Regarding Fe^{2+} build up, non-significant variations were observed in the different genotypes during d 22 and d 102 (Table 2). For this micronutrient, only leaves of "Neg" exhibited greater concentration than those of "Cas" at d 76. The contents of this mineral measured in "Qarah Shani" and "Thompson Seedless" were around 308 ppm and 296 ppm, respectively [48]. In productive plants, the determined foliar level of this nutrient range 38–39 ppm in "Flame Seedless" [49], and 171.6 ppm in "Thompson Seedless" grapevines [50]. In previous studies using 12-year-old espalier plants of "Kyoho Early" (*V.*

vinifera × *V. labrusca*) leaf Fe^{2+} concentrations are around 183.5 ppm [51]. As reported in the distinct cvs., the levels of Fe^{2+} in the leaves seem to vary based on the genotype and growth stage.

Concerning Mn^{2+} variation, the greatest concentrations of this element were measured in "Cas" at d 76 (89 ppm), and then was around 56% higher than those in leaves of "Neg" and "Tin" (Table 2). Afterwards, Mn^{2+} content was kept higher in "Cas" compared to "Neg"; nevertheless, it did not show statistical differences in comparison with "Tin". It has been reported that the average content of leaf Mn^{2+} in productive plants of "Thompson Seedless" grapevines is around 36.2 ppm [50], and in "Kyoho Early" 33.7 ppm [51]. Mn^{2+} is a crucial micronutrient for plant growth and development, involved in various metabolic functions within different plant cell compartments. It acts as an essential cofactor for the oxygen-evolving complex in photosynthetic machinery [52].

Copper showed the lowest concentrations detected in the foliar analysis of the three genotypes during the trial (Table 2). Minimum levels were measured during the d 22 and d 76 (1.15–2.5 ppm) independently of the cvs. followed by a slight increase reaching about 8.5 ppm in "Neg" at d 102. This genotype accumulated similar levels of Cu^{2+} at each sample date; however, its contents were higher than those of "Tin" by approximately 49% and 58% at d 22 and d 102, respectively. These levels are lower than those determined in adult "Kyoho Early" plants, which display a leaf Cu^{2+} concentration around 15.8 ppm the leaves contain a Cu^{2+} concentration around 15.8 ppm [51].

Zinc displayed the second-lowest foliar contents (4.8–10.2 ppm) among the three genotypes throughout the experimental period, following Cu^{2+} (Table 2). A notable variation in zinc accumulation was observed at d 22, with "Cas" and "Neg" exhibiting concentrations 21% and 33% higher, respectively, than those of "Tin". However, in other experimental systems involving productive plants, the leaf content of this mineral ranges 11 to 15 ppm in "Flame Seedless" grapevines [49], around 48 ppm and 57 ppm in "Qarah Shani" and "Thompson Seedless", respectively [48], 35.7 ppm in "Thompson Seedless" [50], and 39.4 ppm in "Kyoho Early" [51]. These reported concentrations are higher than those observed in our study, which may be attributed to several agronomical factors, including soil compositions, fertilizer application rates, genotype traits, and plant age.

Boron contents were similar in the three genotypes mostly over the whole period of study (13.6–28.6 ppm), excepting a higher accumulation of this micronutrient (23%) was noted in the leaves of "Cas" compared to "Neg" (Table 2). In the same range, other experimental systems reported foliar concentrations of this nutrient (21–22 ppm) in adult plants of "Flame Seedless" [49].

Overall, it is worth mentioning that plant nutrient uptake may differ depending on the genotype of *Vitis vinifera*. For instance, with the use of five cvs. ("Airén", "Cencibel", "Garnacha", "Cabernet Sauvignon", and "Chardonnay"), it has been found that the buildup of N, P and Zn^{2+} shows only slight variations; however, there are substantial changes in the accumulation of K^{+}, Ca^{2+}, Mg^{2+}, and Cl^{-}. In addition, the uptake of the following nutrients can vary according to the cultivar: Ca^{2+} (high in "Cencibel", 3.26%), P (low in "Airén", 0.09%), and K^{+} significantly differ in almost all the cvs. ranging from 0.47% to 0.8% in "Cencibel" and "Chardonnay", respectively [31].

In summary, data presented here showed specific similarities and differences in terms of physiological processes. For instance, the evaluation of plant growth by means of functional leaf number revealed a similar tendency between "Neg" and "Tin" with a progressive increase parallelly to plant age, and although "Cas" exhibited the same growth pattern, its development was slower, and subsequently the emerged leaves were lower than the other genotypes. Regarding stomatal movement, gs measurements were similar between the three genotypes except for "Neg" which demonstrated a lower value at d 76 compared to "Cas" and "Neg". Such decrease of gs coincides with the highest leaf temperature which might suggest that "Neg" seems to be sensitive to temperatures higher than 30 °C. Concerning the hormonal changes, "Cas" accumulated the highest concentration of SA compared to "Neg" and "Tin", which both maintained similar contents throughout the experimental period. The data may indicate a possible connection between the elevated SA levels and the fewer leaves in "Cas" plants. To prove this hypothesis, a long-term experiment could be conducted to evaluate the progression of both variables. Relating to ABA, "Cas" and "Neg" accumulated similar levels; however, "Tin" exhibited slightly and temporary reduced concentrations, particularly during the early growth stage (d 22), compared to the other genotypes. JA contents, in turn, fluctuated throughout the stages of growth, displaying a non-uniform pattern of change, and the most apparent observation was likely its high level and the variations among the three genotypes at the initial date. Lastly, IAA contents generally showed similar trends between all the genotypes. In relation to mineral nutrients, similar foliar concentrations of N, P, K^+ and Na^+ were accumulated between the distinct genotypes, while the levels of Mg^{2+} were higher in "Tin" than in "Cas" and "Neg" at the beginning and the end of the experiment. Ca^{2+} content was higher in "Tin" compared to "Cas" and 'Ne' at d 22, and afterward its content was maintained higher than "Neg" but similar to "Cas" at d 102. The levels of micronutrients fluctuated temporarily between cvs.; however, no constant trend was observed, as the plants did not exhibit any deficiencies in any of the elements.

4. Conclusions

The data presented here illustrates the pattern of change and concentration ranges of phytohormones and mineral nutrients, along with leaf growth and stomatal conductance from the beginning of sprouting to 102 days later in three genotypes of *Vitis vinifera*. The findings suggest a relationship between high levels of SA and a reduced number of leaves in "Cas" compared to "Neg" and "Tin" throughout the growth stage. Moreover, an antagonistic change was observed in the accumulation of SA and JA, as well as between ABA and JA particularly at the onset and the end of the trial. Consequently, long-term experiments could further explore the involvement of SA and other hormones, and physiological markers in the growth and development of these genotypes under environmental conditions.

Author Contributions: Conceptualization, J.M.; methodology, J.M., A.U.-G., M.F.L.-C. and A.G.-C.; formal analysis, J.M., A.U.-G., M.F.L.-C. and A.G.-C.; investigation, J.M., A.U.-G., M.F.L.-C. and A.G.-C.; data curation, J.M., A.U.-G., M.F.L.-C. and A.G.-C.; writing—original draft preparation, J.M.; writing—review and editing, J.M.; supervision, J.M.; project administration, J.M.; funding acquisition, J.M. All authors have read and agreed to the published version of the manuscript.

Funding: This research received no external funding.

Data Availability Statement: The original contributions presented in the study are included in the article, further inquiries can be directed to the corresponding author.

Acknowledgments: This research was supported by funding of the Departamento de Ingeniería Agraria y del Medio Natural and the Vicerrectorado de Investigación y Transferencia, Universidad de La Laguna. UPLC-MS equipment used for hormonal analyses was provided by the SCIC of the Universidad Jaume I.

Conflicts of Interest: The authors declare no conflicts of interest.

References

1. Rafique, R.; Ahmad, T.; Ahmed, M.; Khan, M.A. Exploring key physiological attributes of grapevine cultivars under the influence of seasonal environmental variability. *OENO One* **2023**, *57*, 381–397. [CrossRef]
2. Rafique, R.; Ahmad, T.; Ahmed, M.; Khan, M.A.; Wilkerson, C.J.; Hoogenboom, G. Seasonal variability in the effect of temperature on key phenological stages of four table grapes cultivars. *Int. J. Biometeorol.* **2023**, *67*, 745–759. [CrossRef] [PubMed]
3. Salvucci, M.E.; Crafts-Brandner, S.J. Relationship between the heat tolerance of photosynthesis and the thermal stability of rubisco activase in plants from contrasting thermal environments. *Plant Physiol.* **2004**, *134*, 1460–1470. [CrossRef] [PubMed]
4. Liu, G.T.; Wang, J.F.; Cramer, G.; Dai, Z.W.; Duan, W.; Xu, H.G.; Wu, B.H.; Fan, P.G.; Wang, L.J.; Li, S.H. Transcriptomic analysis of grape (*Vitis vinifera* L.) leaves during and after recovery from heat stress. *BMC Plant Biol.* **2012**, *12*, 174. [CrossRef]
5. Canton, M.; Borghezan, M.; Silva, T.C.; Welter, J.F.; Villar, L.; Rosa, D.J.; Silva, A.L.; Pescador, R. Chlorophyll evaluation on leaves of 'Sauvignon Blanc' during vegetative growth in São Joaquim, Santa Catarina, Brazil. *Acta Hortic.* **2017**, *1188*, 15–20. [CrossRef]

6. Carvalho, L.C.; Coito, J.L.; Colaço, S.; Sangiogo, M.; Amâncio, S. Heat stress in grapevine: The pros and cons of acclimation. *Plant Cell Env.* **2015**, *38*, 777–789. [CrossRef]
7. Higgins, S.S.; Larsen, F.E.; Bendel, R.B.; Radamaker, G.K.; Bassman, J.H.; Bidlake, W.R.; Wir, A.A. Comparative gas exchange characteristics of potted, glasshouse-grown almond, apple, fig, grape, olive, peach and Asian pear. *Sci. Hortic.* **1992**, *52*, 313–329. [CrossRef]
8. Harris, J. Abscisic acid: Hidden architect of root system structure. *Plants* **2015**, *4*, 548–572. [CrossRef]
9. Sah, S.K.; Reddym, K.R.; Li, J. Abscisic Acid and Abiotic Stress Tolerance in Crop Plants. *Front. Plant Sci.* **2016**, *7*, 571. [CrossRef]
10. Khandani, Y.; Sarikhani, H.; Gholami, M.; Rad, A.C.; Yousefi, S.; Sodini, M.; Sivilotti, P. Exogenous Auxin Improves the Growth of Grapevine (*Vitis vinifera* L.) under Drought Stress by Mediating Physiological, Biochemical and Hormonal Modifications. *J. Soil Sci. Plant Nutr.* **2024**, *24*, 3422–3440. [CrossRef]
11. Lovisolo, C.; Hartung, W.; Schubert, A. Whole-plant hydraulic conductance and root-to-shoot flow of abscisic acid are independently affected by water stress in grapevines. *Funct. Plant Biol.* **2002**, *29*, 1349–1356. [CrossRef] [PubMed]
12. Perrone, I.; Gambino, G.; Chitarra, W.; Vitali, M.; Pagliarani, C.; Riccomagno, N.; Balestrini, R.; Kaldenhoff, R.; Uehlein, N.; Gribaudo, I.; et al. The grapevine root-specific aquaporin VvPIP2;4N controls root hydraulic conductance and leaf gas exchange under well-watered conditions but not under water stress. *Plant Physiol.* **2012**, *160*, 965–977. [CrossRef] [PubMed]
13. Perrone, I.; Pagliarani, C.; Lovisolo, C.; Chitarra, W.; Roman, F.; Schubert, A. Recovery from water stress affects grape leaf petiole transcriptome. *Planta* **2012**, *235*, 1383–1396. [CrossRef] [PubMed]
14. Ferrandino, A.; Lovisolo, C. Abiotic stress effects on grapevine (*Vitis vinifera* L.): Focus on abscisic acid-mediated consequences on secondary metabolism and berry quality. *Environ. Exp. Bot.* **2010**, *103*, 38–147. [CrossRef]
15. Janda, T.; Szalai, G.; Pál, M. Salicylic acid signalling in plants. *Int. J. Mol. Sci.* **2020**, *21*, 2655. [CrossRef]
16. Li, Q.; Guan, C.; Zhao, Y.; Duan, X.; Yang, Z.; Zhu, J. Salicylic acid alleviates Zn-induced inhibition of growth via enhancing antioxidant system and glutathione metabolism in alfalfa. *Ecotoxicol. Environ. Saf.* **2023**, *265*, 115500. [CrossRef]
17. Wen, P.F.; Chen, J.Y.; Wan, S.B.; Kong, W.F.; Zhang, P.; Wang, W.; Zhan, J.C.; Pan, Q.H.; Huang, W.D. Salicylic acid activates phenylalanine ammonia-lyasein grape berry in response to high temperature stress. *Plant Grow. Regul.* **2008**, *55*, 1–10. [CrossRef]
18. Jiang, J.; Ma, S.; Ye, N.; Jiang, M.; Cao, J.; Zhang, J. WRKY transcription factors in plant responses to stresses. *J. Integr. Plant Biol.* **2017**, *59*, 86–101. [CrossRef]
19. Shen, X.; Guo, X.; Guo, X.; Zhao, D.; Zhao, W.; Chen, J.; Li, T. PacMYBA, a sweet cherry R2R3-MYB transcription factor, is a positive regulator of salt stress tolerance and pathogen resistance. *Plant Physiol. Biochem.* **2017**, *112*, 302–311. [CrossRef]
20. Yoon, Y.; Seo, D.H.; Shin, H.; Kim, H.J.; Kim, C.M.; Jang, G. The Role of stress-responsive transcription factors in modulating abiotic stress tolerance in plants. *Agronomy* **2020**, *10*, 788. [CrossRef]
21. Kim, H.; Seomun, S.; Yoon, Y.; Jang, G. Jasmonic Acid in Plant Abiotic Stress Tolerance and Interaction with Abscisic Acid. *Agronomy* **2021**, *11*, 1886. [CrossRef]
22. De Ollas, C.; Hernando, B.; Arbona, V.; Gómez-Cadenas, A. Jasmonic acid transient accumulation is needed for abscisic acid increase in citrus roots under drought stress conditions. *Physiol. Plant.* **2013**, *147*, 296–306. [CrossRef] [PubMed]
23. Ju, L.; Jing, Y.; Shi, P.; Liu, J.; Chen, J.; Yan, J.; Chu, J.; Chen, K.M.; Sun, J. JAZ proteins modulate seed germination through interaction with ABI 5 in bread wheat and Arabidopsis. *New Phytol.* **2019**, *223*, 246–260. [CrossRef] [PubMed]
24. Verma, V.; Ravindran, P.; Kumar, P.P. Plant hormone-mediated regulation of stress responses. *BMC Plant Biol.* **2016**, *16*, 86. [CrossRef] [PubMed]
25. De Ollas, C.; Dodd, I.C. Physiological impacts of ABA–JA interactions under water limitation. *Plant Mol. Biol.* **2016**, *91*, 641–650. [CrossRef] [PubMed]
26. Deytieux-Belleau, C.; Gagne, S.; L'Hyvernay, A.; Donèche, B.; Geny, L. Possible roles of both absicisic acid and indol-acetic acid in controlling grape berry ripening process. *J. Int. Sci. Vigne. Vin.* **2007**, *41*, 141–148. [CrossRef]
27. Böttcher, C.; Burbidge, C.A.; Boss, P.K.; Davies, C. Interactions between ethylene and auxin are crucial to the control of grape (*Vitis vinifera* L.) berry ripening. *BMC Plant Biol.* **2013**, *13*, 222. [CrossRef]
28. Dinis, L.T.; Bernardo, S.; Conde, A.; Pimentel, D.; Ferreira, H.; Félix, L.; Gerós, H.; Correia, C.M.; Moutinho-Pereira, J. Kaolin exogenous application boosts antioxidant capacity and phenolic content in berries and leaves of grapevine under summer stress. *J. Plant Physiol.* **2016**, *191*, 45–53. [CrossRef]
29. Dinis, L.T.; Malheiro, A.C.; Luzio, A.; Fraga, H.; Ferreira, H.; Gonçalves, I.; Pinto, G.; Correia, C.M.; Moutinho-Pereira, J. Improvement of grapevine physiology and yield under summer stress by kaolin-foliar application: Water relations, photosynthesis and oxidative damage. *Photosynthetica* **2017**, *56*, 641–651. [CrossRef]
30. Dinis, L.T.; Bernardo, S.; Luzio, A.; Pinto, G.; Meijón, M.; Pintó-Marijuan, M.; Cotado, A.; Correia, C.; Moutinho-Pereira, J. Kaolin modulates ABA and IAA dynamics and physiology of grapevine under Mediterranean summer stress. *J. Plant Physiol.* **2018**, *220*, 181–192. [CrossRef]
31. Amorós, J.A.; Pérez-de-los Reyes, C.; García Navarro, F.J.; Bravo, S.; Chacón, J.L.; Martínez, J.; Jiménez Ballesta, R. Bioaccumulation of mineral elements in grapevine varieties cultivated in "La Mancha". *J. Plant Nutr. Soil Sci.* **2013**, *176*, 843–850. [CrossRef]
32. Toumi, M.; Nedjimi, B.; Halitim, A.; Garcia, M. Effects of K-Mg ratio on growth and cation nutrition of *Vitis vinifera* L. cv. "Dattier de Beiruth" grafted on SO4 rootstock. *J. Plant Nutr.* **2016**, *39*, 904–911. [CrossRef]

33. Tagliavini, M.; Scandellari, F. Methodologies and concepts in the study of nutrient uptake requirements and partitioning in fruit trees. Proceedings of the Seventh International Symposium on Mineral Nutrition of Fruit Crops. *Acta Horticul.* **2013**, *984*, 47–56. [CrossRef]
34. Salisbury, F.B.; Ross, C.W. *Plant Physiology*, 4th ed.; Salisbury, F.B., Ross, C.W., Eds.; Wadsworth Publishing Company: Belmont, CA, USA, 1992; 682p, ISBN 0-534-15162-0.
35. Proffitt, T.; Campbell-Clause, J. Managing Grapevine Nutrition and Vineyard Soil Health. 2012. Available online: https://winewa.asn.au/wp-content/uploads/2020/04/Grapevine_Nutrition_LR_2.pdf (accessed on 18 June 2024).
36. Durgbanshi, A.; Arbona, V.; Pozo, O.; Miersch, O.; Sancho, J.V.; Gómez-Cadenas, A. Simultaneous determination of multiple phytohormones in plant extracts by liquid chromatography-electrospray tandem mass spectrometry. *J. Agric. Food Chem.* **2005**, *53*, 8437–8442. [CrossRef]
37. Sáez-Plaza, P.; Michałowski, T.; Navas, M.J.; Asuero, A.G.; Wybraniec, S. An overview of the Kjeldahl method of nitrogen determination. Part I. Early history, chemistry of the procedure, and titrimetric finish. *Crit. Rev. Anal. Chem.* **2013**, *43*, 178–223. [CrossRef]
38. Sáez-Plaza, P.; Michałowski, T.; Navas, M.J.; Asuero, A.G.; Wybraniec, S. An overview of the Kjeldahl method of nitrogen determination. Part II. Sample preparation, working scale, instrumental finish, and quality control. *Crit. Rev. Anal. Chem.* **2013**, *43*, 224–272. [CrossRef]
39. Barton, C.J. Photometric analysis of phosphate rock. *Anal. Chem.* **1948**, *20*, 1068–1073. [CrossRef]
40. Gaines, T.P.; Mitchell, A. Boron determination in plant tissue by the azomethine H method. *Commun. Soil Sci. Plant Anal.* **1979**, *10*, 1099–1108. [CrossRef]
41. Rodríguez-Torres, I. *Variedades de Vid Cultivadas en Canarias: Descriptores Morfológicos, Caracterización Morfológica, Molecular, Agronómica y Enológica*, 2nd ed.; Instituto Canario de Investigaciones Agrarias: Canary Islands, Spain, 2017; p. 194.
42. Greer, D.H. Responses of biomass accumulation, photosynthesis and the net carbon budget to high canopy temperatures of *Vitis vinifera* L. cv. Semillon vines grown in field conditions. *Environ. Exp. Bot.* **2017**, *138*, 10–20. [CrossRef]
43. Zsófi, Z.; Váradi, G.; Bálo, B.; Marschall, M.; Nagy, Z. Heat acclimation of grapevine leaf photosynthesis: Mezzo- and macroclimatic aspects. *Funct. Plant Biol.* **2009**, *36*, 310–322. [CrossRef]
44. Greer, D.H.; Weedon, M.M. Modelling photosynthetic responses to temperature of grapevine (*Vitis vinifera* cv. Semillon) leaves on vines grown in a hot climate. *Plant Cell Environ.* **2012**, *35*, 1050–1064. [CrossRef] [PubMed]
45. Greer, D.H.; Weston, C. Heat stress affects flowering, berry growth, sugar accumulation and photosynthesis of *Vitis vinifera* cv. Semillon grapevines grown in a controlled environment. *Funct. Plant Biol.* **2010**, *37*, 206–214. [CrossRef]
46. Greer, D.H. Stomatal and non-stomatal limitations at different leaf temperatures to the photosynthetic process during the post-harvest period for *Vitis vinifera* cv. Chardonnay vines. *N. Z. J. Crop Hortic. Sci.* **2020**, *48*, 1–21. [CrossRef]
47. Pantin, F.; Simonneau, T.; Muller, B. Coming of leaf age: Control of growth by hydraulics and metabolics during leaf ontogeny. *New Phytol.* **2012**, *196*, 349–366. [CrossRef] [PubMed]
48. Amiri, J.; Eshghi, S.; Tafazoli, E.; Kholdebarin, B.; Abbaspour, N. Ameliorative effects of salicylic acid on mineral concentrations in roots and leaves of two grapevine (*Vitis vinifera* L.) cultivars under salt stress. *Vitis* **2014**, *53*, 181–188. [CrossRef]
49. Ali, I.; Wang, X.; Abbas, W.M.; Hassan, M.U.; Shafique, M.; Tareen, M.J.; Fiaz, S.; Ahmed, W.; Qayyum, A. Quality Responses of Table Grapes "Flame Seedless" as Effected by Foliarly Applied Micronutrients. *Horticulturae* **2021**, *7*, 462. [CrossRef]
50. Mohamed, A.A. Impact of foliar application of nanomicronutrient fertilizers on some quantitative and qualitative traits of "Thompson seedless" grapevine. *Middle East J. Appl. Sci.* **2020**, *10*, 435–441. [CrossRef]
51. Ma, J.; Zhang, M.; Liu, Z.; Chenb, H.; Li, Y.C.; Sun, Y.; Ma, Q.; Zhao, C. Effects of foliar application of the mixture of copper and chelated iron on the yield, quality, photosynthesis, and microelement concentration of table grape (*Vitis vinifera* L.). *Sci. Hortic.* **2019**, *254*, 106–115. [CrossRef]
52. Alejandro, S.; Höller, S.; Meier, B.; Peiter, E. Manganese in Plants: From Acquisition to Subcellular Allocation. *Front. Plant Sci.* **2020**, *11*, 300. [CrossRef]

horticulturae

MDPI

Article

Effects of Light Spectra on Growth, Physiological Responses, and Antioxidant Capacity in Five Radish Varieties in an Indoor Vertical Farming System

Panita Chutimanukul [1], Pakin Piew-ondee [2], Thanyaluk Dangsamer [2], Akira Thongtip [1], Supattana Janta [1], Praderm Wanichananan [1], Ornprapa Thepsilvisut [2], Hiroshi Ehara [3] and Preuk Chutimanukul [2,*]

[1] National Center for Genetic Engineering and Biotechnology (BIOTEC), National Science and Technology Development Agency, Khlong Luang, Pathum Thani 12120, Thailand
[2] Department of Agricultural Technology, Faculty of Science and Technology, Thammasat University, Pathum Thani 12120, Thailand
[3] International Center for Research and Education in Agriculture, Nagoya University, Nagoya 464-8601, Japan
* Correspondence: preuk59@tu.ac.th

Abstract: Radish (*Raphanus sativus* L.) is highly nutritious and contains antioxidants that help reduce the risk of diseases. Light is a crucial factor in their growth and the stimulation of secondary metabolite production. Therefore, this study aimed to investigate the effects of light spectra on the development, physiological responses, and antioxidant capacity of radish varieties including cherry belle (CB), black Spanish (BS), hailstone white (HW), Malaga violet (MV), and sparkler white tip (SW) under a controlled environment. Various spectra of red (R), green (G), and blue (B) light were used. The study found that using a combination of red and blue light (3R:1B) resulted in the highest growth in root diameter, fresh weight, and dry weight across all five radish varieties, with values ranging from 1.83 to 4.63 cm, 13.58 to 89.33 g, and 1.20 to 4.64 g, respectively. In terms of physiological responses, the CB and BS varieties showed a higher photosynthetic rate after exposure to mixed red and blue light (1R:3B, 3R:1B). Additionally, adding green light to the red and blue light also enhanced the photosynthetic rate, with statistically significant differences ranging from 3.31 to 3.99 μmol m^{-2} s^{-1}. The SW variety of radish exhibited an increase in phenolic compounds, flavonoids, and anthocyanins when exposed to light spectra of 1R:1G:1B, 1R:2G:1B, 1R:3G:1B, and 1R:3B. The highest levels of phenolic compounds were 4.67–5.14 mg GAE/g DW, flavonoids were 1.62–1.96 mg Rutin/g DW, and anthocyanins were 1.20–1.58 μg/g DW. However, the antioxidant capacity of five radish varieties under different light spectra did not show significant differences. Thus, the growth, photosynthesis, and antioxidant capacity depend on the optimal light spectrum for each radish variety.

Keywords: *Raphanus sativus* L.; radish; controlled environment; light spectrum; antioxidant capacity

Citation: Chutimanukul, P.; Piew-ondee, P.; Dangsamer, T.; Thongtip, A.; Janta, S.; Wanichananan, P.; Thepsilvisut, O.; Ehara, H.; Chutimanukul, P. Effects of Light Spectra on Growth, Physiological Responses, and Antioxidant Capacity in Five Radish Varieties in an Indoor Vertical Farming System. *Horticulturae* **2024**, *10*, 1059. https://doi.org/10.3390/horticulturae10101059

Academic Editors: Changxia Li and Yue Wu

Received: 26 August 2024
Revised: 1 October 2024
Accepted: 2 October 2024
Published: 3 October 2024

1. Introduction

Radish (*Raphanus sativus* L.) is a small, erect plant classified in the *Brassicaceae* family [1]. It is highly valued and widely consumed worldwide [2], particularly in Asia, where it is a crucial crop in countries like China, Japan, and Korea. Various radish varieties display unique traits, such as differences in root color, shape, and usage for nutrient storage or utilization [3]. Radish cultivation constitutes about 2% of the world's vegetable production, yielding 7 million tons annually [4]. It is commonly consumed fresh, incorporated into dishes, or juiced [5]. Radishes are rich in nutrients, vitamins, and various minerals. Additionally, they contain bioactive compounds such as flavonoids, phenolic acids, anthocyanins [6], and antioxidants [7]. Radishes are known for their high nutritional value. They are considered an option in treating various diseases, including heart disease, cancer, diabetes, and respiratory diseases, helping to reduce blood fat and high blood pressure [8,9]. Consuming vegetables rich in antioxidants can reduce the risk of human diseases by protecting or mitigating damage to cells caused by free radicals [10]. The primary role of antioxidants is to prevent the formation of free radicals [11]. Antioxidants effectively inhibit and treat chronic diseases, blocking or slowing down the reaction of biomolecules with free radicals by transferring electrons to neutralize free radicals and inhibiting oxidation [12]. Their mechanism is based on the capture or neutralization of free radicals, thereby reducing the oxidative damage caused by free radicals to cells and molecules [13–15]. Thus, antioxidant compounds found in natural plant sources can be used as dietary supplements and play a role in disease prevention [16].

Cultivating plants under an indoor vertical farming system using plant factory technology involves applying various technologies from different disciplines to cultivate plants in a controlled environment. This controlled environment regulates factors such as light spectrum, intensity, duration, temperature, humidity, mineral content, carbon dioxide levels, and even artificial light sources, such as light-emitting diodes (LEDs) or fluorescent lamps. The selection of light color and wavelength is crucial as it significantly influences plant growth, photosynthesis, and the production of secondary metabolites [17,18]. Plant factory technology plays a crucial role in food and pharmaceutical production, focusing on increasing quantity, improving quality, and efficiently utilizing resources in an environmentally friendly manner [19,20]. In addition, in plant factories, light is a critical factor for plants because they need it for photosynthesis to support growth and development at various stages. It also stimulates the production of secondary metabolites. It is essential to consider the type of plant being cultivated since different plant species respond differently to light spectrum and wavelength [21]. Cultivating plants in closed systems relies on artificial light sources, particularly light-emitting diodes (LEDs), known for their efficient light emission. LEDs can effectively replace solar energy and promote the growth of plants [22]. Furthermore, LED light bulbs consume less energy and exhibit durability. They allow control over light intensity and wavelength [23]. In general, when plants undergo an increase in photosynthesis, it leads to a higher accumulation of primary metabolites (carbohydrates and amino acids). These substances serve as precursors in producing secondary metabolites with antioxidant properties [24]. Studies have investigated using red and blue LED bulbs, which can stimulate the production of free radicals in plants [25,26]. Therefore, this research aims to study the effects of artificial light on the growth, physiological responses, and secondary metabolite production of five varieties of radish under controlled conditions in a plant factory.

2. Materials and Methods

2.1. Plant Material and Growth Condition

Five varieties of radish seeds were selected from commercial sources, including cherry belle (CB), black Spanish (BS), hailstone white (HW), Malaga violet (MV), and sparkler white tip (SW). All seeds were sown onto germination sponges and germinated under controlled environmental conditions. These conditions included a photosynthetic photon flux density (PPFD) of 150 μmol m^{-2} s^{-1}, a 16 hd^{-1} photoperiod, an air temperature of 25 ± 1.5 °C, and a relative humidity of 70 ± 5%. After 7 days, the seedlings with true leaves and roots were transplanted into the DFT (deep flow technique) system. This transplanting process involved using a modified Enshi nutrient solution with an electrical conductivity (EC) of 1.3 mS/cm and a pH of 6.5. The plants were then provided with a PPFD of 150 μmol m^{-2} s^{-1} from LEDs for a 16 hd^{-1} photoperiod, a temperature of 25 ± 1 °C, a relative humidity of 70 ± 1%, and a CO_2 concentration of 1000 ± 100 ppm, conducted in an indoor vertical farming system at the National Science and Technology Development Agency (NSTDA), Thailand.

2.2. Spectral Light Conditions

In the experiment, the light spectrum was modified based on the method of Chutimanukul et al. [27] using LEDs (AGRI-OPTECH Co., Ltd., Taiwan) that emit red light (R) (λ = 600–700 nm, peak at 660 nm), blue light (B) (λ = 400–500 nm, peak at 450 nm), and green light (G) (λ = 500–600 nm, peak at 525 nm). Then, five varieties of radish were transplanted and subjected to different spectral light treatments in five different ratios of R, G, and B, including (L1) 1R:1G:1B = 33.33:33.34:33.33, (L2) 1R:2G:1B = 25:50:25, (L3) 1R:3G:1B = 20:60:20, (L4) 1R:0G:3B = 25:0:75, and (L5) 3R:0G:1B = 75:0:25. The LEDs were installed horizontally at 40 cm above the cultivation bench. R, B, and G photosynthetic photon flux density (PPFD) was measured using a spectrometer (LI-180, LI-COR Inc., Lincoln, NE, USA). The spectral compositions of the light treatments were recorded using a light meter (Sekonic C-7000, Sekonic, Tokyo, Japan), with a PPFD maintained at 150 μmol m^{-2} s^{-1} for all treatments (Figure 1).

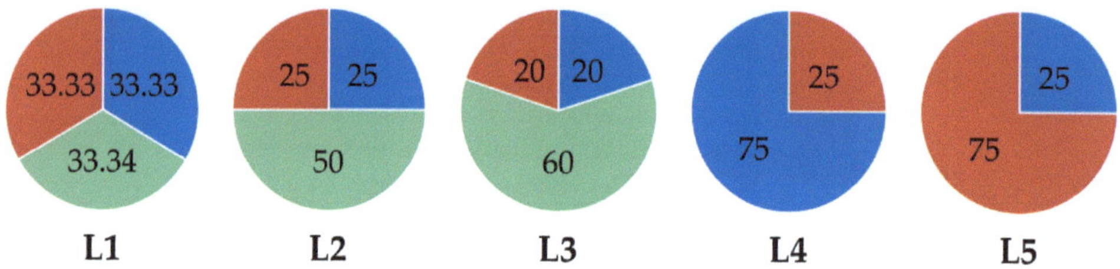

Figure 1. Five different light spectrum treatments were applied to the five radish cultivars.

2.3. Plant Growth Measurements

A total of 28 days after transplanting (DAT) was used for measuring the growth parameters. A ruler and digital caliper were used to measure the root diameter (cm) and root length (cm) (excluding enlarged hypocotyl). After a 72 h drying period at 50 °C in an oven, the fresh weight and dry weight of the roots were measured. Plant samples were collected from each treatment with 4 replications and 3 plants per replication.

2.4. Physiological Measurements

The physiological parameters of radishes were measured 28 days after transplanting, including the photosynthetic rate (*A*), stomatal conductance (g_{sw}), transpiration rate (*E*), and intercellular CO_2 concentration (*Ci*), using the Portable Photosynthesis System (LI-6800, LICOR Inc., Lincoln, NE, USA) with a standard chamber of 3 cm^2 round aperture. The chamber conditions were as follows: 500 mmol m^{-2} s^{-1} of airflow per unit leaf area, a leaf temperature of 25 °C, 70% relative humidity (RH), and a CO_2 concentration of 1000 μmol mol^{-1}. Light intensity measurements were recorded at 150 μmol m^{-2} s^{-1} of Photosynthetic Photon Flux Density (PPFD) on a fully expanded leaf of the plant.

2.5. Secondary Metabolite Quantification

2.5.1. Sample Extraction

After harvest at 28 DAT, all plant samples were dried in an oven at 50 °C for 72 h and ground with mortar and pestle until a fine powder was formed. Plant extraction followed a modified method from Chutimanukul et al. [28] A 10 mg sample of the powdered plant material was mixed with 5 mL of methanol containing 1% HCl in a 15 mL tube. The mixture was vortexed to ensure homogenization and then incubated at room temperature for 3 h. After incubation, the solution was centrifuged at 8500 rpm for 8 min using an Eppendorf Centrifuge 5810R with rotor F-34 6–38 (6 × 125 g). The supernatant solution was then transferred to a 2 mL microcentrifuge tube to analyze phenolics, total flavonoids, anthocyanin, and DPPH radical scavenging activity.

2.5.2. Determination of Total Phenolic Compounds (TPCs)

TPC was determined by a modified Folin-Ciocalteu assay using the modified methods of Chutimanukul et al. [28]. The 200 μL extracted solution was mixed with 200 μL of 1 N Folin-Ciocalteu reagent. After vortexing, it was centrifuged at 10,000 rpm for 2 min and then incubated for 15 min. Following this, 600 µL of 7.5% sodium carbonate (Na_2CO_3) was added, vortexed again, and centrifuged at 10,000 rpm for 2 min. After the 1 h incubation, the absorbance of the mixed solution was measured at 730 nm using a microplate reader (Multiskan Sky, Thermo Scientific, Waltham, MA, USA). TPC was calculated using standard gallic acid with concentrations ranging from 0 to 250 µg/mL (250, 125, 62.5, 31.25, 15.63, 7.8, 3.9, 1.9, and 0.97 µg/mL), prepared from a gallic acid solution dissolved in water to establish calibration curves for TPC concentration calculation. The result was presented as milligrams of gallic acid equivalent (mg of GAE) per gram of sample DW.

2.5.3. Determination of Flavonoid Content

The quantification of flavonoid content was modified from the method of Chutimanukul et al. [28], using a mix of 350 µL of the extracted solution with 75 µL of 5% sodium nitrite ($NaNO_2$) in a 1.5 mL microcentrifuge tube, then centrifuging at 10,000 rpm for 2 min. After incubating at 25 °C for 5 min, 75 µL of 10% aluminum chloride ($AlCl_3{\cdot}6H_2O$) was added and then centrifuged at 10,000 rpm for 2 min. The mixture was then incubated at 25 °C for 5 min. Subsequently, 500 µL of 1 M sodium hydroxide (NaOH) was added and vortexed to homogenize the solution before being centrifuged at 10,000 rpm for 2 min. After 15 min, the absorbance of the mixed solution was measured at 515 nm using a microplate reader (Multiskan Sky, Thermo Scientific, Waltham, MA, USA). Total flavonoid content was calculated from a standard curve of rutin solution dissolved in dimethyl sulfoxide (DMSO). The result was presented as milligrams of rutin equivalents per gram of DW.

2.5.4. Determination of Anthocyanin Content

The method for determining anthocyanin content was adapted from the description by Chutimanukul et al. [28]. The extracted solution with 500 µL was mixed with 400 µL of deionized water and 400 µL of chloroform, followed by centrifugation at 10,000 rpm for 5 min at 25 °C. Then, the supernatant solution was separated, and the absorbance was measured at 515 and 657 nm using a microplate reader (Multiskan Sky, Thermo Scientific, Waltham, MA, USA). The anthocyanin content was calculated using the formula A530 − (0.33 × A657).

2.5.5. DPPH Radical Scavenging Activity

Radishes' free radical scavenging activity was examined using a slightly modified method by Chutimanukul et al. [28], using 2,2-diphenyl-1-picrylhydrazyl (DPPH) as a free radical. A volume of 100 µL of the extracted solution was pipetted into a 1.5 mL microcentrifuge tube and then supplemented with 900 µL of 0.1 mM DPPH. The mixture was vortexed to homogenize the solution, followed by centrifugation at 10,000 rpm for 2 min. After being kept in the darkness at 25 °C for 3 h (during which the solution changed from purple to yellow), the absorbance was measured at 515 nm using a microplate reader (Multiskan Sky, Thermo Scientific, Waltham, MA, USA). Subsequently, the DPPH scavenging activity was calculated using the following equation:

$$\%\text{DPPH inhibition} = \frac{\text{absorbance of control} - \text{absorbance of sample}}{\text{absorbance of control}} \times 100$$

2.6. Statistical Analysis

The experimental data were evaluated by one-way analysis of variance with Duncan's multiple range test (DMRT) at a significance tested level $p < 0.05$. The statistical analyses were conducted using IBM SPSS Statistics 21 (IBM Corporation; Armonk, NY, USA).

3. Results

3.1. Biomass Accumulation and Plant Growth

The development of radish roots in five cultivars under different light spectra revealed significant differences in root diameter, shown in Figures 2 and 3. CB, HW, MV, and SW under 3R:1B light spectra had the highest root diameters, measuring 4.63, 4.29, 3.88, and

3.72 cm, respectively. Conversely, the 1R:2G:1B spectrum resulted in BS having the smallest root diameter at 1.11 cm. Additionally, under the 3R:1B spectrum, SW had the longest root length at 5.85 cm, while HW had the shortest at 2.73 cm under 1R:1G:1B. Differences in light spectra did not significantly affect plant root length in CB, BS, and MV. Regarding weight, significant differences were observed in both fresh weights and dry weights under the 3R:1B spectrum. SW, CB, MV, and HW had the highest fresh root weights at 89.37, 87.53, 83.37, and 77.7 g, respectively, while MV, HW, and SW had the highest dry weights at 4.64, 4.03, and 3.65 g, respectively. Conversely, BS had the lowest fresh and dry weights at 5.73 and 0.47 g under the 1R:2G:1B spectrum. Significant differences were not found in the root dry weight of CB.

Figure 2. The phenotype of cherry belle (**A**), black Spanish (**B**), hailstone white (**C**), Malaga violet (**D**), and sparkler white tip (**E**) under five light treatments (1R:1G:1B, 1R:2G:1B, 1R:3G:1B, 1R:3B, and 3R:1B). R, G, and B represent the red, green, and blue light spectra, respectively, and the number before the letter stands for the ratio of each light spectrum.

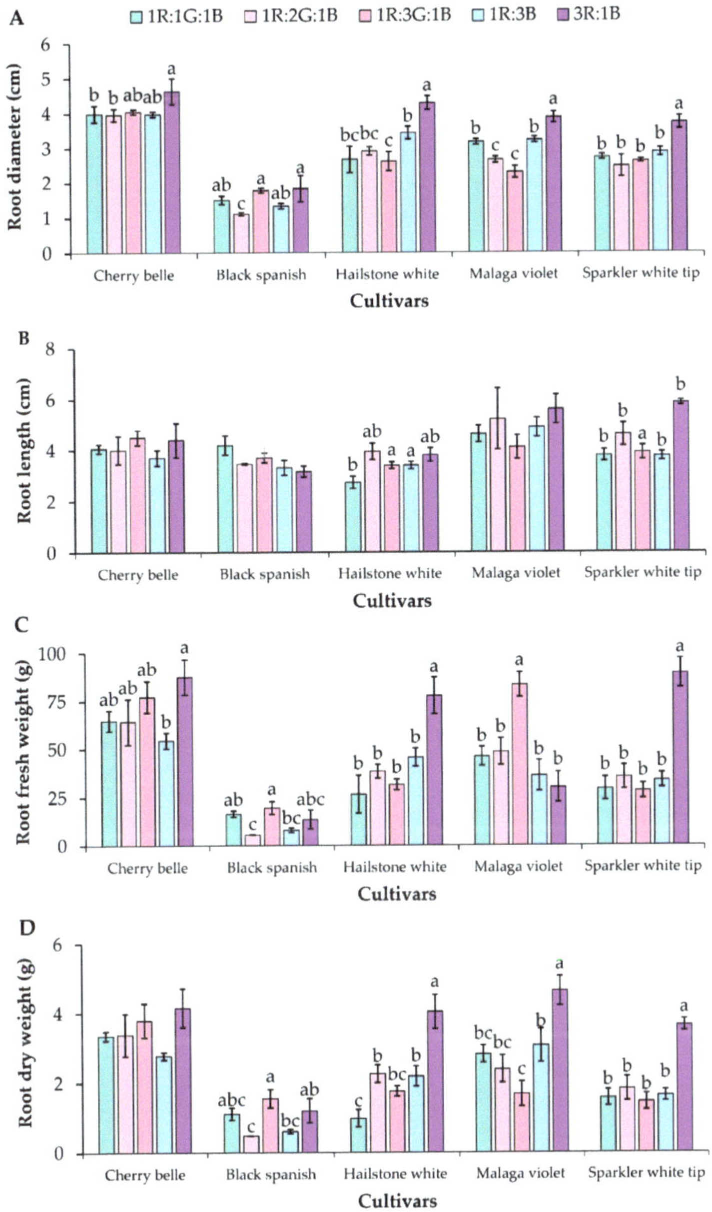

Figure 3. Biomass and plant growth of five radish varieties on different light spectrum treatments on root diameter (**A**), root length (**B**), root fresh weight (**C**), and root dry weight (**D**). R, G, and B represent the red, green, and blue light spectra, respectively, and the number before the letters stands for the ratio of each light spectrum. Values are means with standard deviations ($n = 4$). The letters above the bars indicate statistically significant differences by Duncan's multiple range tests ($p < 0.05$). The absence of letters indicates no significant difference.

3.2. Physiological Responses

The experiment revealed that different light spectra significantly affected the physiological response of radish, particularly in terms of the photosynthetic rate (*A*) (Figure 4A). For the CB variety, significant differences were observed, where the 3R:1B, 1R:2G:1B, and 1R:3B light spectra promoted the highest photosynthetic rates at 3.99, 3.41, and 3.31 μmol m^{-2} s^{-1}, respectively. The highest photosynthetic rates were observed for the BS variety with the 3R:1B and 1R:3B light spectra at 3.56 and 3.44 μmol m^{-2} s^{-1}, respectively. However, no statistical differences were observed in the HW, MV, and SW cultivars.

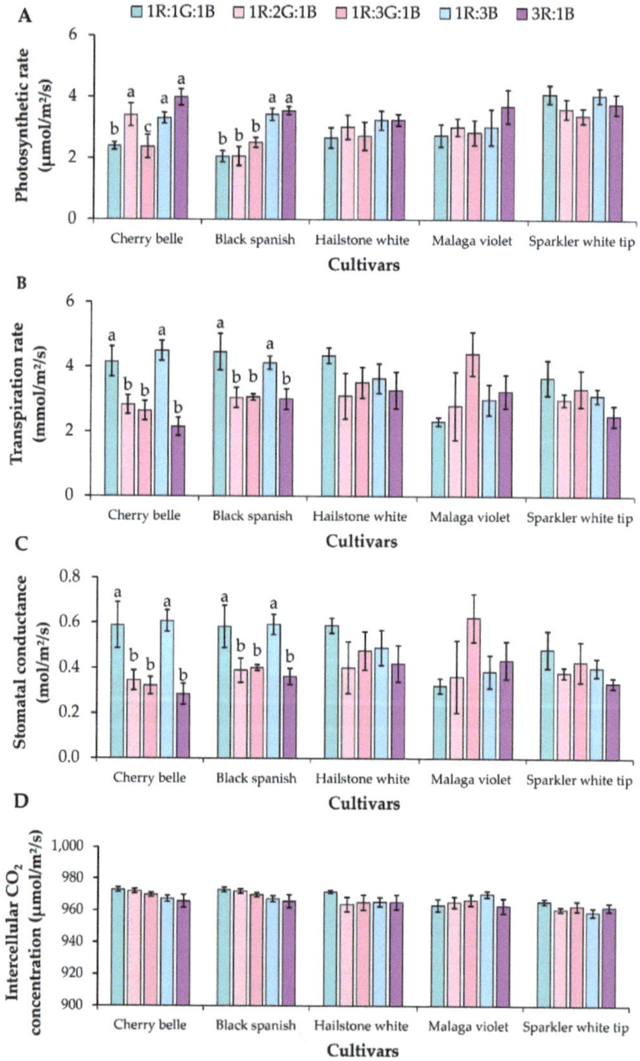

Figure 4. Physiological responses of five radish varieties on different light spectrum treatments on photosynthetic rate (**A**), transpiration rate (**B**), stomatal conductance (**C**), and intercellular CO_2 concentration (**D**). R, G, and B represent the red, green, and blue light spectra, respectively, and the number before the letters stands for the ratio of each light spectrum. Values are means with standard deviations ($n = 4$). The letters above the bars indicate statistically significant differences by Duncan's multiple range tests ($p < 0.05$). The absence of letters indicates no significant difference.

The transpiration rate (E) was highest in the CB variety under the 1R:1G:1B and 3R:1B light spectra, resulting in values of 4.15 and 4.49 mol m^{-2} s^{-1}, respectively. Similarly, in the BS variety, under the 1R:1G:1B and 3R:1B light spectra, the transpiration rates were 4.35 and 3.64 mol m^{-2} s^{-1}, respectively (Figure 4B). Stomatal conductance (g_{sw}) showed significant increases in both CB and BS radish varieties under the 3R:1B and 1R:1G:1B light spectra, with values reaching 0.61 and 0.59 mol m^{-2} s^{-1} for CB and 0.59 and 0.58 mol m^{-2} s^{-1} for BS, respectively. However, significant differences were not observed in E and g_{sw} in the HW, MV, and SW varieties (Figure 4C). CO_2 concentrations (Ci) in all radish cultivars ranged from 959.08 to 973.02 μmol m^{-2} s^{-1}, with no statistically significant effect observed from the different light spectra. (Figure 4D).

3.3. Secondary Metabolite Quantification

The study of phenolic compound quantities in radish of five cultivars subjected to different light spectra revealed that radish cultivars CB and SW under light spectra of 1R:1G:1B, 1R:3B, 1R:2G:1B, and 1R:3G:1B exhibited the highest phenolic compound quantities. For cultivar CB, the quantities were 3.52, 3.52, 3.97, and 3.77 mgGAE/g DW, respectively, while for cultivar SW, they were 5.11, 4.67, 4.74, and 5.14 mgGAE/g DW, respectively. Radish cultivar MV under the spectrum of 1R:1G:1B had the highest phenolic compound quantity at 5.22 mgGAE/g DW. Cultivars BS and HW under different light spectra showed no statistically significant difference in phenolic compound quantities (Figure 5A).

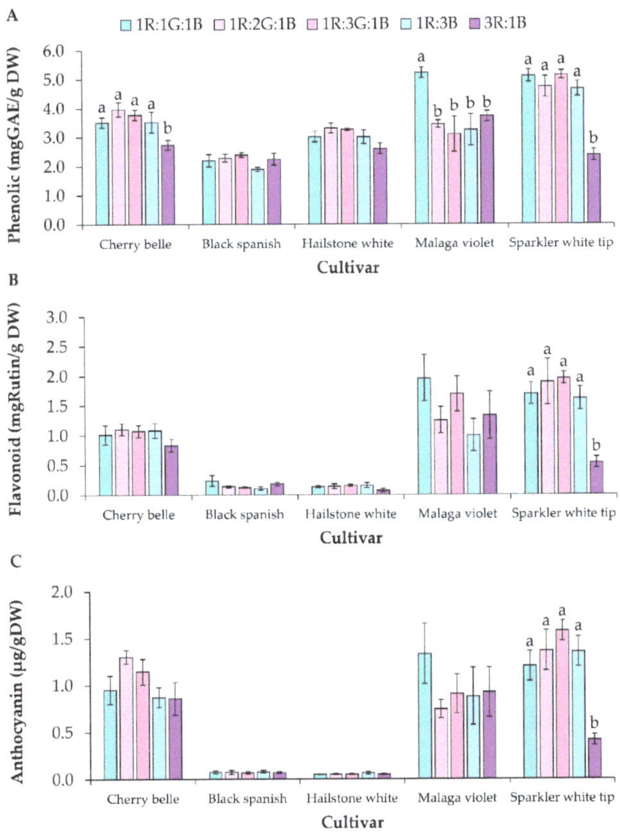

Figure 5. *Cont.*

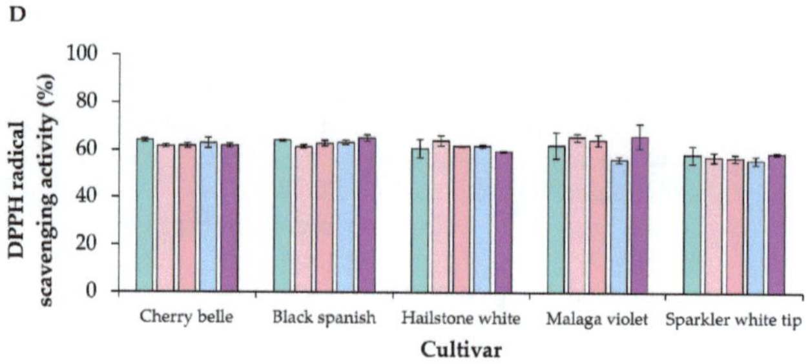

Figure 5. The quantities of secondary metabolites and antioxidant capacity of five radish varieties on different light spectrum treatments on total phenolic (**A**), flavonoid (**B**), anthocyanin (**C**), and DPPH scavenging activity (**D**). R, G, and B represent the red, green, and blue light spectra, respectively, and the number before the letters stands for the ratio of each light spectrum. Values are means with standard deviations ($n = 4$). The letters above the bars indicate statistically significant differences by Duncan's multiple range tests ($p < 0.05$). The absence of letters indicates no significant difference.

The study of flavonoid quantities in radish of five cultivars subjected to different light spectra found that radish cultivar SW under spectra of 1R:1G:1B, 1R:3B, 1R:2G:1B, and 1R:3G:1B exhibited the highest flavonoid quantities. The quantities were 1.70, 1.62, 1.89, and 1.96 mgRutin/g DW, respectively. Conversely, the cultivar under the spectrum of 3R:1B showed the lowest flavonoid quantity at 0.54 mgRutin/g DW. Cultivars CB, BS, HW, and MV under different light spectra showed no statistically significant difference in flavonoid quantities (Figure 5B).

The study of anthocyanin quantities in radish of five cultivars subjected to different light spectra found that radish cultivar SW under spectra of 1R:1G:1B, 1R:3B, 1R:2G:1B, and 1R:3G:1B exhibited the highest anthocyanin quantities. The quantities were 1.20, 1.35, 1.37, and 1.58 µg/gDW, respectively. Conversely, the cultivar under the spectrum of 3R:1B showed the lowest anthocyanin quantity at 0.42 µg/gDW. Cultivars CB, BS, HW, and MV under different light spectra showed no statistically significant difference in anthocyanin quantities (Figure 5C).

The study of DPPH radical scavenging activity in radish of five cultivars subjected to different light spectra found that all five cultivars showed no statistically significant difference in antioxidant capacities (Figure 5D).

4. Discussion

The experimental results showed that the growth response of radishes in all cultivars, including root length, root fresh weight, and root dry weight, tended to increase when exposed to light spectra with high ratios of red light, indicating that different light spectra had a clear effect on plant growth. A study by Yorio et al. [29] investigated the promotion of growth factors in radishes by providing different light spectra. It was found that providing only red light did not promote radish growth as effectively as combining red and blue light. Specifically, providing red light alone resulted in reduced growth compared to providing fluorescent light, and red light combined with blue light. Blue light was found to be essential for plant growth. Therefore, providing a combination of red and blue light promoted radish growth, increasing the dry weight of the roots. In the study of Zha and Liu [30], which demonstrated the efficiency factors of light, including light quality, intensity, and duration on cherry belle radish cultivars, it was found that providing red light alone was a factor that led to the low development of radish roots. Moreover, blue light promoted root. However, providing blue light in high ratios also adversely affected radish growth. Thus, light spectrum ratios need to be managed appropriately. It was found that providing red and blue light in a 2:1 ratio (2R:1B) for 16 hd^{-1} promoted radish growth in terms of root length, diameter, fresh weight, and dry weight compared to growth in stem aspects when compared to providing red and blue light in a 1:1 ratio (1R:1B). Additionally, increasing light intensity led to higher radish growth. This intensity has a relationship with photosynthesis to produce plant energy. Similarly, in a study by Gam et al. [31], which studied light provision from LED light sources, it was found that providing red and blue light with a higher ratio, such as 4:1 (4R:1B), resulted in the highest growth in terms of stem diameter, fresh weight, dry weight, and highest biomass production, and stimulated higher production of phytochemicals in plants compared to other light spectra from LED or fluorescent light sources. A report by Pritchard et al. [32] found that when plants received high levels of carbon dioxide, it stimulated growth in the underground parts, such as roots, more than the above-ground parts, such as stems and leaves. About 50% of the nutrients from photosynthesis are sent to the roots underground for plant nutrition. In addition, light intensity is a factor because light intensity plays an important role in the growth of radishes, especially in root development, affecting plant photosynthesis and growth, and the accumulation of nutrients in radish roots. If plants receive light with high intensity, they can photosynthesize more, but if the light intensity decreases, growth will decrease accordingly [30]. Blue light is crucial in promoting photosynthesis, chlorophyll production, and regulating plant growth. Cryptochromes (CRYs) help control plant developmental processes in response to blue light [33], with optimal growth occurring at appropriate blue light proportions. It enhances photosynthetic efficiency and nutrient absorption [34].

Red and blue light spectra significantly enhance the efficiency of photosynthesis in plants, as they fall within the visible light range that plants utilize for this process. Samuoliene et al. [35] reported that red and blue light spectra affect the photosynthetic efficiency of radish. When exposed solely to red light, radishes exhibited a reduced photosynthetic rate due to the accumulation of pigments used in photosynthesis and low carbohydrate levels in the leaves. Red light alone was insufficient for optimal photosynthesis, leading to a lower photosynthetic rate. However, when combined with blue light, radishes demonstrated a higher photosynthetic rate. This finding aligns with research by Liu and Lersel [36], which examined the effects of red, green, and blue light spectra on plant photosynthesis. Their study revealed that red light had the highest photosynthetic rate compared to blue and green light. Different wavelengths affect light absorption and energy utilization in photosynthesis. Excessive wavelengths that plants do not use effectively, such as yellow or green, can reduce photosynthetic efficiency. Choosing light spectra close to plants' wavelengths in photosynthesis enhances this process. Red light plays a crucial role in photosynthesis as one of the primary wavelengths plants use for energy production. In the photosynthetic process, chlorophyll absorbs red light, particularly in Photosystem I, initiating reactions that produce ATP and NADPH necessary for carbohydrate synthesis in the Calvin cycle [37]. Additionally, red light affects stomatal opening, facilitating gas exchange and the absorption of carbon dioxide essential for photosynthesis [36]. Red and blue light are absorbed more efficiently by photosynthetic pigments compared to green light. This means that red and blue light primarily excite chloroplasts in the upper layers of leaves, while green light can penetrate deeper, reaching chloroplasts in lower layers [36].

Stomatal conductance is a measure of the opening and closing of stomata, which correlates with the water potential indicating the plant's water status, and it reflects the width of stomatal pores when they are open per unit of time. If the stomata are widely open, it indicates a high value. It is measured in mmol m^{-2} s^{-1} and indicates efficiency and gas exchange in plants, affecting photosynthesis and transpiration [38]. Therefore, if a plant can open its stomata widely and keep them open for longer, it will enhance photosynthetic efficiency because plants can exchange carbon dioxide from the atmosphere more effectively. However, a decrease in stomatal conductance, resulting from narrower stomatal openings, can help prevent water loss in plants [37]. Stomatal opening is primarily influenced by blue light and red light [39]. Blue light signals stomatal opening, while red light provides the energy necessary for photosynthesis and stomatal function. In many vascular plants, the presence of BL is necessary for stomatal opening, especially when combined with red light. Weak blue light alone does not significantly induce stomatal opening; however, when it is superimposed on strong red light, it can lead to a rapid and substantial opening of the stomata [39].

When considering the quantity of phenolic compounds in radish cultivars CB and SW under the light spectra of 1R:1G:1B, 1R:3B, 1R:2G:1B, and 1R:3G:1B, it resulted in the highest quantity of phenolic compounds. Under the same light spectra, the radish cultivar SW exhibited the highest quantity of flavonoids. Radish cultivar SW when subjected to light spectra of 1R:1G:1B, 1R:3B, 1R:2G:1B, and 1R:3G:1B resulted in the highest quantity of anthocyanins. In general, when plants undergo photosynthesis with higher light intensity, it leads to an accumulation of primary metabolites (carbohydrates and amino acids), which are precursors in producing antioxidant compounds [24]. Additionally, both red and blue light spectra affect the production of antioxidant compounds [40]. It is observed that blue and red light promoted antioxidant activities and secondary compounds, such as phenolic compounds, pinitol accumulation, and betacyanin [41]. The quantity of phenolic compounds serves as an important index of seedling quality, and the accumulation of phenolic compounds can be stimulated by cultivating under different wavelengths of LED light [42], especially blue light spectra, which induce phenolic compound synthesis, as light stimulates certain key enzymes that act as intermediaries in phenolic compound synthesis [43]. Batista et al. [44] reported that blue light plays a role in regulating microtubule synthesis-related genes and secondary metabolite synthesis-related genes, and that red light and green light are also closely associated with the expression of defense-related genes. Moreover, in lettuce, blue light has been demonstrated to enhance the content of antioxidant polyphenols, anthocyanins, and carotenoids [45,46]. Jang et al. [47] reported that camellia callus produces larger amounts of phenolics when exposed to RB and RGB because of defense-related mechanisms. Kondo [48] states that blue light plays a crucial role in stimulating the accumulation of anthocyanins. Stutte et al. [49] reported that blue light is essential for producing purple pigments associated with anthocyanin synthesis. It is known that green light, the same as blue and UV, is also absorbed by the cryptochromes, although the specific photoreceptor of such light remains to be identified in higher plants. The green light can be more efficiently absorbed by the outer leaves of the canopy and stimulate photosynthesis at lower leaf levels [42,50,51]. Few accessible reports showed that green light used as monochromatic or as part of a broader light combination frequently had no effect or reduced accumulation of phenolic compounds and can reverse the positive impact of monochromatic blue light [42,50]. The green light had positive results in only a few cases. For example, it enhanced the total phenolic and total flavonoid production of *Prunella vulgaris* callus cultures [52]. Tang et al. [53] reported that blue and red in the RBG light and the RBW light increased the antioxidant activities of three vegetable species (lettuce, radish, and tomato) compared to the W light. However, contrasting examples in the literature indicate that the amount of secondary metabolism under artificial lighting is light and species dependent. For example, no essential difference was found in pigment content in *Dieffenbachia amoena, Ficus elastica*, and Boston lettuce [54,55]. Consequently, the production of phenolic compounds depends concurrently on the light environment and the physiological and biochemical factors. Additionally, phenolic compounds and flavonoids are phytochemicals with antioxidant properties found in natural plants and are the primary antioxidants involved in inhibiting oxidation processes in plants [56,57]. However, it is possible that in this experiment, phenolic compounds and flavonoids did not affect the antioxidant activity in radish. This is because non-enzymatic antioxidants can be classified into several groups, including ascorbic acid, phenolic compounds such as phenols, phenolics, flavonoids, tannins, glycosides, alkaloids, and lignin [58], which may include other substances that influence the antioxidant activity in radish. Therefore, further in-depth studies are required.

5. Conclusions

This study demonstrated the effects of light spectra on growth, secondary metabolites, and antioxidant capacity. It was found that radishes of all five varieties exposed to the 3R:1B light spectrum showed tendencies for increased growth in terms of diameter, fresh weight, and dry weight. The mixed red and blue light spectrum (3R:1B) effectively promoted plant growth. Similarly, physiological responses in the CB and BS varieties indicated a higher photosynthetic rate after exposure to the mixed red and blue light (1R:3B, 3R:1B). Additionally, adding green light to the red and blue light spectrum also enhanced the photosynthetic rate. For the SW variety, the quantity of phenolic compounds, flavonoids, and anthocyanins tended to increase when exposed to the 1R:1G:1B, 1R:2G:1B, 1R:3G:1B, and 1R:3B light spectra. The antioxidant capacity of five radish varieties under different light spectra did not show significant differences. Thus, the growth, photosynthesis, and antioxidant capacity depend on the optimal light spectrum for each radish variety. However, this system faces limitations in terms of high energy costs. Nevertheless, the demand for vertical farming systems is expected to increase in the future, particularly for herbal plants and functional foods, contributing to food security and sustainability.

Author Contributions: Conceptualization, P.C. (Panita Chutimanuku), A.T. and P.C. (Preuk Chutimanukul); methodology, P.C. (Panita Chutimanuku), A.T. and P.C. (Preuk Chutimanukul); validation, P.C. (Panita Chutimanuku), P.P.-o., T.D., O.T. and P.C. (Preuk Chutimanukul); formal analysis, S.J. and P.C. (Preuk Chutimanukul); investigation, P.P.-o., T.D. and P.C. (Preuk Chutimanukul); resources, P.C. (Panita Chutimanuku), S.J., P.W. and P.C. (Preuk Chutimanukul); data curation, P.C. (Panita Chutimanuku), H.E. and P.C. (Preuk Chutimanukul); writing—original draft preparation, P.C. (Panita Chutimanuku), P.P.-o., T.D. and P.C. (Preuk Chutimanukul); writing—review and editing, P.C. (Panita Chutimanuku) and P.C. (Preuk Chutimanukul); supervision, P.C. (Preuk Chutimanukul); funding acquisition, P.C. (Preuk Chutimanukul) All authors have read and agreed to the published version of the manuscript.

Funding: This study was supported by Thammasat University Research Fund (contract No. TUFT 8/2567) and the National Center for Genetic Engineering and Biotechnology (BIOTEC), Thailand (basic research fund: fiscal year 2023 with contract No. 4709540).

Data Availability Statement: The datasets used and/or analyzed during the current study are included in the article; further inquiries can be directed to the corresponding author.

Acknowledgments: The authors would like to thank Thammasat University Center of Excellence in Agriculture Innovation Center through Supply Chain and Value Chain, the Faculty of Science and Technology, Thammasat University, the National Center for Genetic Engineering and Biotechnology (BIOTEC), and the National Science and Technology Development Agency for providing technical supports and instrument.

Conflicts of Interest: The authors declare no conflicts of interest.

References

1. Nishio, T. Economic and Academic Importance of Radish. In *The Radish Genome*; Springer: Berlin/Heidelberg, Germany, 2017; pp. 1–10.
2. Kitashiba, H.; Li, F.; Hirakawa, H.; Kawanabe, T.; Zou, Z.; Hasegawa, Y.; Tonosaki, K.; Shirasawa, S.; Fukushima, A.; Yokoi, S.; et al. Draft sequences of the radish (*Raphanus sativus* L.) genome. *DNA Res.* **2014**, *21*, 481–490. [CrossRef] [PubMed]
3. Lu, G.; Cao, J.; Yu, X.; Xiang, X.; Chen, H. Mapping QTLs for root morphological traits in *Brassica rapa* L. based on AFLP and RAPD markers. *J. Appl. Genet.* **2008**, *49*, 23–31. [CrossRef] [PubMed]
4. Kopta, T.; Pokluda, R. Yields, quality and nutritional parameters of radish (*Raphanus sativus*) cultivars when grown organically in the Czech Republic. *Hortic. Sci.* **2013**, *40*, 16–21. [CrossRef]
5. Banihani, S.A. Radish (*Raphanus sativus*) and Diabetes. *Nutrients* **2017**, *9*, 1014. [CrossRef]
6. Papetti, A.; Milanese, C.; Zanchi, C.; Gazzani, G. HPLC–DAD–ESI/MSn characterization of environmentally friendly polyphenolic extract from *Raphanus sativus* L. var. "Cherry Belle" skin and stability of its red components. *Food Res. Int.* **2014**, *65*, 238–246. [CrossRef]
7. Rani, A.; Arfat, Y.; Aziz, R.S.; Ali, L.; Ahmed, H.; Asim, S.; Rashid, M.; Hocart, C.H. Enzymatically assisted extraction of antioxidant and anti-mutagenic compounds from radish (*Raphanus sativus*). *Environ. Technol. Innov.* **2021**, *23*, 101620. [CrossRef]
8. Rakhmawati, R. Potency of Lobak Leaves (*Raphanus sativus* L. var. hortensis Back) as Anticancer and Antimicrobial Candidates. *Biodiversitas J. Biol. Divers.* **2009**, *10*, 158–162. [CrossRef]
9. Curtis, I. The noble radish: Past, present and future. *Trends Plant Sci.* **2003**, *8*, 305–307. [CrossRef]

10. Kumar, S. The importance of antioxidant and their role in pharmaceutical science—A review. *Asian J. Res. Chem. Pharm. Sci.* **2014**, *1*, 27–44.
11. Hu, Y.; Zheng, Z.; Zhai, D.; Liang, P.; Wang, Z.; Jiang, C.; Guo, Y.; Chen, H.; Shen, C.; Wu, Y.; et al. A Mini-Review: Exploring the Application Prospects of the Three Major Rules in the Field of Antioxidants. *J. Mol. Struct.* **2024**, *1304*, 137746. [CrossRef]
12. Losada-Barreiro, S.; Sezgin-Bayindir, Z.; Paiva-Martins, F.; Bravo-Díaz, C. Biochemistry of Antioxidants: Mechanisms and Pharmaceutical Applications. *Biomedicines* **2022**, *10*, 3051. [CrossRef] [PubMed]
13. Kurutas, E. The importance of antioxidants which play the role in cellular response against oxidative/nitrosative stress: Current state. *Nutr. J.* **2016**, *15*, 71. [CrossRef] [PubMed]
14. Niki, E. Antioxidants: Basic Principles, Emerging Concepts, and Problems. *Biomed. J.* **2014**, *37*, 106–111. [CrossRef] [PubMed]
15. Nimse, S.B.; Pal, D. Free radicals, natural antioxidants, and their reaction mechanisms. *RSC Adv.* **2015**, *5*, 27986–28006. [CrossRef]
16. Goyeneche, R.; Roura, S.; Ponce, A.G.; Vega-Galvez, A.; Quispe, I.; Uribe, E.; Di Scala, K. Chemical characterization and antioxidant capacity of red radish (*Raphanus sativus* L.) leaves and roots. *J. Funct. Foods* **2015**, *16*, 256–264. [CrossRef]
17. Kozai, T. Propagation, grafting and transplant production in closed systems with artificial lighting for commercialization in Japan. *Propag. Ornam. Plants* **2007**, *7*, 145–149.
18. Kozai, T.; Kubota, C.; Chun, C.; Afreen, F.; Ohyama, K. Necessity and Concept of the Closed Transplant Production System. In *Transplant Production in the 21st Century: Proceedings of the International Symposium on Transplant Production in Closed System for Solving the Global Issues on Environmental Conservation, Food, Resources and Energy*; Springer: Dordrecht, The Netherlands, 2000.
19. Morimoto, T.; Torii, T.; Hashimoto, Y. Optimal control of physiological processes of plants in a green plant factory. *Control Eng. Pract.* **1995**, *3*, 505–511. [CrossRef]
20. Nicole, C.C.s.; Charalambous, F.; Martinakos, S.; Voort, S.; Li, Z.; Verhoog, M.; Krijn, M. Lettuce growth and quality optimization in a plant factory. *Acta Hortic.* **2016**, *1134*, 231–238. [CrossRef]
21. Kang, J.; Sugumaran, K.; Atulba, S.L.; Jeong, B.R.; Hwang, S. Light intensity and photoperiod influence the growth and development of hydroponically grown leaf lettuce in a closed-type plant factory system. *Horticulture* **2013**, *54*, 501–509. [CrossRef]
22. Shimizu, H. Light Environment Optimization for Lettuce Growth in Plant Factory. *IFAC Proc. Vol.* **2011**, *18*, 605–609. [CrossRef]
23. Kozai, T. Resource use efficiency of closed plant production system with artificial light: Concept, estimation and application to plant factory. *Proc. Jpn. Acad. Ser. B Phys. Biol. Sci.* **2013**, *89*, 447–461. [CrossRef] [PubMed]
24. Bantis, F.; Ouzounis, T.; Radoglou, K. Artificial LED lighting enhances growth characteristics and total phenolic content of Ocimum basilicum, but variably affects transplant success. *Sci. Hortic.* **2016**, *198*, 277–283. [CrossRef]
25. Shao, L.; Fu, Y.; Liu, H.; Liu, H. Changes of the antioxidant capacity in Gynura bicolor DC under different light sources. *Sci. Hortic.* **2015**, *184*, 40–45. [CrossRef]
26. Shiga, T.; Shoji, K.; Shimada, H.; Hashida, S.-n.; Goto, F.; Yoshihara, T. Effect of light quality on rosmarinic acid content and antioxidant activity of sweet basil, *Ocimum basilicum* L. *Plant Biotechnol.* **2009**, *26*, 255–259. [CrossRef]
27. Chutimanukul, P.; Wanichananan, P.; Janta, S.; Toojinda, T.; Darwell, C.T.; Mosaleeyanon, K. The influence of different light spectra on physiological responses, antioxidant capacity and chemical compositions in two holy basil cultivars. *Sci. Rep.* **2022**, *12*, 588. [CrossRef]
28. Chutimanukul, P.; Jindamol, H.; Thongtip, A.; Korinsak, S.; Romyanon, K.; Toojinda, T.; Darwell, C.T.; Wanichananan, P.; Panya, A.; Kaewsri, W.; et al. Physiological responses and variation in secondary metabolite content among Thai holy basil cultivars (*Ocimum tenuiflorum* L.) grown under controlled environmental conditions in a plant factory. *Front. Plant Sci.* **2022**, *13*, 1008917. [CrossRef]
29. Yorio, N.C.; Goins, G.D.; Kagie, H.R.; Wheeler, R.M.; Sager, J.C. Improving spinach, radish, and lettuce growth under red light-emitting diodes (LEDs) with blue light supplementation. *HortScience* **2001**, *36*, 380–383. [CrossRef]
30. Zha, L.; Liu, W. Effects of light quality, light intensity, and photoperiod on growth and yield of cherry radish grown under red plus blue LEDs. *Hortic. Environ. Biotechnol.* **2018**, *59*, 511–518. [CrossRef]
31. Gam, D.T.; Khoi, P.H.; Ngoc, P.B.; Linh, L.K.; Hung, N.K.; Anh, P.T.L.; Thu, N.T.; Hien, N.T.T.; Khanh, T.D.; Ha, C.H. LED Lights Promote Growth and Flavonoid Accumulation of Anoectochilus roxburghii and Are Linked to the Enhanced Expression of Several Related Genes. *Plants* **2020**, *9*, 1344. [CrossRef]
32. Pritchard, S.G.; Prior, S.A.; Rogers, H.H.; Davis, M.A.; Runion, G.B.; Popham, T.W. Effects of elevated atmospheric CO_2 on root dynamics and productivity of sorghum grown under conventional and conservation agricultural management practices. *Agric. Ecosyst. Environ.* **2006**, *113*, 175–183. [CrossRef]
33. Yang, Z.; Liu, B.; Su, J.; Liao, J.; Lin, C.; Oka, Y. Cryptochromes Orchestrate Transcription Regulation of Diverse Blue Light Responses in Plants. *Photochem. Photobiol.* **2017**, *93*, 112–127. [CrossRef] [PubMed]
34. Liang, Y.; Kang, C.; Kaiser, E.; Kuang, Y.; Yang, Q.; Li, T. Red/blue light ratios induce morphology and physiology alterations differently in cucumber and tomato. *Sci. Hortic.* **2021**, *281*, 109995. [CrossRef]
35. Samuoliene, G.; Sirtautas, R.; Brazaitytė, A.; Miliauskienė, J.; Sakalauskienė, S.; Duchovskis, P. The impact of red and blue light-emitting diode illumination on radish physiological indices. *Cent. Eur. J. Biol.* **2011**, *6*, 821–828. [CrossRef]
36. Liu, J.; van Iersel, M.W. Photosynthetic Physiology of Blue, Green, and Red Light: Light Intensity Effects and Underlying Mechanisms. *Front. Plant Sci.* **2021**, *12*, 619987. [CrossRef]
37. Stirbet, A.; Lazár, D.; Guo, Y.; Govindjee, G. Photosynthesis: Basics, history and modelling. *Ann. Bot.* **2019**, *126*, 511–537. [CrossRef]

38. Engineer, C.B.; Hashimoto-Sugimoto, M.; Negi, J.; Israelsson-Nordström, M.; Azoulay-Shemer, T.; Rappel, W.-J.; Iba, K.; Schroeder, J.I. CO_2 Sensing and CO_2 Regulation of Stomatal Conductance: Advances and Open Questions. *Trends Plant Sci.* **2016**, *21*, 16–30. [CrossRef]
39. Doi, M.; Kitagawa, Y.; Shimazaki, K. Stomatal Blue Light Response Is Present in Early Vascular Plants. *Plant Physiol.* **2015**, *169*, 1205–1213. [CrossRef]
40. Lattanzio, V.; Lattanzino, V.M.T.; Cardinali, A. Role of phenolics in the resistance mechanisms of plants against fungal pathogens and insects. *Phytochem. Adv. Res.* **2006**, *661*, 23–67.
41. Weeplian, T.; Yen, T.-B.; Ho, Y.-S. Growth, Development, and Chemical Constituents of Edible Ice Plant (*Mesembryanthemum crystallinum* L.) Produced under Combinations of Light-emitting Diode Lights. *HortScience* **2018**, *53*, 865–874. [CrossRef]
42. Zhang, X.; Bian, Z.; Yuan, X.; Chen, X.; Lu, C. A review on the effects of light-emitting diode (LED) light on the nutrients of sprouts and microgreens. *Trends Food Sci. Technol.* **2020**, *99*, 203–216.
43. Taulavuori, K.; Pyysalo, A.; Taulavuori, E.; Julkunen-Tiitto, R. Responses of phenolic acid and flavonoid synthesis to blue and blue-violet light depends on plant species. *Environ. Exp. Bot.* **2018**, *150*, 183–187. [CrossRef]
44. Batista, D.S.; Felipe, S.H.S.; Silva, T.D.; Castro, K.M.; Mamedes-Rodrigues, T.C.; Miranda, N.A.; Ríos-Ríos, A.M.; Faria, D.V.; Fortini, E.A.; Chagas, K.; et al. Light quality in plant tissue culture: Does it matter? *In Vitro Cell. Dev. Biol. Plant* **2018**, *54*, 195–215.
45. Li, Q.; Kubota, C. Effects of supplemental light quality on growth and phytochemicals of baby leaf lettuce. *Environ. Exp. Bot.* **2009**, *67*, 59–64. [CrossRef]
46. Johkan, M.; Shoji, K.; Goto, F.; Hashida, S.-N.; Yoshihara, T. Blue Light-emitting Diode Light Irradiation of Seedlings Improves Seedling Quality and Growth after Transplanting in Red Leaf Lettuce. *HortScience* **2010**, *45*, 1809–1814. [CrossRef]
47. Jang, E.B.; Ho, T.-T.; Park, S.-Y. Effect of light quality and tissue origin on phenolic compound accumulation and antioxidant activity in Camellia japonica calli. *In Vitro Cell. Dev. Biol. Plant* **2020**, *56*, 567–577. [CrossRef]
48. Kondo, S.; Tomiyama, H.; Rodyoung, A.; Okawa, K.; Ohara, H.; Sugaya, S.; Terahara, N.; Hirai, N. Abscisic acid metabolism and anthocyanin synthesis in grape skin are affected by light emitting diode (LED) irradiation at night. *J. Plant Physiol.* **2014**, *171*, 823–829. [CrossRef]
49. Stutte, G.W. Light-emitting Diodes for Manipulating the Phytochrome Apparatus. *HortScience* **2009**, *44*, 231–234. [CrossRef]
50. Wong, C.E.; Teo, Z.W.N.; Shen, L.; Yu, H. Seeing the lights for leafy greens in indoor vertical farming. *Trends Food Sci. Technol.* **2020**, *106*, 48–63. [CrossRef]
51. Landi, M.; Zivcak, M.; Sytar, O.; Brestic, M.; Allakhverdiev, S.I. Plasticity of photosynthetic processes and the accumulation of secondary metabolites in plants in response to monochromatic light environments: A review. *Biochim. Biophys. Acta Bioenerg.* **2020**, *1861*, 148131. [CrossRef]
52. Fazal, H.; Abbasi, B.H.; Ahmad, N.; Ali, S.S.; Akbar, F.; Kanwal, F. Correlation of different spectral lights with biomass accumulation and production of antioxidant secondary metabolites in callus cultures of medicinally important *Prunella vulgaris* L. *J. Photochem. Photobiol. B Biol.* **2016**, *159*, 1–7. [CrossRef]
53. Tang, Y.; Mao, R.; Guo, S. Effects of LED spectra on growth, gas exchange, antioxidant activity and nutritional quality of vegetable species. *Life Sci. Space Res.* **2020**, *26*, 77–84. [CrossRef]
54. Heo, J.; Lee, Y.; Kim, D.; Chang, Y.; Chun, C. Effects of Supplementary LED Lighting on Growth and Biochemical Parametets in Dieffenbachia amoena 'Camella' and Ficus elastica 'Melany'. *J. Korean Hort. Sci. Tech.* **2010**, *28*, 51–58.
55. Martineau, V.; Lefsrud, M.; Naznin, T.; Kopsell, D. Comparison of Light-emitting Diode and High-pressure Sodium Light Treatments for Hydroponics Growth of Boston Lettuce. *HortScience* **2012**, *47*, 477–482. [CrossRef]
56. Barros, L.; Ferreira, M.-J.; Queirós, B.; Ferreira, I.; Baptista, P. Total phenols, ascorbic acid, B-Carotene and lycopene in Portuguese wild edible mushrooms and their antioxidant activities. *Food Chem.* **2007**, *103*, 413–419. [CrossRef]
57. Murakami, M.; Yamaguchi, T.; Takamura, H.; Matoba, T. Effects of Ascorbic Acid and α-Tocopherol on Antioxidant Activity of Polyphenolic Compounds. *J. Food Sci.* **2006**, *68*, 1622–1625. [CrossRef]
58. Lobo, V.; Patil, A.; Phatak, A.; Chandra, N. Free radicals, antioxidants and functional foods: Impact on human health. *Pharmacogn. Rev.* **2010**, *4*, 118–126. [CrossRef]

horticulturae

MDPI

Article

Molecular Responses to Drought and Waterlogging Stresses of Kiwifruit (*Actinidia chinensis* var. *deliciosa*) Potted Vines

Elena Baldi, Chiara Pastore *, Giacomo Chiarelli, Maurizio Quartieri, Francesco Spinelli and Moreno Toselli

Department of Agricultural and Food Sciences, University of Bologna, 40126 Bologna, Italy; elena.baldi7@unibo.it (E.B.); giacomo.chiarelli3@unibo.it (G.C.); maurizio.quartieri@unibo.it (M.Q.); francesco.spinelli3@unibo.it (F.S.); moreno.toselli@unibo.it (M.T.)

* Correspondence: chiara.pastore@unibo.it

Abstract: Environmental extremes, such as drought and flooding, are becoming increasingly common, resulting in significant crop losses. The aim of the present study was to understand the molecular response induced by drought and waterlogging conditions, and to link these responses to the physiological adaptation of plants. For this purpose, leaf RNA expression was analyzed in potted kiwifruit plants by Illumina Next Generation Sequences. Stressed plants showed an impairment of all physiological parameters (leaf-gas exchange and stem-water potential) with a more evident effect in waterlogging condition than in drought condition. However, the impact on the transcriptome in waterlogged plants was less intense than in drought stress. Drought affected several metabolic pathways, among which "plant hormone signal transduction", "protein processing in endoplasmic reticulum", and "mitogen-activated protein kinase signaling pathway" were the most representative in terms of number of genes involved. The genes involved in the biosynthesis of phenylpropanoids were positively influenced by both drought and waterlogging. Finally, waterlogging stimulated secondary metabolisms by upregulating the genes responsible for the biosynthesis of terpenoids and flavonoids, such as stilbenoids. The obtained results show that the two contrasting stress conditions share several common physiological responses and molecular mechanisms.

Keywords: water stress; RNA-seq; gene expression; stem-water potential; leaf-gas exchange; metabolic pathway

Citation: Baldi, E.; Pastore, C.; Chiarelli, G.; Quartieri, M.; Spinelli, F.; Toselli, M. Molecular Responses to Drought and Waterlogging Stresses of Kiwifruit (*Actinidia chinensis* var. *deliciosa*) Potted Vines. *Horticulturae* **2024**, *10*, 834. https://doi.org/10.3390/horticulturae10080834

Academic Editors: Changxia Li and Yue Wu

Received: 8 July 2024
Revised: 26 July 2024
Accepted: 30 July 2024
Published: 7 August 2024

1. Introduction

Climate change, mainly due to global warming, is increasing the frequency and intensity of extreme weather events leading to conditions of drought or waterlogging in orchards. Both these conditions are detrimental to plants, resulting in a reduction of plant performance and yield [1]. Drought stress occurs in plants when there is low soil-water availability for a prolonged period. On the other hand, an excess of water in the soil leads to the saturation of soil pores, reducing oxygen availability and impairing gas exchange in tissues. In response to water deficit, roots' anatomy and architecture change to enhance water- and nutrient-uptake, and to provide mechanical strength to the root system [2]. In addition, plants reduce stomatal and cuticular conductance and evaporative surfaces [3], leading to a substantial reduction of net photosynthesis [4]. Moreover, drought stress is associated with the reduction of water potential that is lowered by osmotic adjustment and synthesis of low molecular-weight proteins, amino acids [5], and compatible sugars [6] that protect plants from damage caused by water deficiency [7]. Phytohormones play a crucial role in modulating the negative effects of drought stress in fruit crops [8], namely abscisic acid (ABA), a key regulator of plant adaptation to drought stress. The accumulation of ABA in *Arabidopsis thaliana* (L.) Heynh. activates downstream signals leading to stomatal closure, thereby limiting excessive water loss in plants [9]. Excessive water in the soil may lower oxygen to 1–5% percent, resulting in hypoxia conditions for roots [10], which negatively affects plant growth and fruit production [11].

At the whole-plant level, hypoxia causes a reduction in photosynthesis, stomatal conductance, transpiration and leaf-water potential [12]. Moreover, leaf-chlorophyll concentration, hormonal imbalances, and impairment in the uptake of water and minerals have also been observed [13]. These responses can change energy consumption, cellular metabolism, and gene expression with effects on apricot growth and development [14]. Symptoms of root hypoxia stress often do not occur until several days or weeks after the onset of waterlogging [11], and comprise leaf chlorosis, curling, defoliation, leaf abscission, marginal browning, necrosis, yellowing, root rotting, wilting, and fruit abscission [15].

In temporary stress conditions, plants are able to recover with a timing that depends on species, soil [16], and severity of the stress, otherwise prolonged waterlogging, as well as drought, result in the death of most cultivated plants. However, kiwifruit plants are extremely sensitive to waterlogging stress, and plant growth is reduced even after one day of exposure to anoxic conditions [17]. When roots were exposed to more than four days of waterlogging, vine growth did not recover even after aeration was restored to the roots [17]. In kiwifruit, stomata closure is one of the first symptoms of waterlogging [17]. In vines exposed to waterlogging for more than three days, no recovery of stomatal activity was observed [17]. Other studies indicate that in *Actinidia chinensis* var. *deliciosa* waterlogging damage became irreversible between 7 to 11 days of continuous exposure to an excess of water [18].

Symptoms of waterlogging and drought stress in leaves are similar, and thus they can mislead growers that often keep watering waterlogged plants to recover them from an incorrectly identified drought stress. In addition to the study of the physiological response to water stress, the analysis of gene expression in response to stress shed light on how plants adapt to the environment, providing useful information for developing tools to mitigate the stress in orchard conditions, or to induce plants with tolerance. Since roots are the first organ exposed to waterlogging and drought stress, several experiments focused on the study of gene expression in kiwifruit rootstocks [19,20]. However, elucidating the leaf response may provide a further level of understanding for adaptation to water stress (e.g., stomata closure and photo-dissipation) occurring in the canopy. Furthermore, since leaves are easier and faster to sample and process than roots, the transcriptome study at leaf-level may open new possibilities for an early detection of stress by identifying gene markers activated prior to symptoms' occurrence, and allow us to discriminate drought from waterlogging response directly in field conditions. The results obtained by transcriptome sequencing in *Prunus* spp. [21] and in *Citrus* sp. [13] in response to root hypoxia gave valuable information on species and genotype tolerance to drought and waterlogging, even highlighting common mechanisms of response mainly involving carbon metabolism, nutrient uptake and transport, and hormone synthesis/signaling. Despite several studies on kiwifruit vines, responses to waterlogging and drought stresses have focused on the transcriptional changes in roots [19,20], and knowledge of the responses in leaves is limited.

The aim of the present experiment was to associate drought and waterlogging physiological responses to transcriptional changes, highlighting the molecular mechanism underling kiwifruit plants' responses. We hypothesized that genes responding to waterlogging are different from the genes affected by drought stress in comparison to control leaves.

2. Materials and Methods

2.1. Plant Material and Treatments

The study was carried out in summer 2022 at the experimental station of the Department of Agricultural and Food Science of the University of Bologna (Italy), on the north side of the metropolitan area of the city, on 12, one-year-old, potted, self-rooted 'Hayward' (*Actinidia chinensis* var. *deliciosa* (A.Chev.)) plants. The plants were transplanted on March 2022 into 70 L pots filled with clay–loam soil and trained to a single cane.

Irrigation started at the beginning of May in order to maintain soils at field-water capacity. Starting from 17th June, four pots were inserted into plastic containers (40 cm large, 60 cm long, and 30 cm high) filled with tap water (waterlogging = WL), so that plants were immersed in 30 cm of water. Plants were regularly irrigated and maintained in the waterlogging status during the entire vegetative season. On 25th July, four of the remaining plants were stressed by stopping irrigation for 2 days (drought = D); control pots were regularly irrigated by returning the water lost by evapotranspiration.

Leaf-gas exchange measurements and transcriptome analyses of leaf samples were conducted on 27th July (10 days after WL and 2 days after D).

2.2. Plant Physiological Measurements

Leaf-gas exchange was measured on one well-lit, mature, fully expanded leaf per plant, using an open circuit gas-exchange system equipped with an LED light source (LI-6400, LI-COR, Lincoln, NE, USA). In detail, the leaf's net photosynthesis rate (Pn), stomatal conductance (gs), intercellular CO_2 concentration (Ci), and transpiration rate (E) were measured. Light intensity inside the chamber was maintained at the natural PAR (photosynthetically active radiation) values experienced by the leaves immediately before the measurements.

Stem water potential (Ψw) was measured with a pressure chamber on one leaf per plant, previously wrapped with aluminum foil and enclosed in a plastic bag at least one hour before excision to allow moisture equilibration with the soil [22,23]. Pressure was gently applied using a pump-up pressure chamber (PMS Instrument Company, Albany, OR, USA).

Physiological data were statistically analyzed in a completely randomized experimental design with 4 biological replicates per treatment; analysis of variance was performed for these data, and when they showed statistically significant ($p \leq 0.05$) effects, means were separated by the Student Newman–Keuls (SNK) test.

2.3. RNA Isolation and Sequencing

The same leaf used for gas-exchange measurements was sampled, immediately frozen in liquid nitrogen, and stored at −80 °C until RNA extraction. Leaves were ground in liquid nitrogen, and RNA was extracted for each biological replication using the Spectrum Plant Total Rna Kit (Merck KGaA, Darmstadt, Germany) following the manufacturer's instructions. Illumina sequencing (San Diego, CA, USA) and statistical analysis of RNA-seq data were performed by Novogene Co., Ltd. (Cambridge, UK) after checking total RNA quantity and quality on an Agilent 5400 Fragment Analyzer System (Santa Clara, CA, USA).

Sequencing libraries were constructed with the Novogene NGS RNA Library Prep Set (PT042)* for Illumina and sequenced on an Illumina NovaSeq 6000 platform with PE150. RNA-seq raw reads were processed to obtain high-quality clean reads by removing the adapter, reads with "N" > 10%, and reads when low quality nucleotides (base quality less than 5) constituted more than 50 percent of the read.

Clean reads were then mapped onto the kiwifruit reference genome and the manually curated *A. chinensis* var. *chinensis* Red5 genome [24] with the HISAT2 algorithm (v 2.0.5 [25]). Transcripts were assembled by StringTie software (v1.3.3) and differentially expressed gene (DEG) analysis was performed using the DESeq2 R package [26]. The resulting *p*-values were adjusted using the Benjamini and Hochberg's approach for controlling the false discovery rate. Genes with a corrected *p*-value (padj) below 0.05 were considered differentially expressed.

The enrichment analysis of DEGs was performed to find out which biological functions or pathways were significantly associated with them. ClusterProfiler 3.8.1 [27] was used for enrichment analysis, including GO (Gene Ontology, http://www.geneontology.org/, accessed on 31 March 2023) and KEGG (Kyoto Encyclopedia of Genes and Genomes, (http://www.genome.jp/kegg/, accessed on 31 March 2023)) enrichment. The GO terms and the KEGG pathways with padj < 0.05 were considered significantly enriched.

Transcriptome data were deposited in NCBI databases under the accession number PRJNA1076811.

3. Results

3.1. Physiological Response to Drought and Waterlogging

After 10 days of waterlogging or 2 days of drought, leaves showed a strong reduction of net photosynthesis with a more evident effect of waterlogging than drought (Table 1). Both types of water stress induced a reduction in stomatal conductance and transpiration in comparison to the control, while intercellular CO_2 concentration was only significantly increased by waterlogging plants in comparison to drought, and the control showed similar values (Table 1). Leaves for both treatments had higher stem-water potential than control leaves (Table 1).

Table 1. Effect of soil water availability on the leaf's net photosynthesis rate (P_n-µmol CO_2 m^2 s^{-1}), stomatal conductance (g_s-µmol H_2O m^2 s^{-1}), intercellular CO_2 concentration (C_i-µmol CO_2 m^2 s^{-1}), transpiration rate (E-µmol H_2O m^2 s^{-1}), and stem-water potential (Ψw-MPa).

TREATMENT	Pn	gs	C_i	E	Stem Ψ_w
Control	13.2 a	0.174 a	244 b	4.09 a	−0.438 a
Waterlogging	0.155 c	0.063 b	480 a	0.787 b	−0.563 a
Drought	3.64 b	0.036 b	266 b	1.27 b	−1.18 b
Significance	***	**	*	***	***

*, **, ***: effect significant at $p \leq 0.05$, $p \leq 0.01$ and $p \leq 0.001$, respectively. Values in the same column followed by the same letter are not statistically different according to the Student Newman–Keuls test ($p \leq 0.05$).

3.2. Overview of Sequencing Data and Statistical Summary of DEGs

Illumina RNA-Seq generated 73.6 million raw reads and 72.8 million clean reads on average for each leaf sample. The Q30 percentages ranged from 93.5 to 94%, while the GC contents were around 45.4% and 46.2%. The mapping of the clean reads on the reference genome using HISAT2 v2.0.5 software allowed us to map 82% of the clean reads, and 78.5% of them mapped onto a unique position in the genome. In total, 33,044 genes were analyzed, and the gene expression was estimated by Fragments Per Kilobase of transcript sequence per Millions of base pairs sequenced (FPKM). In a first analysis of gene expression (FPKM > 1 as the threshold to determine whether one gene is expressed; padj < 0.05), it was observed that more than 66% of annotated genes were commonly expressed among the three treatments (Figure 1); the number of shared expressed genes between drought and the control was low, while it was four-fold higher between the control and waterlogging. Drought induced the highest number of specifically expressed genes, while in waterlogging a lower number of unique, differentially expressed genes was detected (Figure 1), suggesting a less differential molecular response for this treatment.

This result was further confirmed when differentially expressed genes (DEG) were analyzed. Only 260 genes (0.8% of all annotated transcripts present in the reference genome) were differentially expressed between waterlogging and the control (Figure 2; log2Fold Change ≥ 2), while a stronger effect was detected following drought, with 5195 (15.7% of all annotated transcripts) DEGs compared to the control (Figure 2). The number of downregulated DEGs between the control and waterlogging treatments was lower than those upregulated (Figure 2). Conversely, in the comparison between the control and drought, a higher number of downregulated than upregulated DEGs (Figure 2) were in evidence.

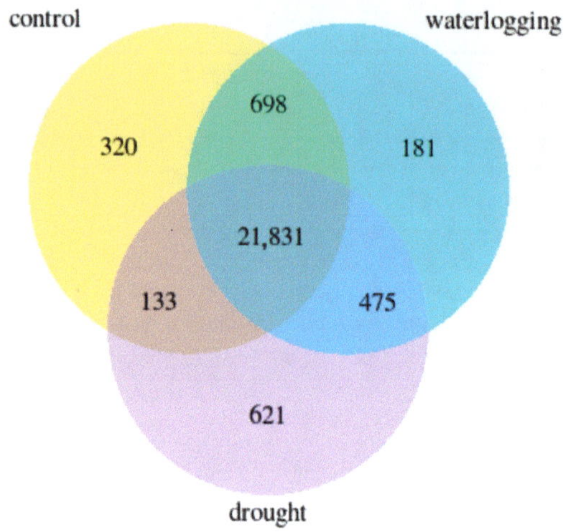

Figure 1. Venn diagram of leaves' expressed genes for each treatment (FPKM > 1).

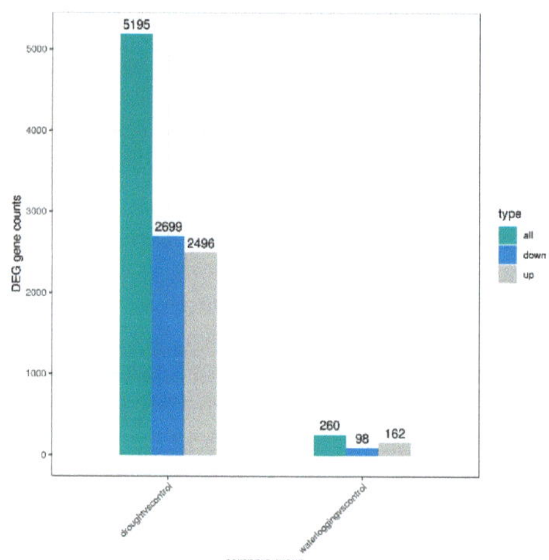

Figure 2. Differentially expressed genes (DEGs) between control and drought (left), and waterlogging and control (right). All = number of all DEGs; down = number of DEGs that were downregulated; up = number of DEGs that were upregulated (padj < 0.05).

3.3. *Gene Ontology and Kyoto Encyclopaedia of Genes and Genome-Enrichment Analyses*

In the Gene Ontology (GO), the number of main functional categories of genes affected by waterlogging (Figure 3a) were lower in comparison to those observed in drought (Figure 3b). Among the shared GO classes, the "extracellular region", which included genes that were not attached to the cell surface, was significantly affected by both stresses and resulted in the greatest representation in the waterlogging vs. the control (12 genes differentially expressed: 6 up- and 6 downregulated), and was one of the most represented in the drought vs. control comparison (125 genes differentially expressed: 55 up- and 70 downregulated).

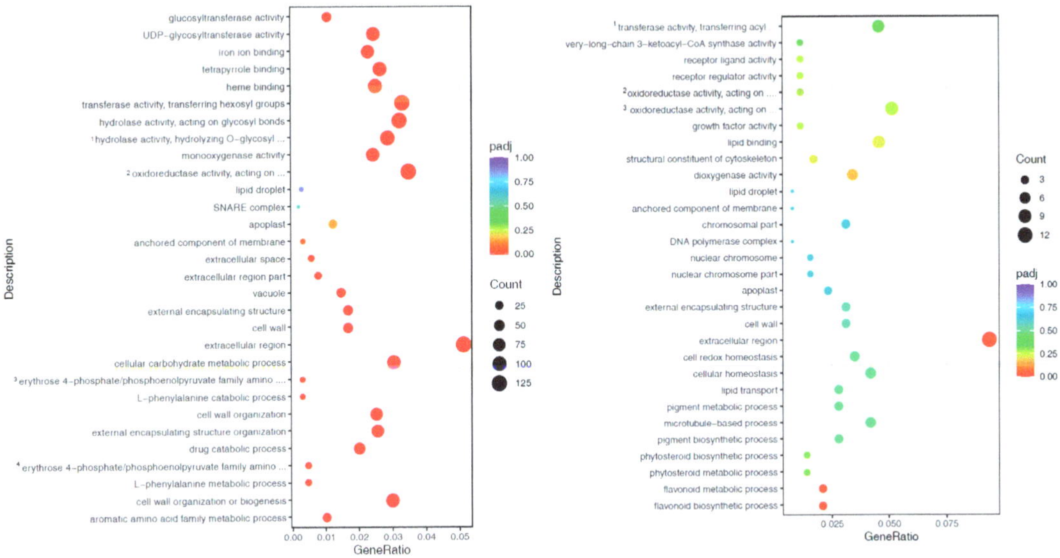

(**a**) (**b**)

Figure 3. Gene Ontology (GO) enrichment scatter plot of top 30 significantly enriched terms in the GO enrichment analysis. (**a**). drought vs. control ([1] hydrolase activity, hydrolyzing O-glycosyl compounds; [2] oxidoreductase activity, acting on paired donors, with incorporation or reduction of molecular oxygen; [3] erythrose 4-phosphate/phosphoenolpyruvate family amino acid catabolic process; [4] erythrose 4-phosphate/phosphoenolpyruvate family amino acid metabolic process); (**b**). waterlogging vs. control ([1] transferase activity, transferring acyl groups other than amino-acyl groups; [2] oxidoreductase activity, acting on paired donors, with oxidation of a pair of donors resulting in the reduction of molecular oxygen to two molecules of water; [3] oxidoreductase activity, acting on paired donors, with incorporation or reduction of molecular oxygen). Gene ratio = ratio between the number of differentially expressed genes in each GO term and all differentially expressed genes that can be found in the GO database; padj = adjusted p-values ≤ 0.05 are significant; count = the number of differentially expressed genes concerning this term.

Related to the "cellular component", other functional categories linked to the cellular integrity, such as "cell wall" and "external encapsulating structure", were represented in both stress conditions, but were differentially expressed in comparison to the control only in drought, where 41 genes were differentially expressed (7 up- and 34 downregulated, for both cellular components). Unlike drought stress, waterlogging significantly affected the biological process involving flavonoid metabolism and biosynthesis, with three genes differentially expressed (Figure 3b). Conversely, drought was able to modify several genes involved in carbohydrate and amino acid metabolisms, but also genes involved both in the biological process and molecular function. The highest proportion related to the photosynthetic process were iron, tetrapyrrole, and heme, binding with 84 (42 up and 42 down), 97 (51 up and 46 down), and 92 (52 up and 41 down) DEGs, respectively (Figure 3a). Moreover, drought stress also showed a different gene expression for the cellular carbohydrate metabolic process that was mainly downregulated (62 genes versus 27 upregulated) and for several biological processes that induce changes in cellular organization (cell wall organization and biogenesis, vacuole, etc.).

Kyoto Encyclopedia of Genes and Genomes (KEGG) enrichment analysis allows a detailed identification of significantly enriched metabolic pathways or signal transduction pathways associated with differentially expressed genes between waterlogging and drought, both in comparison to the control. Several metabolic pathways were upregulated by drought (Figure 4a), among which "plant hormone signal transduction", "protein processing in endoplasmic reticulum", and "mitogen-activated protein kinase (MAPK) signaling pathway–plant" were the most representative in terms of number of genes involved (63, 51 and 32, respectively). The biosynthesis of phenylpropanoid was positively affected from both drought and waterlogging, with a more evident effect as consequence of water deficit (43 genes) than excess (94 genes). In addition, waterlogging stimulated secondary metabolisms by upregulating the genes responsible for the biosynthesis of terpenoids, stilbenoids, and triterpenoids (Figure 4b).

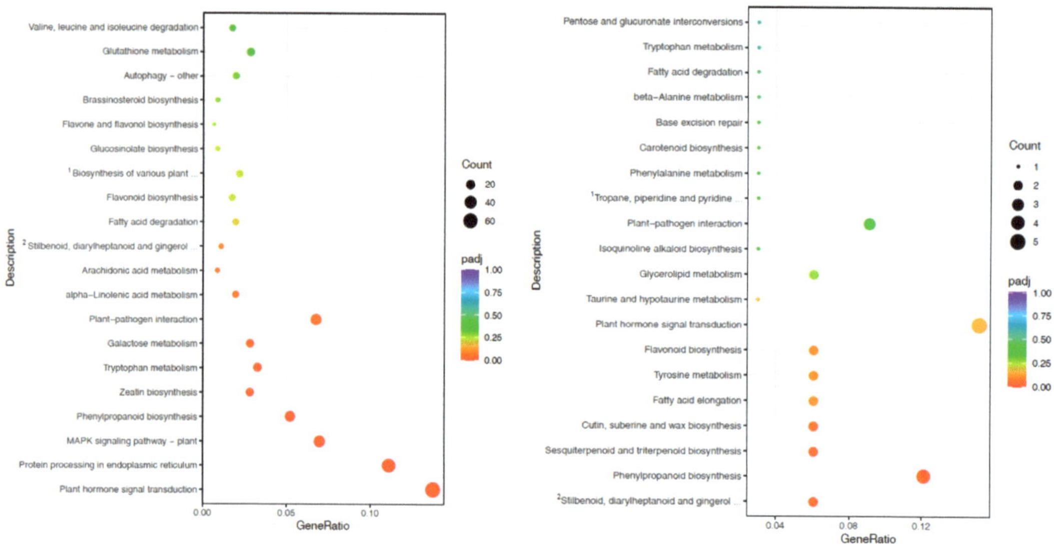

(a) (b)

Figure 4. Kyoto Encyclopedia of Genes and Genomes (KEGG) enrichment scatter plot of top 20 significantly upregulated enriched terms in the KEGG enrichment analysis. (**a**). drought vs. control ([1] biosynthesis of various plant secondary metabolites; [2] stilbenoid, diarylheptanoid and gingerol biosynthesis) (**b**). waterlogging vs. control ([1] tropane, piperidine and pyridine alkaloid biosynthesis; [2] stilbenoid, diarylheptanoid and gingerol biosynthesis). Gene ratio = ratio between the number of differentially expressed genes in each pathway and all differentially expressed genes that can be found in the KEGG database; padj = adjusted p-values ≤ 0.05 are significant; count = the number of differentially expressed genes concerning this term.

Both drought and waterlogging induced the downregulation of genes involved in the biosynthesis of flavonoids (Figure 5). Drought stress also led to the down regulation of genes responsible for starch and sucrose metabolism (37), carbon fixation (20), co-factor biosynthesis (49), and amino sugar and nucleotide sugar metabolism (27). Similarly, the biosynthesis of amino acids (57), in particular of phenylalanine, tyrosine, and tryptophan (18), was negatively affected by drought (Figure 5a). In waterlogging only, the downregulation of the biosynthesis of unsaturated fatty acids was observed, and it was limited to a small number of genes involved (29); no more genes were differentially downregulated for this treatment (Figure 5).

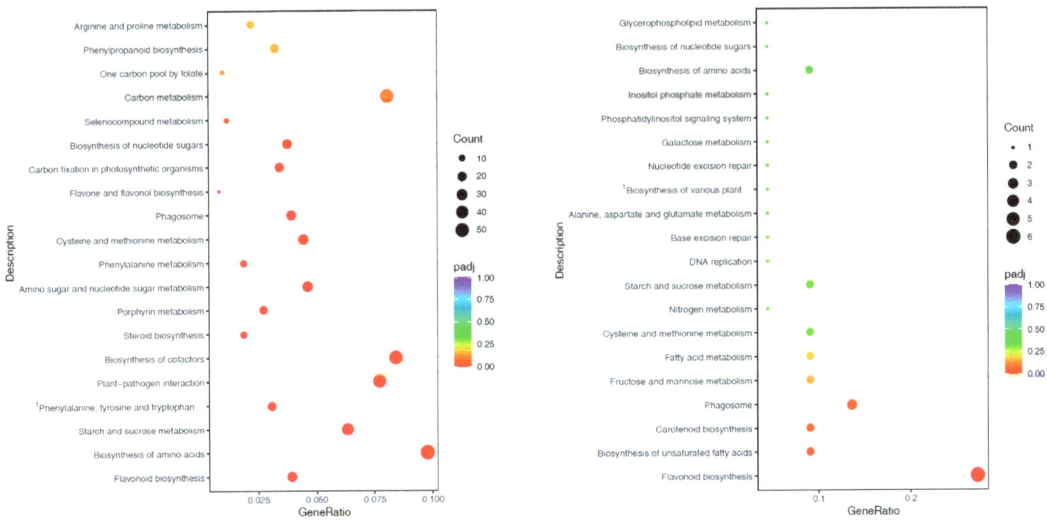

(**a**) (**b**)

Figure 5. Kyoto Encyclopedia of Genes and Genomes (KEGG) enrichment scatter plot of top 20 significantly downregulated enriched terms in the KEGG enrichment analysis. (**a**). drought vs. control ([1] phenylalanine, tyrosine and tryptophan biosynthesis); (**b**). waterlogging vs. control ([1] biosynthesis of various plant secondary metabolites). Gene ratio = ratio between the number of differentially expressed genes in each pathway and all differentially expressed genes that can be found in the KEGG database; padj = adjusted *p*-values ≤ 0.05 are significant; count = the number of differentially expressed genes concerning this term.

3.4. Pathways Involved in Drought and Waterlogging Stress Response

As reported above, several biological process, cellular components, and molecular functions were influenced by the water stress conditions considered. However, in order to shed light on the gene changes that led to the physiological response (Table 1), we focused on the following metabolic pathways: carbon and sugar metabolism, carbon fixation, photosynthesis, hormone signaling, and the secondary metabolites pathway.

3.4.1. Effect of Drought and Waterlogging on Genes Involved in Secondary Metabolites Pathways

Several pathways involved in secondary metabolite production were differentially regulated under drought and waterlogging. The biosynthesis of flavonoids showed several differentially expressed genes both in drought (31) and waterlogged plants (8; Figure 6). In both water conditions, the number of detected upregulated genes was lower in comparison to the downregulated ones. In particular, both stresses caused the upregulation of two genes involved in stilbene biosynthesis (two isoforms of spermidine hydroxy cinnamoyl transferase, CEY00_Acc29568, CEY00_Acc10082, and CEY00_Acc00879) and the downregulation of several genes involved in the synthesis of different flavonoid compounds, namely chalcone isomerase (CEY00_Acc03848, CEY00_Acc27670, and CEY00_Acc03638), dihydroflavonol 4-reductase (CEY00_Acc19353 and CEY00_Acc01005), anthocyanidin synthase (CEY00_Acc28876 and CEY00_Acc16762), and anthocyanidin reductase (CEY00_Acc09639 and CEY00_Acc17426; Figure 6); while only in drought-stressed plants, there was found a down-regulation in chalcone synthase (CEY00_Acc00260, CEY00_Acc24966, CEY00_Acc02004, and CEY00_Acc08970), and flavonol synthase (CEY00_Acc24372 and CEY00_Acc11493) genes. Similar to what was observed for flavonoids, phenylpropanoid biosynthesis (Figure S1A) was also affected by drought, with 43 different transcripts in comparison to the control, while only four differentially expressed genes were detected following waterlogging (Figure S1B). In particular, all the genes detected in waterlogging were shared with drought and were upregulated—these were spermidine hydroxycinnamoyl transferase, 5-O-(4-coumaroyl)-D-quinate 3′-monooxygenase, and two members of the peroxidase family, which together led to the stimulation of the production of lignin. Moreover, several differentially expressed genes in comparison to the control and involved in other secondary metabolite biosynthesis were found following drought, namely "stilbenoid, diarylheptanoid and gingerol biosynthesis" (7 genes), "biosynthesis of various plant secondary metabolites" (18 genes), and "flavone and flavonol biosynthesis" (8 genes). On the other hand, waterlogging induced the upregulation of two genes involved in the biosynthesis of sesquiterpenes and triterpenoids (Figure S2).

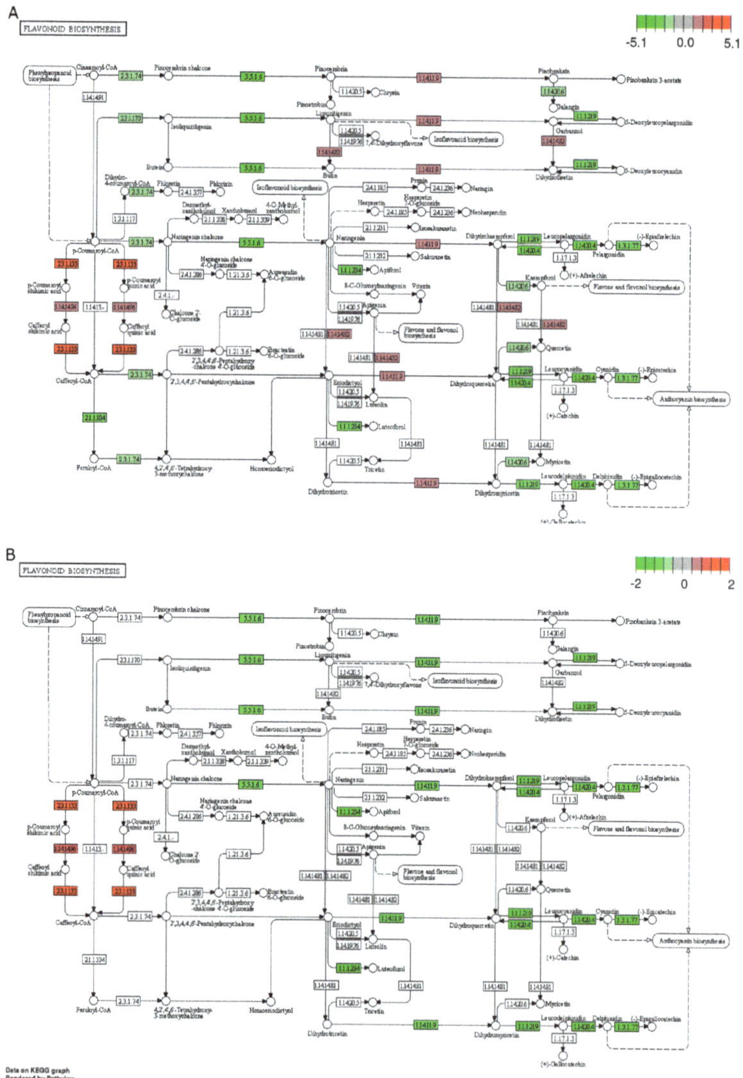

Figure 6. Differentiated genes in control vs. drought stress (**A**) and control vs. waterlogging (**B**) in flavonoid metabolism. The green box represents downregulated genes, the red box upregulated genes, and the white box not differentially expressed; padj < 0.05.

3.4.2. Effect of Drought and Waterlogging on Genes Involved in Carbon and Sugar Metabolism, Carbon Fixation, and Photosynthesis

In waterlogging conditions, no differentially expressed genes were found in comparison to the control. In relation to drought stress, we focused on the starch and sucrose metabolism (Figure 7) that showed 51 differentially expressed genes, among which 37 were down- and 14 upregulated. Most of the downregulated genes were related to enzymes that are involved in the metabolism of glucose, like glucose-6-phosphate isomerase (CEY00_Acc04867) and hexokinase (CEY00_Acc32432 and CEY00_Acc10665), or that contribute to starch production, like the 1,4-alpha-glucan branching enzyme (CEY00_Acc16922) and granule-bound starch synthase (CEY00_Acc03251). In addition, amylase (β-CEY00_Acc15000, CEY00_Acc28118, CEY00_Acc28117, CEY00_Acc09124, α - CEY00_Acc31412, CEY00_Acc06307, and CEY00_Acc23514), some transferase as 4-alpha-glucanotransferase (CEY00_Acc13256), sucrose phosphate synthase (CEY00_Acc11414 and CEY00_Acc15447), glucose-1-phosphate adenylyl transferases §(CEY00_Acc03134), as well as trehalose-phosphatase/synthase (CEY00_Acc02730, CEY00_Acc01464, CEY00_Acc, CEY00_Acc21372, CEY00_Acc06522, CEY00_Acc20715, CEY00_Acc25163, CEY00_Acc29021, CEY00_Acc10556, CEY00_Acc19227, and CEY00_Acc06511) were downregulated (Figure 7). On the other hand, those upregulated were mainly linked to the increased transcription of β-glucosidase (CEY00_Acc23018, CEY00_Acc13317, and CEY00_Acc09335) and β-fructofuranosidase (CEY00_Acc27954 and CEY00_Acc22611; Figure 7).

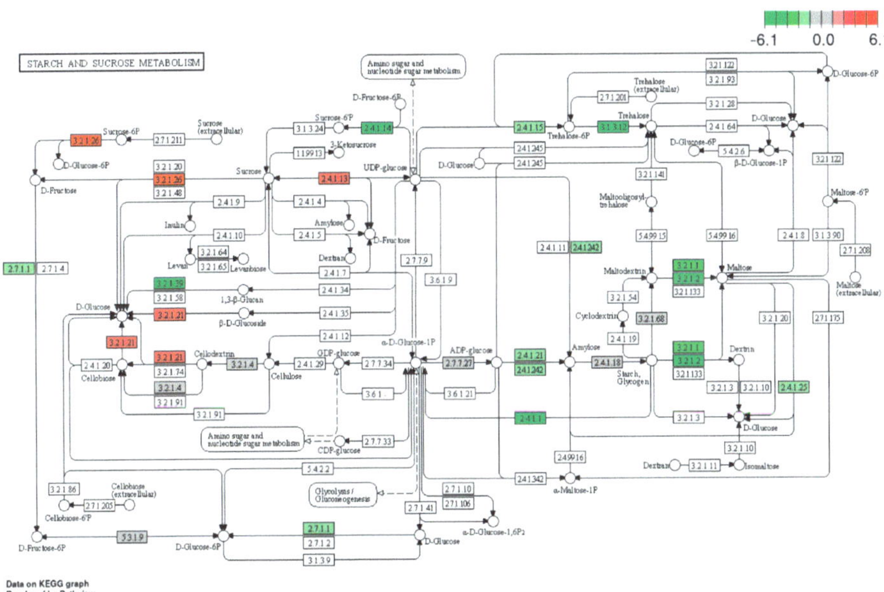

Figure 7. Differentially expressed genes in the control vs. drought stress conditions relating to starch and sucrose metabolism. Green boxes represent downregulated genes, red boxes upregulated genes. Genes in grey and white boxes were not differentially expressed; padj < 0.05.

Moreover, in the pathway related to "carbon fixation in photosynthetic organisms", a downregulation of several genes like phosphoenolpyruvate carboxylase (CEY00_Acc26501, CEY00_Acc19136, and CEY00_Acc28561), malate dehydrogenase (CEY00_Acc27472 and CEY00_Acc19122), fructose-bisphosphate aldolase (CEY00_Acc27551), ribose 5-phosphate isomerase (CEY00_Acc28440, CEY00_Acc09423, and CEY00_Acc17197), and ribulose bisphosphate carboxylase (CEY00_Acc28506) was observed (Figure 8).

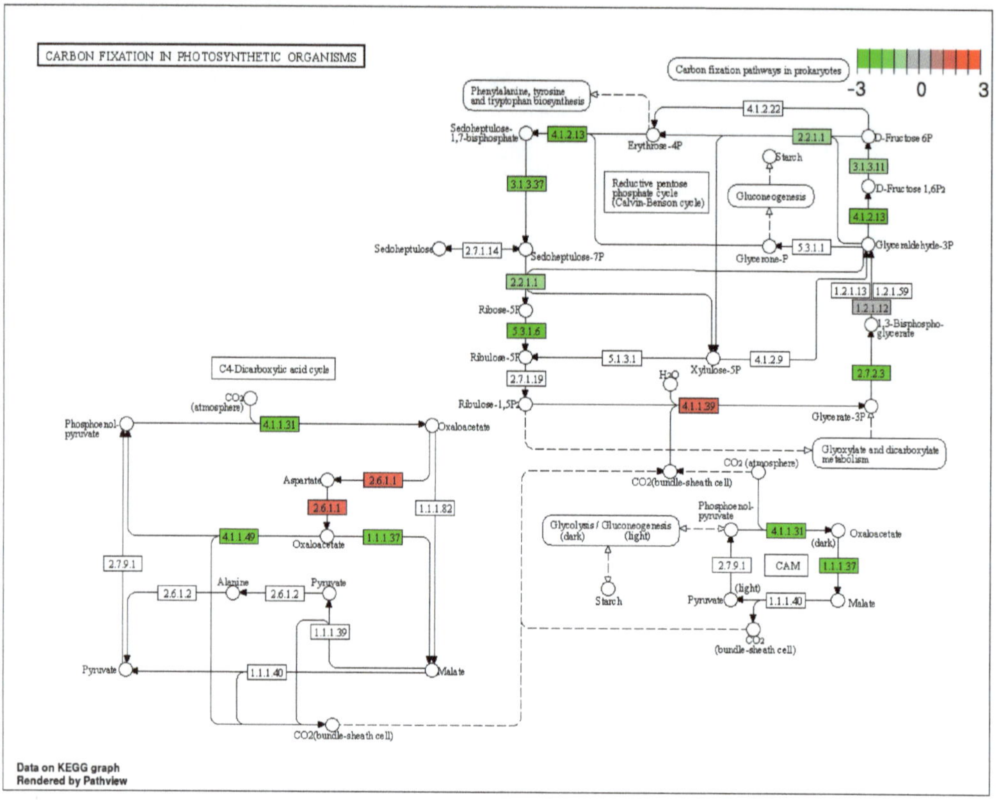

Figure 8. Differentially expressed genes in control vs. drought stress conditions relating to carbon fixation in photosynthetic organisms. Green boxes represent downregulated genes, red boxes upregulated. Genes in white boxes were not differentially expressed; padj < 0.05.

The observed downregulation of genes involved in carbon fixation following drought can be considered a direct effect of the general downregulation of the photosynthetic process. Indeed, several components of the light-dependent reactions of photosynthesis (photosystem I and II, cytochrome b6/f, electron transport chain, and ATP-ase complex) were downregulated in comparison to the control (Figure 9).

3.4.3. Effect of Drought and Waterlogging on Genes Involved in Hormone Metabolism

The responsiveness of the plant to the main plant hormones (auxins, cytokines, gibberellins, abscisic acid, ethylene, brassinosteroids, jasmonic acid, and salicylic acid) was affected by drought and waterlogging starting from several pathways (diterpenoid, cysteine and methionine, $\alpha-$linolenic acid, phenylalanine metabolism, and brassinosteroid biosynthesis). In particular, water deficiency caused the differential expression of 97 genes (63 up- and 34 downregulated) in comparison to the control, while in waterlogging only 5 upregulated genes were detected. As previously seen for phenylpropanoid biosynthesis, even in this case, the five genes detected for waterlogging were common to drought, and represented the genes involved.

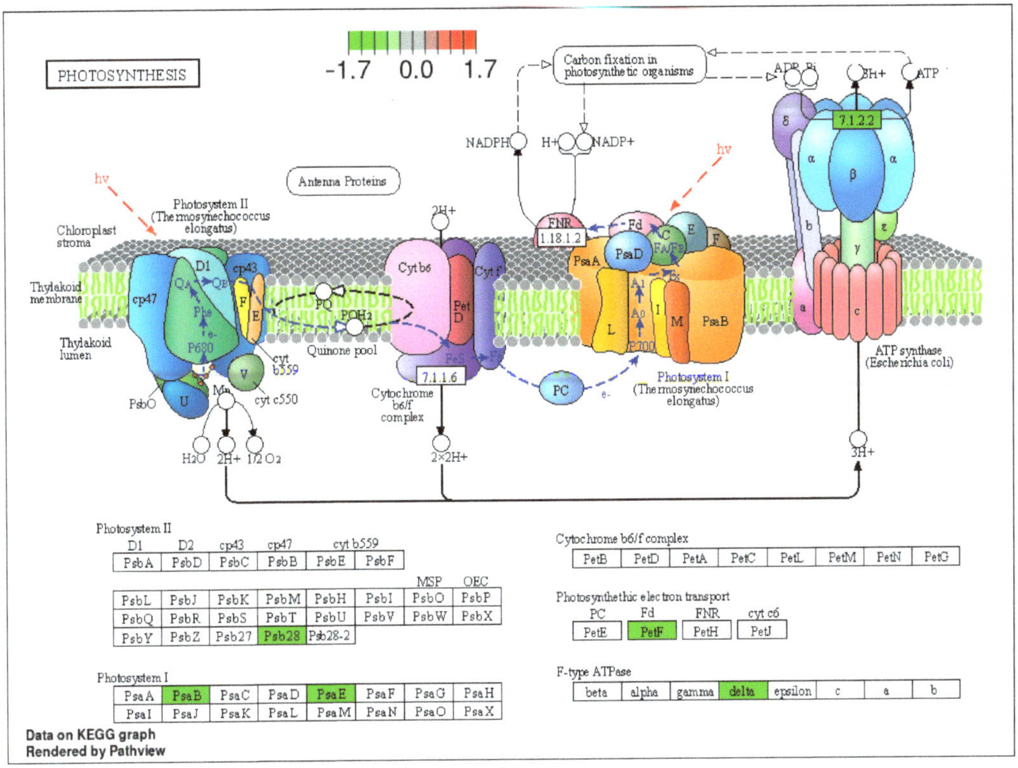

Figure 9. Differentially expressed genes in control vs. drought stress conditions relating to photosynthesis. Green boxes represent downregulated genes. Genes in white boxes were not differentially expressed; padj < 0.05.

Both stresses induced the upregulation of two genes involved in the auxin pathway—the auxin responsive GH3 gene (GH3) and the SAUR-like auxin-responsive gene (SAUR)—even with a higher significance in waterlogging than in drought (Figure 10). Moreover, drought induced the upregulation of one auxin influx carrier (belonging to the AUX1-LAX family) and of two members of the auxin-responsive proteins (AUX/IAA; Figure 10). In the pathways of signal transduction induced by abscisic acid (ABA), both stressing conditions led to the upregulation of the protein phosphatase 2C (PP2C) and of two abscisic acid (ABA) responsive element binding factors (ABF). Moreover, following drought, all the genes involved in the signal transduction induced by ABA were upregulated, except one member of the abscisic acid receptor PYR/PYL (Figure 10).

Concerning the other hormones, as previously described, no effect of waterlogging was detected (Figure S3B). Following drought, a general downregulation of the genes involved in cytokinin and jasmonic acid responses were, instead, detected (Figure S3A), while a general induction was found for ethylene response (with the upregulation of the ethylene response sensor 2 (ETR), EIN3-binding F box protein 1, and ethylene response factor 1), and for the salicylic acid-mediated response (with the upregulation of the pathogenesis-related protein 1, PR1; Figure S3A). A less clear behavior was detected for the brassinosteroid-induced response, with the upregulation of some genes at the beginning and the down regulation of others at the end of the hormone-signal transduction (Figure S3A).

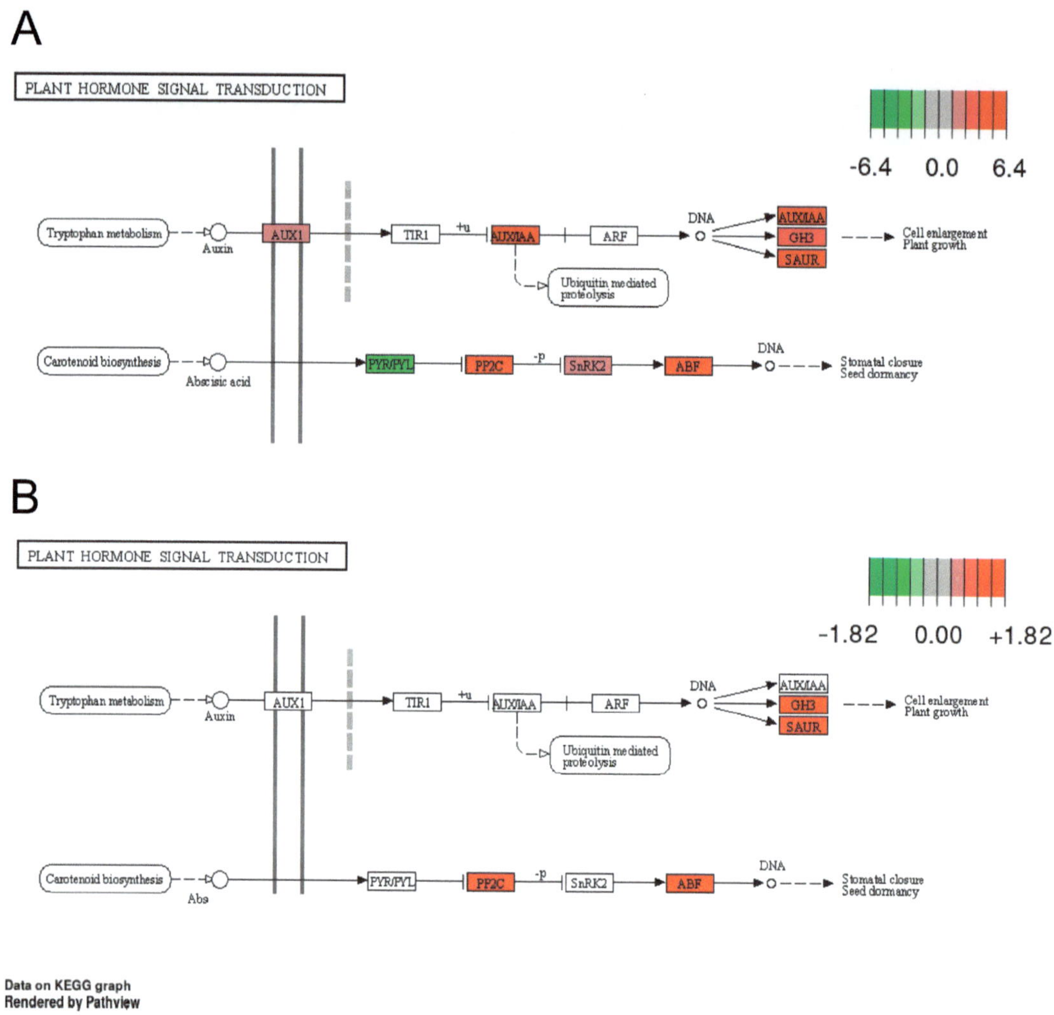

Figure 10. Differentiated genes in control vs. drought stress (**A**) and control vs. waterlogging (**B**) in plant hormone signal transduction. The green box represents downregulated genes, the red box upregulated genes, and the white box not differentially expressed genes; padj < 0.05.

4. Discussion

The results demonstrate the unique and differential response of kiwifruit leaf under two extreme and contrasting water regimes, drought and waterlogging. Several mechanisms, such as hormonal signaling, photosynthesis, respiration, and secondary metabolite production were implicated in plant responses to water excess or deficit showing different impacts on the transcriptome, with nonetheless similar physiological behavior. In the present experiment, we observed a higher impact on differential gene expression under drought as compared to under waterlogging as also evidenced on soybeans [28,29].

In general, drought stress induced the upregulation of several genes linked to the GO categories "plant hormone" and "MAPK signaling transduction", as well as "protein processing in endoplasmatic reticulum" which, in our experimental conditions, could together allow the plant to respond to water deficiency. The MAP kinases are involved in a wide variety of physiological processes mainly linked to signal transduction caused by wounding, pathogens, and abiotic stresses [30], and their upregulation could consequently play a critical role in enhancing stress resistance [31]. Similarly, genes belonging to the category "protein processing in endoplasmatic reticulum" are involved in the processing of misfolded proteins, which are retained within the endoplasmic reticulum and are directed toward degradation through the proteasome. The accumulation of misfolded proteins in the endoplasmatic reticulum is, in general, linked to severe stress conditions and, in these situations, the protective mechanisms activated by the unfolded protein response (UPR) is not sufficient to restore normal endoplasmatic reticulum function and cells could die by apoptosis. It was reported that drought stress causes accumulation of misfolded or unfolded proteins that induce endoplasmic reticulum homeostasis imbalance, resulting in a higher load on secretory proteins [32]. In similar situations, the overexpression of genes involved in the UPR signaling can lead to the stimulation of the expression of specific genes, which regulate the response to stress [33].

On the other hand, a reduced number of GO categories were significantly influenced by waterlogging. In detail, phenylpropanoids, terpenoids, stilbenoids, and triterpenoids biosynthesis genes were upregulated while, as also for drought, the flavonoid pathway genes were downregulated.

Secondary metabolites are required for protection, stress tolerance, competition, and species interaction [34,35]. Among them, flavonoids were those that showed differentially expressed genes both in drought and waterlogging conditions. Flavonoids are one of the most important plant secondary metabolites with multiple functions in response to stress, and they are composed of six subgroups: chalcones, flavones, flavan-3-ols, flavandiols, anthocyanins, and proanthocyanidins or condensed tannins [36,37].

Chalcone isomerase is one of the genes involved in the first-step biosynthetic pathway of flavonoids; its down regulation, as a consequence of both the stressing conditions, together with the reduced expression of flavanone 4-reductase, dihydroflavonol 4-reductase, anthocyanidin synthase, and anthocyanidin reductase could bring about reduced activation of the flavonoid branch of the phenylpropanoid pathway that led to flavan-3-ols production [38,39]. Even if in other species, as in sorghum and corn, the role of flavan-3-ols as antioxidants under drought stress has been well described [40], other secondary metabolite compounds probably play a key role in kiwifruit leaves' response to drought.

Shikimate O-hydroxycinnamoyl transferase and spermidine hydroxycinnamoyl transferase, upregulated by both stresses, take part in the monolignol pathway [41] that leads to the production of the stress-induced lignin polymer [42]. It was previously demonstrated ([43] and literature here cited) that drought stress affects both the xylem anatomy and lignification of trees due to the modification by stressed cells of their pattern of lignin deposition by strengthening the cell wall. Both drought and waterlogging can lead to wilting and increased lignification that would contribute to improved mechanical support of the plant aerial structure, as well as water transport [43]. Lignification could also help to reduce cell expansion and plant growth during this phase, thereby favoring reallocation of carbon resources to other defense mechanisms [43].

Waterlogging also induced the upregulation of several genes involved in terpenes and terpenoids biosynthesis, consistent with the role in mitigating the effects of oxidative stress by modulating the oxidative status of plants [44].

According to several researchers, [45–47] under drought stress the reduction of photosynthetic activity leads to changes in carbohydrate metabolism in leaves [48]. Drought stress resulted in the downregulation of several genes involved in photosynthesis, carbohydrate metabolism and, in particular, of the one related to starch accumulation. According to their form, carbohydrates serve diverse functions. For example, starch is the major source of energy reserves, whereas sucrose, fructose, and glucose are solutes for cell osmotic potential. In addition, sucrose also has a role in stabilizing proteins and membranes after stress damage [49–51]. Plants can regulate cells' osmotic potential by fine-tuning the rate of starch accumulation and its ratio with soluble sugars that are osmotically active [52]. In our experiment, drought induced the downregulation of genes involved in starch production

(1,4-alpha-glucan branching enzyme, granule-bound starch synthase) and the metabolism of glucose (glucose-6-phosphate isomerase; hexokinase). This was also followed by a reduction in the expression of genes involved in starch degradation such as β and α amylase, and sucrose phosphate synthase.

Moreover, in the pathway related to carbon fixation in photosynthetic organisms, there was a downregulation of several genes like phosphoenolpyruvate carboxylase, malate dehydrogenase, aldolase superfamily proteins, ribose 5-phosphate isomerase, glyceraldehyde-3-phosphate dehydrogenase, and ribulose bisphosphate carboxylase. All these enzymes take part in the Calvin cycle that is the most common CO_2 fixation process in C3 plants. Drought stress, by inducing stomatal closure, leads to a decrease in CO_2 availability, thus impairing the efficiency of the Calvin cycle, which is crucial for photosynthetic responses and metabolic pathways.

On the other hand, upregulated genes were mainly linked to the increased expression of β-glucosidase. Previous studies on beans [53], wheat [54], and tobacco [55] evidenced an increase in β-glucosidase abundance under water deficit. This protein is present in all living cells and plays fundamental biological roles in several processes such as degradation of cellulose and other carbohydrates [56]. It is suggested that glycosyl hydrolases enzymes utilize cell-wall polysaccharides as an alternate carbon source under sugar depletion in the cytoplasm during dehydration [57,58]. These results, consequently, fit with the general concept that drought stress might affect biochemical metabolism in the cell wall. Moreover, in previous experiments [59], it was found that β-glucosidase is responsible for the production of the free form of abscisic acid. The data presented by Dietz and co-authors [59] provide evidence that β-glucosidase of barley leaves has a specific function in ABA-dependent root:shoot signaling in plants. The phytohormone ABA has a critical role in regulating the plant response to water stress [60,61]. Its concentration increases in roots in response to drought stress [61,62], resulting in stomatal closure and a consequent reduction of photosynthetic rates and the accumulation of protective proteins and metabolites to maintain cellular water status [60,61]. On the other hand, in response to flooding, ABA levels decrease in roots and increase in the aerial organs [63]. A transcriptome study performed on Arabidopsis revealed an upregulation of ABA biosynthesis genes in leaves, and a downregulation in roots under flooding conditions [64]. It was thus suggested that ABA may accumulate in leaves of flooded plants because of reduced translocation of photoassimilates out of leaves, and roots do not act as the source of ABA because most of them die within the first few days of flooding [65]. Thanks to xylem sap analysis, other authors [63] have revealed that an increase in leaves' ABA concentrations was not due to its translocation from roots to shoots, but to an increased production in old leaves and transportation to younger leaves [66]. As for citrus and also grapes, waterlogging can induce substantial ABA accumulation in leaves, which is mainly responsible for the severe stomatal closure following waterlogging in plants [67].

In the present experiment, the highest impact on photosynthesis was found in plants exposed to waterlogging. The reduction in photosynthesis was primarily connected to a significantly lower stomatal conductance, which resulted in lower intercellular CO_2 concentration. Both stress conditions induced an upregulation of abscisic acid (ABA) responsive element binding factor (ABF). ABA is the key regulator of stomata opening, and thus the rise in ABF induction suggests that the plants respond to stress by a prompt stomata closure. Stomatal conductance in waterlogged plants was 24-fold lower than in drought-exposed ones. This result corroborates the finding that kiwifruit responds to water excess by a prompt stomata closure within 3 h after the exposure to stress [17]. Moreover, stomata closure impairs evapotranspiration, and thus the plant is unable to dissipate water excess. This phenomenon makes kiwifruit vines extremely sensitive to waterlogging, and permanent damage to the root system occurs within three to five days of exposition [68].

In our experiment, we did not observe a differentiated expression of genes directly related to ABA biosynthesis, however, we found some effects on the carotenoid signaling pathways induced by ABA [69]. Specifically, we measured a downregulation of PYR/PYR receptors and an upregulation of PPC2, ABF, and SnRK2 (only in response to drought). According to previous research [70–72], the interactions between PYR/PYL/RCAR proteins and their target PP2Cs induce the activation of downstream targets of the group A PP2Cs, including SnRK2 protein kinases that play key roles in the regulation of transcriptional response and stomatal closure, thereby confirming our findings.

Beside ABA, auxin signal transduction was also altered by stress, with a trend of upregulation of genes encoding auxin influx carrier (AUX1), auxin-responsive protein IAA family (AUX/IAA), and SAUR, highlighting the crucial roles of auxin-signaling genes in drought stress responses. Comparable results were previously found in rice [73], *Arabidopsis* [74], and *Davidia involucrata* [75]. In recent years, several researchers demonstrated the potential link between auxin response and abiotic stress [76–78] due to the interaction with other plant hormones like salicylic acid and ABA [79]. In a study on transgenic *Arabidopsis* [80], the protective role of endogenous and exogenous auxin in drought stress resistance through the regulation of root architecture, ABA-responsive genes expression, reactive oxygen species metabolism, and metabolic homeostasis was demonstrated. In addition, in a study on white clover [81], it was demonstrated that exogenous auxin application mitigated plant dehydration symptoms by improving stem dry weight, relative water content, and total chlorophyll content in leaves. This was mainly because exogenous auxins induced the production of other hormones, suggesting that auxins increasing production at leaf-level could have a positive effect on improving the drought tolerance of white clover. In a grape experiment [67], some auxin-induced genes such as GH3 (auxin-conjugating enzyme) were upregulated under waterlogging stress, confirming our findings. Consequently, as also for drought, auxins, by interacting with other hormones, could play a positive role as signaling molecules in response to waterlogging stress.

5. Conclusions

Economic returns from kiwifruit cultivation strongly depend on individual fruit sizes and quality, which are directly affected by water availability. For this reason, kiwifruits are usually irrigated and, in some areas, this practice has led to demands for water that exceed local resources, even if more frequently water scarcity is experienced due to climate change. Plant responses toward abiotic stress are complex and involve several different regulatory networks between genes related to hormonal signaling, photosynthesis, respiration, and secondary metabolite production. Comparative transcriptome profiling performed in this study under drought and waterlogging has provided an opportunity to evaluate and differentiate the categorized molecular responses. Surprisingly, despite similar physiological behavior, the two contrasting stress conditions have a quite different impact on the transcriptome, suggesting the strongest effect came from the drought stress in comparison to the flooding. Even if some shared molecular mechanisms were observed, our results suggest that, while the plant actively reacts to water deficit, the response to water excess seems to be passive, with the plant reducing at minimum its metabolic activity. Further studies are required to understand whether, in the short-term, even an excess of water can induce a specific response, in order to identify genes as early markers of stress. Concerning drought, the massive transcriptome reprogramming that emerged for the leaves in the present study suggests that several effects may also be present in fruit, and it would be interesting to ascertain whether, considering climate change and agricultural sustainability, a controlled reduction of the water use for kiwifruit irrigation could yield fruit equally appreciated by consumers.

Supplementary Materials: The following supporting information can be downloaded at: https://www.mdpi.com/article/10.3390/horticulturae10080834/s1, Figure S1: differentially expressed genes involved in phenylpropanoid biosynthesis in control vs. drought stress and control vs. waterlogging; Figure S2: differentially expressed genes in control vs. waterlogging in sesquiterpenoid and triterpenoid biosynthesis; Figure S3: differentially expressed genes in control vs. drought stress and control vs. waterlogging in plant hormone signal transduction.

Author Contributions: Conceptualization, E.B., C.P. and M.T.; data curation, E.B. and C.P.; formal analysis, E.B., C.P., G.C. and M.Q.; investigation, E.B., C.P., G.C. and M.Q.; methodology, E.B. and C.P.; resources, M.T.; supervision, F.S. and M.T.; validation, E.B., C.P. and M.T.; writing–original draft, E.B. and C.P.; writing–review and editing, F.S. and M.T. All authors have read and agreed to the published version of the manuscript.

Funding: This research received no external funding.

Data Availability Statement: All data were uploaded to the National Center for Biotechnology Information (NCBI) with the following BioProject ID: PRJNA1076811.

Conflicts of Interest: The authors declare no conflicts of interest.

References

1. Zandalinas, S.I.; Sengupta, S.; Fritschi, F.B.; Azad, R.K.; Nechushtai, R.; Mittler, R. The impact of multifactorial stress combination on plant growth and survival. *New Phytol.* **2021**, *230*, 1034–1048. [CrossRef] [PubMed]
2. Morris, E.C.; Griffiths, M.; Golebiowska, A.; Mairhofer, S.; Burr-Hersey, J.; Goh, T.; von Wangenheim, D.; Atkinson, B.; Sturrock, C.J.; Lynch, J.P.; et al. Shaping 3D root system architecture. *Curr. Biol.* **2017**, *27*, R919–R930. [CrossRef] [PubMed]
3. Jarzyniak, K.M.; Jasiński, M. Membrane transporters and drought resistance—A complex issue. *Front. Plant Sci.* **2014**, *5*, 687. [CrossRef] [PubMed]
4. Daszkowska-Golec, A.; Szarejko, I. Open or close the gate–stomata action under the control of phytohormones in drought stress conditions. *Front. Plant Sci.* **2013**, *4*, 138. [CrossRef] [PubMed]
5. Good, A.G.; Zaplachinski, S.T. The effects of drought stress on free amino acid accumulation and protein synthesis in *Brassica napus*. *Physiol. Plant.* **1994**, *90*, 9–14. [CrossRef]
6. Maruyama, K.; Urano, K.; Yoshiwara, K.; Morishita, Y.; Sakurai, N.; Suzuki, H.; Kojima, M.; Sakakibara, H.; Shibata, D.; Saito, K.; et al. Integrated analysis of the effects of cold and dehydration on rice metabolites, phytohormones, and gene transcripts. *Plant Physiol.* **2014**, *164*, 1759–1771. [CrossRef] [PubMed]
7. Bacelar, E.L.; Moutinho-Pereira, J.M.; Gonçalves, B.M.; Brito, C.V.; Gomes-Laranjo, J.; Ferreira, H.M.; Correia, C.M. Water use strategies of plants under drought conditions. In *Plant Responses to Drought Stress: From Morphological to Molecular Features*; Aroca, R., Ed.; Springer: Berlin/Heidelberg, Germany, 2012; pp. 145–170. [CrossRef]
8. Liu, X.; Gao, T.; Liu, C.; Mao, K.; Gong, X.; Li, C.; Ma, F. Fruit crops combating drought: Physiological responses and regulatory pathways. *Plant Physiol.* **2023**, *192*, 1768–1784. [CrossRef] [PubMed]
9. Kalladan, R.; Lasky, J.R.; Chang, T.Z.; Sharma, S.; Juenger, T.E.; Verslues, P.E. Natural variation identifies genes affecting drought-induced abscisic acid accumulation in *Arabidopsis thaliana*. *Proc. Natl. Acad. Sci. USA* **2017**, *114*, 11536–11541. [CrossRef] [PubMed]
10. Sasidharan, R.; Bailey-Serres, J.; Ashikari, M.; Atwell, B.J.; Colmer, T.D.; Fagerstedt, K.; Fukao, T.; Geigenberger, P.; Hebelstrup, K.H.; Hill, R.D.; et al. Community recommendations on terminology and procedures used in flooding and low oxygen stress research. *New Phytol.* **2017**, *214*, 1403–1407. [CrossRef]
11. Salvatierra, A.; Toro, G.; Mateluna, P.; Opazo, I.; Ortiz, M.; Pimentel, P. Keep calm and survive: Adaptation strategies to energy crisis in fruit trees under root hypoxia. *Plants* **2020**, *9*, 1108. [CrossRef]
12. Habibi, F.; Liu, T.; Shahid, M.A.; Schaffer, B.; Sarkhosh, A. Physiological, biochemical, and molecular responses of fruit trees to root zone hypoxia. *Environ. Exp. Bot.* **2023**, *206*, 105179. [CrossRef]
13. Xie, R.; Zheng, L.; Jiao, Y.; Huang, X. Understanding physiological and molecular mechanisms of citrus rootstock seedlings in response to root zone hypoxia by RNA-Seq. *Environ. Exp. Bot.* **2021**, *192*, 104647. [CrossRef]
14. Domingo, R.; Pérez-Pastor, A.; Ruiz-Sánchez, M.C. Physiological responses of apricot plants grafted on two different rootstocks to flooding conditions. *J. Plant Physiol.* **2002**, *159*, 725–732. [CrossRef]
15. McGee, T.; Shahid, M.A.; Beckman, T.G.; Chaparro, J.X.; Schaffer, B.; Sarkhosh, A. Physiological and biochemical characterization of six *Prunus* rootstocks in response to flooding. *Environ. Exp. Bot.* **2021**, *183*, 104368. [CrossRef]
16. Baldi, E.; Quartieri, M.; Chiarelli, G.; Larocca, G.N.; Masroleo, M.; Xylogiannis, E.; Falsone, G.; Toselli, M. Water retention potentials of Italian soils and physiological responses of potted golden kiwifruit. *Sci. Hort.* **2024**, *329*, 113013. [CrossRef]
17. Smith, G.S.; Judd, M.J.; Miller, S.A.; Buwalda, J.G. Recovery of kiwifruit vines from transient waterlogging of the root system. *New Phytol.* **1990**, *115*, 325–333. [CrossRef] [PubMed]
18. Li, Z.; Bai, D.; Zhong, Y.; Abid, M.; Qi, X.; Hu, C.; Fang, J. Physiological responses of two contrasting kiwifruit (*Actinidia* spp.) rootstocks against waterlogging stress. *Plants* **2021**, *10*, 2586. [CrossRef]
19. Zhang, J.Y.; Huang, S.N.; Mo, Z.H.; Xuan, J.P.; Jia, X.D.; Wang, G.; Guo, Z.R. De novo transcriptome sequencing and comparative analysis of differentially expressed genes in kiwifruit under waterlogging stress. *Mol. Breed.* **2015**, *35*, 208. [CrossRef]
20. Li, Z.; Bai, D.; Zhong, Y.; Lin, M.; Sun, L.; Qi, X.; Hu, C.; Fang, J. Full-length transcriptome and RNA-seq analyses reveal the mechanisms underlying waterlogging tolerance in kiwifruit (*Actinidia valvata*). *Int. J. Mol. Sci.* **2022**, *23*, 3237. [CrossRef]

21. Arismendi, M.J.; Almada, R.; Pimentel, P.; Bastias, A.; Salvatierra, A.; Rojas, P.; Salvatierra, A.; Rojas, P.; Hinrichsen, P.; Pinto, M.; et al. Transcriptome sequencing of *Prunus* sp. rootstocks roots to identify candidate genes involved in the response to root hypoxia. *Tree Genet. Genomes* **2015**, *11*, 11. [CrossRef]
22. Naor, A.; Klein, I.; Doron, I. Stem water potential and apple fruit size. *J. Am. Soc. Hort. Sci.* **1995**, *120*, 577–582. [CrossRef]
23. Turner, N.C. Measurement of plant water status by the pressure chamber technique. *Irrig. Sci.* **1988**, *9*, 289–308. [CrossRef]
24. Pilkington, S.M.; Crowhurst, R.; Hilario, E.; Nardozza, S.; Fraser, L.; Peng, Y.; Gunaseelan, K.; Simpson, R.; Tahir, J.; Deroles, S.C.; et al. A manually annotated *Actinidia chinensis* var. *chinensis* (kiwifruit) genome highlights the challenges associated with draft genomes and gene prediction in plants. *BMC Genom.* **2018**, *19*, 257. [CrossRef]
25. Pertea, M.; Kim, D.; Pertea, G.; Leek, J.T.; Salzberg, S.L. Transcript-level expression analysis of RNA-seq experiments with HISAT, StringTie and Ballgown. *Nat. Prot.* **2016**, *11*, 1650–1667. [CrossRef] [PubMed]
26. Anders, S.; McCarthy, D.J.; Chen, Y.; Okoniewski, M.; Smyth, G.K.; Huber, W.; Robinson, M.D. Count-based differential expression analysis of RNA sequencing data using R and Bioconductor. *Nat. Protoc.* **2013**, *8*, 1765–1786. [CrossRef] [PubMed]
27. Yu, G.; Wang, L.; Han, Y.; He, Q. clusterProfiler: An R package for comparing biological themes among gene clusters. *OMICS* **2012**, *16*, 284–287. [CrossRef] [PubMed]
28. Chen, W.; Yao, Q.; Patil, G.B.; Agarwal, G.; Deshmukh, R.K.; Lin, L.; Wang, B.; Wang, Y.; Prince, S.J.; Song, L.; et al. Identification and comparative analysis of differential gene expression in soybean leaf tissue under drought and flooding stress revealed by RNA-Seq. *Front. Plant Sci.* **2016**, *7*, 1044. [CrossRef] [PubMed]
29. Nanjo, Y.; Maruyama, K.; Yasue, H.; Yamaguchi-Shinozaki, K.; Shinozaki, K.; Komatsu, S. Transcriptional responses to flooding stress in roots including hypocotyl of soybean seedlings. *Plant Mol. Biol.* **2011**, *77*, 129–144. [CrossRef] [PubMed]
30. Huang, G.T.; Ma, S.L.; Bai, L.P.; Zhang, L.; Ma, H.; Jia, P.; Liu, J.; Zhong, M.; Guo, Z.F. Signal transduction during cold, salt, and drought stresses in plants. *Mol. Biol. Rep.* **2012**, *39*, 969–987. [CrossRef]
31. Hamel, L.P.; Nicole, M.C.; Duplessis, S.; Ellis, B.E. Mitogen-activated protein kinase signaling in plant-interacting fungi: Distinct messages from conserved messengers. *Plant Cell* **2012**, *24*, 1327–1351. [CrossRef]
32. Czékus, Z.; Szalai, G.; Tari, I.; Khan, M.I.R.; Poór, P. Role of ethylene in ER stress and the unfolded protein response in tomato (*Solanum lycopersicum* L.) plants. *Plant Physiol. Biochem.* **2022**, *181*, 1–11. [CrossRef] [PubMed]
33. Manghwar, H.; Li, J. Endoplasmic reticulum stress and unfolded protein response signaling in plants. *Int. J. Mol. Sci.* **2022**, *23*, 828. [CrossRef] [PubMed]
34. Wolfender, J.L.; Rudaz, S.; Hae Choi, Y.; Kyong Kim, H. Plant metabolomics: From holistic data to relevant biomarkers. *Curr. Med. Chem.* **2013**, *20*, 1056–1090. [CrossRef] [PubMed]
35. Parida, A.K.; Panda, A.; Rangani, J. Metabolomics-guided elucidation of abiotic stress tolerance mechanisms in plants. In *Plant Metabolites and Regulation under Environmental Stress*; Academic Press: Cambridge, MA, USA, 2018; pp. 89–131. [CrossRef]
36. Winkel-Shirley, B. Biosynthesis of flavonoids and effects of stress. *Curr. Opin. Plant Biol.* **2002**, *5*, 218–223. [CrossRef] [PubMed]
37. Petrussa, E.; Braidot, E.; Zancani, M.; Peresson, C.; Bertolini, A.; Patui, S.; Vianello, A. Plant flavonoids—Biosynthesis, transport and involvement in stress responses. *Int. J. Mol. Sci.* **2013**, *14*, 14950–14973. [CrossRef] [PubMed]
38. Yin, Y.C.; Zhang, X.D.; Gao, Z.Q.; Hu, T.; Liu, Y. The research progress of chalcone isomerase (CHI) in plants. *Mol. Biotechnol.* **2019**, *61*, 32–52. [CrossRef] [PubMed]
39. Li, Y.; Cui, W.; Qi, X.; Lin, M.; Qiao, C.; Zhong, Y.; Hu, C.; Fang, J. MicroRNA858 negatively regulates anthocyanin biosynthesis by repressing AaMYBC1 expression in kiwifruit (*Actinidia arguta*). *Plant Sci.* **2020**, *296*, 110476. [CrossRef] [PubMed]
40. Gu, H.; Wang, Y.; Xie, H.; Qiu, C.; Zhang, S.; Xiao, J.; Li, H.; Chen, L.; Li, X.; Ding, Z. Drought stress triggers proteomic changes involving lignin, flavonoids and fatty acids in tea plants. *Sci. Rep.* **2020**, *10*, 15504. [CrossRef] [PubMed]
41. Gomez-Cano, L.; Gomez-Cano, F.; Dillon, F.M.; Alers-Velazquez, R.; Doseff, A.I.; Grotewold, E.; Gray, J. Discovery of modules involved in the biosynthesis and regulation of maize phenolic compounds. *Plant Sci.* **2020**, *291*, 110364. [CrossRef]
42. Cesarino, I. Structural features and regulation of lignin deposited upon biotic and abiotic stresses. *Curr. Opin. Biotechnol.* **2019**, *56*, 209–214. [CrossRef]
43. Bhardwaj, R.; Handa, N.; Sharma, R.; Kaur, H.; Kohli, S.; Kumar, V.; Kaur, P. Lignins and abiotic stress: An overview. In *Physiological Mechanisms and Adaptation Strategies in Plants under Changing Environment*; Ahmad, P., Wani, M., Eds.; Springer Science & Business Media: Berlin/Heidelberg, Germany, 2014; Volume 1, pp. 267–296. [CrossRef]
44. Boncan, D.A.T.; Tsang, S.S.; Li, C.; Lee, I.H.; Lam, H.M.; Chan, T.F.; Hui, J.H. Terpenes and terpenoids in plants: Interactions with environment and insects. *Int. J. Mol. Sci.* **2020**, *21*, 7382. [CrossRef] [PubMed]
45. Tabaeizadeh, Z. Drought-induced responses in plant cells. *Int. Rev. Cytol.* **1998**, *182*, 193–247. [CrossRef] [PubMed]
46. Krasensky, J.; Jonak, C. Drought, salt, and temperature stress-induced metabolic rearrangements and regulatory networks. *J. Exp. Bot.* **2012**, *63*, 1593–1608. [CrossRef] [PubMed]
47. Chen, W.; He, S.; Liu, D.; Patil, G.B.; Zhai, H.; Wang, F.; Stephenson, T.J.; Wang, Y.; Wang, B.; Nguyen, H.T.; et al. A sweet potato geranylgeranyl pyrophosphate synthase gene, IbGGPS, increases carotenoid content and enhances osmotic stress tolerance in *Arabidopsis thaliana*. *PLoS ONE* **2015**, *10*, e0137623. [CrossRef]
48. Lemoine, R.; Camera, S.L.; Atanassova, R.; Dédaldéchamp, F.; Allario, T.; Pourtau, N.; Bonnemain, J.L.; Laloi, M.; Coutos-Thvenot, P.; Maurousset, L.; et al. Source-to-sink transport of sugar and regulation by environmental factors. *Front. Plant Sci.* **2013**, *4*, 272. [CrossRef] [PubMed]

49. Livingston, D.P.; Hincha, D.K.; Heyer, A.G. Fructan and its relationship to abiotic stress tolerance in plants. *Cell. Mol. Life Sci.* **2009**, *66*, 2007–2023. [CrossRef]
50. Santarius, K.A. The protective effect of sugars on chloroplast membranes during temperature and water stress and its relationship to frost, desiccation and heat resistance. *Planta* **1973**, *113*, 105–114. [CrossRef] [PubMed]
51. Zhou, Q.; Yu, B. Changes in content of free, conjugated and bound polyamines and osmotic adjustment in adaptation of vetiver grass to water deficit. *Plant Physiol. Biochem.* **2010**, *48*, 417–425. [CrossRef] [PubMed]
52. Thalmann, M.; Pazmino, D.; Seung, D.; Horrer, D.; Nigro, A.; Meier, T.; Kölling, K.; Pfeifhofer, H.W.; Zeeman, S.C.; Santelia, D. Regulation of leaf starch degradation by abscisic acid is important for osmotic stress tolerance in plants. *Plant Cell* **2016**, *28*, 1860–1878. [CrossRef]
53. Zadražnik, T.; Moen, A.; Egge-Jacobsen, W.; Meglič, V.; Šuštar-Vozlič, J. Towards a better understanding of protein changes in common bean under drought: A case study of N-glycoproteins. *Plant Physiol. Biochem.* **2017**, *118*, 400–412. [CrossRef]
54. Gregorova, Z.; Kovacik, J.; Klejdus, B.; Maglovski, M.; Kuna, R.; Hauptvogel, P.; Matusikova, I. Drought-induced responses of physiology, metabolites, and PR proteins in *Triticum aestivum*. *J. Agric. Food Chem.* **2015**, *63*, 8125–8133. [CrossRef] [PubMed]
55. Cardenas-Manríquez, G.; Vega-Muñoz, I.; Villagómez-Aranda, A.L.; León-Galvan, M.F.; Cruz-Hernandez, A.; Torres-Pacheco, I.; Rangel-Cano, R.M.; Rivera-Bustamante, R.F.; Guevara-Gonzalez, R.G. Proteomic and metabolomic profiles in transgenic tobacco (*N. tabacum* Xanthi nc) to CchGLP from *Capsicum chinense* BG-3821 resistant to biotic and abiotic stresses. *Environ. Exp. Bot.* **2016**, *130*, 33–41. [CrossRef]
56. Chen, S.; Ding, G.; Wang, Z.; Cai, H.; Xu, F. Proteomic and comparative genomic analysis reveals adaptability of *Brassica napus* to phosphorus-deficient stress. *J. Proteom.* **2015**, *117*, 106–119. [CrossRef] [PubMed]
57. Bhushan, D.; Pandey, A.; Choudhary, M.K.; Datta, A.; Chakraborty, S.; Chakraborty, N. Comparative proteomics analysis of differentially expressed proteins in chickpea extracellular matrix during dehydration stress. *Mol. Cell. Proteom.* **2007**, *6*, 1868–1884. [CrossRef] [PubMed]
58. Gupta, D.B.; Rai, Y.; Gayali, S.; Chakraborty, S.; Chakraborty, N. Plant organellar proteomics in response to dehydration: Turning protein repertoire into insights. *Front. Plant Sci.* **2016**, *7*, 460. [CrossRef] [PubMed]
59. Dietz, K.J.; Sauter, A.; Wichert, K.; Messdaghi, D.; Hartung, W. Extracellular β-glucosidase activity in barley involved in the hydrolysis of ABA glucose conjugate in leaves. *J. Exp. Bot.* **2000**, *51*, 937–944. [CrossRef] [PubMed]
60. Ali, S.; Hayat, K.; Iqbal, A.; Xie, L. Implications of abscisic acid in the drought stress tolerance of plants. *Agronomy* **2020**, *10*, 1323. [CrossRef]
61. Wu, J.; Wang, J.; Hui, W.; Zhao, F.; Wang, P.; Su, C.; Gong, W. Physiology of plant responses to water stress and related genes: A review. *Forests* **2022**, *13*, 324. [CrossRef]
62. Todaka, D.; Takahashi, F.; Yamaguchi-Shinozaki, K.; Shinozaki, K. ABA-responsive gene expression in response to drought stress: Cellular regulation and long-distance signalling. In *Advances in Botanical Research*; Academic Press: Cambridge, MA, USA, 2019; Volume 92, pp. 83–113. [CrossRef]
63. Arbona, V.; Zandalinas, S.I.; Manzi, M.; González-Guzmán, M.; Rodriguez, P.L.; Gómez-Cadenas, A. Depletion of abscisic acid levels in roots of flooded *Carrizo citrange* (*Poncirus trifoliata* L. Raf.× *Citrus sinensis* L. Osb.) plants is a stress-specific response associated to the differential expression of PYR/PYL/RCAR receptors. *Plant Mol. Biol.* **2017**, *93*, 623–640. [CrossRef]
64. Hsu, F.C.; Chou, M.Y.; Peng, H.P.; Chou, S.J.; Shih, M.C. Insights into hypoxic systemic responses based on analyses of transcriptional regulation in *Arabidopsis*. *PLoS ONE* **2011**, *6*, e28888. [CrossRef]
65. Jackson, M.B.; Young, S.F.; Hall, K.C. Are roots a source of abscisic acid for the shoots of flooded pea plants? *J. Exp. Bot.* **1988**, *39*, 1631–1637. [CrossRef]
66. Rodríguez-Gamir, J.; Ancillo, G.; González-Mas, M.C.; Primo-Millo, E.; Iglesias, D.J.; Forner-Giner, M.A. Root signaling and modulation of stomatal closure in flooded citrus seedlings. *Plant Physiol. Biochem.* **2011**, *49*, 636–645. [CrossRef] [PubMed]
67. Zhu, X.; Li, X.; Jiu, S.; Zhang, K.; Wang, C.; Fang, J. Analysis of the regulation networks in grapevine reveals response to waterlogging stress and candidate gene-marker selection for damage severity. *R. Soc. Open Sci.* **2018**, *5*, 172253. [CrossRef]
68. Smith, G.S.; Miller, S.A. Effects of root anoxia on the physiology of kiwifruit vines. *Acta Hort.* **1991**, *297*, 401–408. [CrossRef]
69. Creelman, R.A.; Gage, D.A.; Stults, J.T.; Zeevaart, J.A. Abscisic acid biosynthesis in leaves and roots of *Xanthium strumarium*. *Plant Physiol.* **1987**, *85*, 726–732. [CrossRef] [PubMed]
70. Lee, S.C.; Lan, W.; Buchanan, B.B.; Luan, S. A protein kinase-phosphatase pair interacts with an ion channel to regulate ABA signaling in plant guard cells. *Proc. Natl. Acad. Sci. USA* **2009**, *106*, 21419–21424. [CrossRef] [PubMed]
71. Lee, S.C.; Lim, C.W.; Lan, W.; He, K.; Luan, S. ABA signaling in guard cells entails a dynamic protein–protein interaction relay from the PYL-RCAR family receptors to ion channels. *Mol. Plant* **2013**, *6*, 528–538. [CrossRef] [PubMed]
72. Lim, C.W.; Baek, W.; Jung, J.; Kim, J.H.; Lee, S.C. Function of ABA in stomatal defense against biotic and drought stresses. *Int. J. Mol. Sci.* **2015**, *16*, 15251–15270. [CrossRef]
73. Jung, H.; Lee, D.K.; Do Choi, Y.; Kim, J.K. OsIAA6, a member of the rice Aux/IAA gene family, is involved in drought tolerance and tiller outgrowth. *Plant Sci.* **2015**, *236*, 304–312. [CrossRef]
74. Salehin, M.; Li, B.; Tang, M.; Katz, E.; Song, L.; Ecker, J.R.; Kliebenstein, D.J.; Estelle, M. Auxin-sensitive Aux/IAA proteins mediate drought tolerance in *Arabidopsis* by regulating glucosinolate levels. *Nat. Commun.* **2019**, *10*, 4021. [CrossRef]
75. Liu, Q.; Feng, Z.; Xu, W.; Vetukuri, R.R.; Xu, X. Exogenous melatonin-stimulated transcriptomic alterations of *Davidia involucrata* seedlings under drought stress. *Trees* **2021**, *35*, 1025–1038. [CrossRef]

76. Du, H.; Wu, N.; Chang, Y.; Li, X.; Xiao, J.; Xiong, L. Carotenoid deficiency impairs ABA and IAA biosynthesis and differentially affects drought and cold tolerance in rice. *Plant Mol. Biol.* **2013**, *83*, 475–488. [CrossRef] [PubMed]
77. Du, H.; Wu, N.; Fu, J.; Wang, S.; Li, X.; Xiao, J.; Xiong, L. A GH3 family member, OsGH3-2, modulates auxin and abscisic acid levels and differentially affects drought and cold tolerance in rice. *J. Exp. Bot.* **2012**, *63*, 6467–6480. [CrossRef] [PubMed]
78. Kim, J.; Baek, D.; Park, H.C.; Chun, H.J.; Oh, D.H.; Lee, M.K.; Cha, Y.J.; Kim, W.Y.; Kim, M.C.; Chung, W.S.; et al. Overexpression of *Arabidopsis* YUCCA6 in potato results in high-auxin developmental phenotypes and enhanced resistance to water deficit. *Mol. Plant* **2013**, *6*, 337–349. [CrossRef] [PubMed]
79. Kazan, K.; Manners, J.M. Linking development to defense: Auxin in plant–pathogen interactions. *Trends Plant Sci.* **2009**, *14*, 373–382. [CrossRef]
80. Shi, H.; Chen, L.; Ye, T.; Liu, X.; Ding, K.; Chan, Z. Modulation of auxin content in *Arabidopsis* confers improved drought stress resistance. *Plant Physiol. Biochem.* **2014**, *82*, 209–217. [CrossRef]
81. Zhang, Y.; Li, Y.; Hassan, M.J.; Li, Z.; Peng, Y. Indole-3-acetic acid improves drought tolerance of white clover via activating auxin, abscisic acid and jasmonic acid related genes and inhibiting senescence genes. *BMC Plant Biol.* **2020**, *20*, 150. [CrossRef]

horticulturae

Article

An Evaluation of the Effect of Fertilizer Rate on Tree Growth and the Detection of Nutrient Stress in Different Irrigation Systems

Krishna Neupane, Anthony Witcher and Fulya Baysal-Gurel *

Department of Agricultural Sciences and Engineering, College of Agriculture, Tennessee State University, Otis L. Floyd Nursery Research Center, 472 Cadillac Lane, McMinnville, TN 37209, USA; kneupane@my.tnstate.edu (K.N.); awitcher@tnstate.edu (A.W.)
* Correspondence: fbaysalg@tnstate.edu

Abstract: Early season monitoring of nutrient stress is important in red maple (*Acer rubrum* L.) and flowering dogwood (*Cornus florida* L.) to optimize management practices and ensure healthy crop production in containers. Two different irrigation systems (drip and overhead irrigation) were used in this study. Two rates (low and high) of controlled-release fertilizer were used with no fertilizer as a control treatment. Data were recorded for plant height, stem diameter, substrate pH and electrical conductivity (EC), chlorophyll content, normalized difference vegetation index (NDVI), visual observation of plant quality, and leaf nutrient content. The results of this study showed that the increase in plant height and stem diameter was greater among the fertilized maple tree, whereas no differences were observed in the flowering dogwoods for an increase in plant height. NDVI was greater for drip irrigation for both fertilizer rates in both red maples and flowering dogwoods. A positive correlation of 73% to 83% was observed for red maples and 79% to 83% was observed for flowering dogwoods between handheld NDVI and unmanned aerial vehicle-mounted NDVI sensors. In red maple, a high fertilizer rate resulted in greater substrate pH, whereas in flowering dogwood, no differences were observed. Varied responses were observed among the treatments for nutrient content; however, both rates of fertilizer application were sufficient for both tree species. Drip-irrigated red maples had higher nitrogen and phosphorous content, whereas nitrogen content was higher in both irrigation systems in flowering dogwoods. This study provides useful insights into understanding the effect of nutrient stress on tree growth and the application of sensing technology for the monitoring and early detection of nutrient stress in container-grown nursery crops.

Keywords: flowering dogwood; nutrient stress; red maple; UAVs; NDVI

Citation: Neupane, K.; Witcher, A.; Baysal-Gurel, F. An Evaluation of the Effect of Fertilizer Rate on Tree Growth and the Detection of Nutrient Stress in Different Irrigation Systems. *Horticulturae* **2024**, *10*, 767. https://doi.org/10.3390/horticulturae10070767

Academic Editors: Changxia Li and Yue Wu

Received: 19 June 2024
Revised: 15 July 2024
Accepted: 15 July 2024
Published: 19 July 2024

1. Introduction

The monitoring of the nutrient status provides an overall condition of health and vigor and is considered an important aspect of crop production. Several researchers and governing bodies have defined plant nutrients and fertilizers differently. Dobermann et al. [1] defined plant nutrients as an important element, the absence of which might restrict plants from completing their lifecycle, whereas Mee et al. [2] highlighted the importance of plant nutrients as essential components for growth and vitality. Kirkby [3] published an article stating that fourteen essential nutrients are required for plants to complete their life cycle. Fertilizers, on the other hand, are defined as important additional substances that help plants increase their growth, maintain their health, and improve productivity. Both inadequate and excessive amounts of nutrients available to the plants are considered nutrient stress or fertilizer stress. Insufficient fertilizer application causes a deficiency of nutrients in the soil, while excessive application becomes detrimental to plants, soil, water, and the environment [4]. Fertilizer stress could result in restricted root and shoot growth, chlorosis, early defoliation of leaves, and decreased biomass [2,5,6]. Thus, it is important to manage the application of fertilizer, providing only the required amount without over-exceeding or restraining nutrient status in plants.

The total amount of nutrients present in the soil may not always be available to the plants. The availability of nutrients to plants depend on several factors, such as soil moisture content, soil temperature, pH, the physical conditions of the soil, and other toxic elements or salts [7]. Li et al. [8] reported that the movement of nutrients from the soil to roots and roots to the above-ground part is greatly influenced by adequate water content. Similarly, Rose et al. [9,10] described that adequate water is required for nutrient uptake in container-grown crops. Low soil water potential causes a restriction in the flow of nutrients to the roots through diffusion and mass flow [9,11]. Additionally, an adequate supply of nutrients to the plants helps them to withstand adverse environmental conditions [6,12–14]. A study conducted by Novair et al. [15] showed that the application of organic amendments like azo-compost and rice straw plays a vital role in increasing nutrient (phosphorus) uptake by plants in low-irrigation conditions. An increase in phosphorus uptake was recorded by Samadi et al. [16] while using biofertilizers in corn fields in drought conditions.

Irrigation systems play a vital role, particularly in terms of maintaining soil water status. They also play a huge factor in the growth of plants. Irrigation systems and fertilizers affect plant growth [17]. Diverse types of irrigation systems have been developed to irrigate containerized crops. Each system has its own pros and cons. Overhead irrigation systems supply water while also washing off the dust and fungal debris attached to the foliage of plants and reducing the temperature of the plants [18]. However, overhead irrigation uses excess water to ensure complete coverage of the production area [19]. On the other hand, drip irrigation reduces leaching and evaporation loss and allows efficient management of the water [20]. A study conducted by Rowe et al. [21] found that overhead-irrigated *Sedum album* L. and *S. floriferum* L. had greater plant height and plant health, while a study conducted by Davies et al. [22] found no significant differences between overhead irrigation and drip irrigation in terms of the vegetative growth of *Cornus alba* L. and *Lonicera periclymenum* L. On the other hand, Yang et al. [20] reported that drip irrigation is a clear winner among other irrigation systems as it is efficient in terms of water productivity.

Additionally, several approaches have been identified for monitoring the biotic and abiotic stress of plants. Remote sensing has been widely used by scientists and researchers to identify and monitor crop stress. Chowhan and Chakraborty [23] explained remote sensing as a dynamic technique that records stress in the crop through the changes in electromagnetic radiation and quantifies the amount of stress. Mee, Siva, and Ahmad [2] explained that specific wavelength measurements could help to identify different types of plant stresses. Cadet and Samson [24] and Gulzar et al. [25] used the fluorescence emission technique to detect nitrogen, potassium, and phosphorus deficiency in sunflowers. An in situ real-time monitoring system was used to monitor nitrate and phosphorous uptake in plants [26]. Mahajan et al. [27] used hyperspectral reflectance techniques, applying eight different vegetative indices, to monitor nitrogen, phosphorus, sulfur, and potassium in wheat. Additionally, remote sensing was used to evaluate the performance of irrigation systems and their management [28]. This shows that the application of remote sensing and UAVs can have an enormous impact on the monitoring of water status and fertilizer stress in woody ornamental production systems.

Red maple and flowering dogwoods are important ornamental landscape crops. However, their water requirements are different. Red maples are moderately tolerant to flooding [29], but flowering dogwoods are intolerant to flooding [30]. Most commercial ornamental nursery growers use overhead sprinklers (small containers) or micro-irrigation systems (large containers) to supply adequate water to the crops [31]. However, only 20–40% of the water from overhead irrigation is retained in the containers [32]. The objective of this study was to evaluate the response of two tree species in a container production system while using different fertilizer rates in two irrigation systems. Also, the objective was to monitor which irrigation system performs well in different fertility conditions to improve the production of these tree species. Similarly, this study was also designed to monitor nutrient stress through aerial imagery and sensing technology. This study will help with understanding the irrigation system, water use efficiency, fertilizer requirements, and optimization of important ornamental crops like red maple and flowering dogwoods.

2. Materials and Methods

2.1. Experimental Site, Planting Materials, and Design of Study

This study was conducted at the Otis L. Floyd Nursery Research Center in McMinnville, TN, during the summer of 2022. Red maple 'Sun Valley' and flowering dogwood 'Cherokee Princess' bare root liners were purchased from a commercial nursery and were transplanted (February 2022) into 11.4 L (3-gallon) nursery containers (C1200; Nursery Supplies, Chambersburg, PA, USA) filled with a standard pine bark substrate (1.6 cm screened; Morton's Horticultural Products, Inc., McMinnville, TN, USA). The transplanted trees were moved to an outdoor gravel pad in the last week of April. Trees appearing similar in shoot size and health conditions visually were selected for this study. Twenty individual trees were arranged randomly into four replications (block of five individuals per replication) for each treatment in a split-plot design. This study was conducted from 9 May to 10 August 2022.

2.2. Fertilizer Rate

Three fertilizer treatments included control (no fertilizer), low rate of fertilizer, and high rate of fertilizer. Controlled-release fertilizer (Florikan 17-7-8 NPK total, 6 months @ 80 °F; Florikan LLC, Sarasota, FL, USA) (composition: total N: 17%; P: 7%; K: 8%; S: 4%; Mg: 1.20%; Fe: 0.20%; Mn: 0.06%; Mo: 0.002%) was applied as a top-dress to each container on 9 May 2022 in different rates for each tree species. The low rates for flowering dogwood and red maple were 18 and 35 g per container, respectively. High rates were 55 and 70 g per container for flowering dogwood and red maple, respectively (manufacturer's recommendation). Lower rates were used for flowering dogwood due to sensitivity to highly soluble salt levels [33].

2.3. Irrigation System

Two distinct types of irrigation systems (drip and overhead) were used. Drip irrigation was provided three times daily (11 a.m., 12 p.m., and 1 p.m.) using a dribble ring (15.2 cm diameter; Dramm Corp., Manitowoc, WI, USA) fitted with a pressure-compensating emitter (4 L h^1; Netafim USA, Fresno, CA, USA). Irrigation application volume for each treatment was adjusted monthly based on replacement of daily water use/loss from the previous 24 h period. A subset of containers (4 replications) was irrigated (hand-watered) to saturation, allowed to drain for 1 h, then weighed. After 24 h, the same containers were weighed, and the difference in weight was calculated and converted to ml, which was then applied daily (equally divided among the three irrigation times). Overhead irrigation was provided once daily using multi-stream nozzles (MP3000 Rotator; Hunter Industries, San Marcos, CA, USA). Overhead irrigation application volume was adjusted periodically to maintain a leaching fraction of 10–30%. A subset of containers (4 replications) was nested inside 9.5 L plastic buckets so there was no gap between the top rim of the bucket and the container sidewall to measure irrigation leach volume of containers. A separate set of empty buckets (4 replications) was also placed between plants to measure irrigation application volume. One hour after the daily irrigation, irrigation application rate and leaching volume were recorded, and leaching fraction was calculated: (leach volume ÷ application volume) × 100 = leaching fraction. Each irrigation system was designated in a separate gravel pad controlled by a single solenoid.

2.4. Data Collection

2.4.1. Plant Growth Data

Changes in growth (plant height and stem diameter) were recorded as plant growth data. Plant height was recorded as total length from the substrate surface to the top growing point of the plant. Stem diameter was measured 5 cm above the base of the tree using a digital caliper (Mitutoyo Corp., Kanagawa, Japan). Plant height and stem diameter were recorded (twenty plants per treatment) at the beginning of this study and monthly for 3 months (June, July, August).

2.4.2. Normalized Difference Vegetation Index (NDVI)

NDVI values were collected from a handheld Field Scout CM 1000 NDVI Meter (Spectrum Technologies Inc., Aurora, IL, USA) and an unmanned aerial vehicle (UAV) (DJI Mavic 2; DJI Technology Co., Ltd., Shenzhen, Guangdong, China) mounted with Sentera single-sensor NDVI camera (Sentera Inc., Saint Paul, MN, USA). The UAV was flown monthly at a height of 20 m in bright, sunny conditions between 9 a.m. and 2 p.m. The aerial images obtained from UAV were processed to obtain NDVI values using commercial Sentera Field Agent software [Version v5.20.0].

2.4.3. Relative Chlorophyll

Relative chlorophyll content was recorded as Soil–Plant Analytical Development (SPAD) values monthly using a 502-leaf chlorophyll meter (Spectrum Technologies Inc., Aurora, IL, USA). The SPAD meter was calibrated each time while collecting data for different treatments to reduce the chances of error [34]. Data were recorded for every plant on a bright, sunny day between 9 a.m. and 2 p.m.

2.4.4. Visual Observation

Plants were visually rated for leaf chlorosis (monthly) to document the severity of any symptoms that could have resulted from nutrient deficiency. Ratings were ranked on a scale of 1–5 (1 = 1–20%; 2 = 21–40%; 3 = 41–60%; 4 = 61–80%; 5 = 81–100%) for the percentage of foliage showing chlorotic symptoms. Data were recorded for every tree used in this study.

2.4.5. Substrate Chemical Properties

Substrate pH and electrical conductivity (EC) were recorded monthly from leachate samples collected using the pour-through method [35]. Briefly, containers were irrigated just to saturation and allowed to drain for 1 h; then, 295 mL water was slowly applied to the substrate surface, and leachate was collected after 5 min and then brought to the lab for analysis. Leachate samples (four replications) were collected from a single plant from each replication and treatment. Leachates were analyzed, as their study helps to understand the salinity and nutrient uptake by the plants [36].

2.4.6. Leaf Tissue Analysis

At the end of this study, leaf samples were collected from representative plants of each treatment (n = 3) to analyze the nutrient content in the leaves. The samples were collected, dried in a forced-air oven at 50 °C, and then sent to a commercial laboratory (Waypoint Analytical Corp., Memphis, TN, USA) for analysis. Leaf tissue was analyzed for nitrogen (N), phosphorus (P), and potassium (K) content. Total nitrogen was analyzed using the automated combustion method (Total Nitrogen Analyzer: Leco, Carl Erba, Elementar, and PerkinElmer with resistance furnace with thermal conductivity detector). Total phosphorus and potassium were analyzed using the microwave digestion/dissolution closed vessel method.

2.5. Statistical Analysis

Data were analyzed using SAS 9.4. Variables were analyzed using two-factor analysis with fertilizer rates and irrigation systems as factors. Plant growth data (height and stem diameter), NDVI (handheld meter), leaf greenness, leaf nutrient content, and substrate pH and EC were analyzed using general linear model, and means were separated using Fisher's LSD test (Proc GLM). Visual observation ratings were analyzed using General

Linear Mixed Model (GLMM) using normal distribution. Data with 100% and 0% rankings were converted to 0.99 and 0.01 to meet the assumptions of Proc GLIMMIX. Brown–Forsythe *t*-test ANOVA was used for data with unequal variances. Data for each month (June, July, and August) were analyzed separately.

3. Results

Among the treatments, significant differences were observed in the increase in plant height, stem diameter, NDVI, chlorophyll content, visual observation of plant quality, pH and EC, and leaf nutrient contents. A strong and positive correlation was observed between the UAV-mounted NDVI and the handheld NDVI.

3.1. Plant Height

For red maple, significant differences were observed among the treatments for fertilizer rate (June: $p \leq 0.0001$, July: $p \leq 0.0001$, August: $p \leq 0.0001$) and irrigation systems (June: $p = 0.0109$, July: $p = 0.0006$, August: $p \leq 0.0001$) throughout the study duration. However, significant interactions between fertilizer rates and irrigation systems were observed in June and July but not in August (June: $p = 0.0028$, July: $p = 0.0026$, August: $p = 0.0597$). In June and July, both rates of fertilizers in both irrigation systems led to a greater increase in plant height compared to the no-fertilizer controls. In August, both rates of fertilizer in overhead irrigation and the low rate of fertilizer in drip irrigation led to greater increases in plant height compared to the no-fertilizer controls and the high rate of fertilizer in drip irrigation (Figure 1).

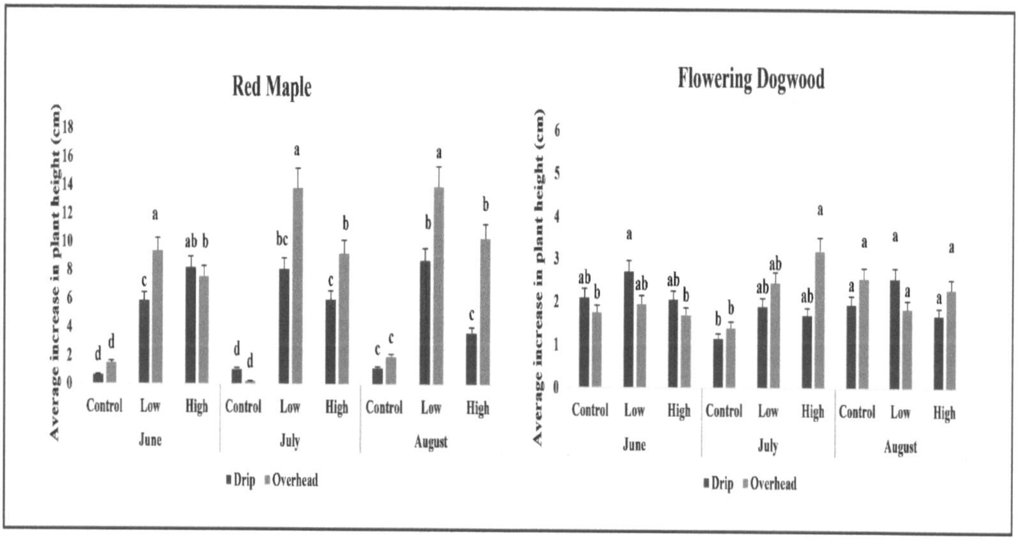

Figure 1. Average increase in plant height (cm) for containerized red maples (**left**) and flowering dogwoods (**right**) over three months using different fertilizer rates in overhead and drip irrigation systems. The increase in height in June was calculated as the difference between height in June and initial height of the plant. Different letters above the error bar (standard error) represent significant differences in increase in plant height (within each month) among the treatments ($\alpha = 0.05$, Fisher's LSD test).

For flowering dogwood, no significant differences were observed among the treatments for fertilizer rate (June: $p = 0.2740$, July: 0.1138; August: $p = 0.9517$), irrigation system (June: $p = 0.0546$, July: 0.1136, August: $p = 0.8086$), and interaction between fertilizer rates and irrigation systems (June: $p = 0.7511$, July: 0.5374, August: $p = 0.6723$) in terms of an increase in plant height throughout the study duration. In June, the low rate of fertilizer in drip irrigation led to a greater increase in plant height, but no significant differences were observed among the treatments within each irrigation system. In July, the high rate of fertilizer in overhead irrigation led to a significantly greater increase in plant height compared to both no-fertilizer controls, whereas no significant differences were observed among the treatments in August (Figure 1).

3.2. Stem Diameter

In red maple, significant differences in fertilizer rate were observed throughout the study period (June: $p \leq 0.0001$; July: $p \leq 0.0001$; August: $p \leq 0.0001$). For the irrigation system, significant differences were observed in July and August but not in June (June: $p = 0.1612$; July: $p \leq 0.0001$; August: $p = 0.0050$). The interaction between the fertilizer rate and the irrigation system was significant in June and July (June: $p = 0.0066$; July: $p \leq 0.0001$; August: $p = 0.1520$). Throughout the study duration, both rates of fertilizers applied on both irrigation systems led to a greater increase in stem diameter compared to the no-fertilizer controls. Controls of both irrigation systems had statistically similar increases in stem diameter (Figure 2).

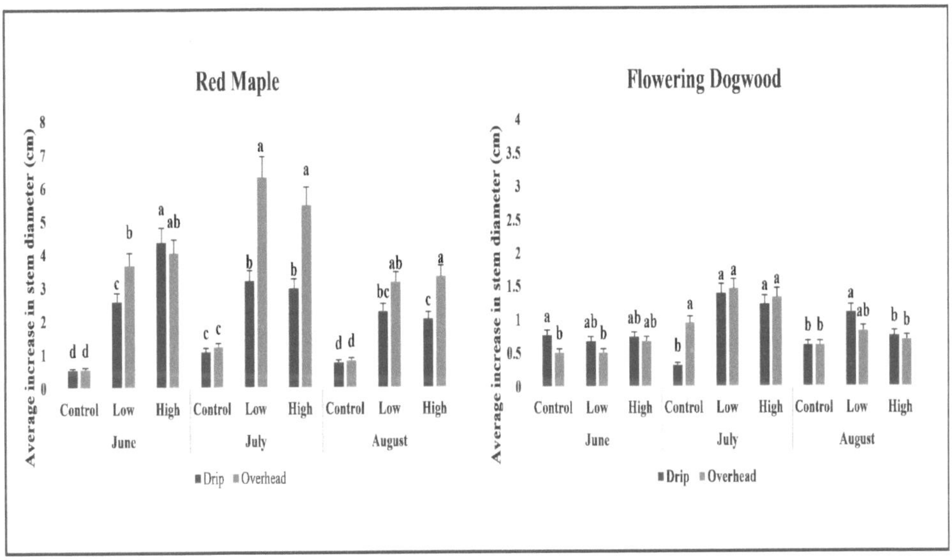

Figure 2. Average increase in stem diameter (cm) for containerized red maple (**left**) and flowering dogwoods (**right**) over three months using different fertilizer rates in overhead and drip irrigation systems. The increase in stem diameter in June was calculated as the differences in height in June and initial stem diameter of the plant. Different letters above the error bar (standard error) represent significant differences in increase in stem diameter (within each month) among the treatments ($\alpha = 0.05$, Fisher's LSD test).

In flowering dogwood, significant differences were only observed in June for the irrigation system (June: $p = 0.0205$; July: $p = 0.1267$; August: $p = 0.2428$) and in July and August for the fertilizer rate (June: $p = 0.4277$; July: $p = 0.0007$; August: $p = 0.0112$). No interactions between fertilizer rates and irrigation systems were observed (June: $p = 0.4903$; July: $p = 0.3460$; August: $p = 0.4875$). In June, drip irrigation led to a greater stem diameter compared to overhead irrigation in the no-fertilizer control, but there were no differences using the two other fertilizer rates. In July, all treatments had greater stem diameters compared to those that underwent drip irrigation with no-fertilizer controls. In August, fertilizer that underwent a low rate of drip irrigation led to greater stem diameter compared to the no-fertilizer and high fertilizer rates (regardless of irrigation system) (Figure 2).

3.3. NDVI (Handheld Meter)

For red maple, significant differences were observed in fertilizer rates throughout the study period (June: $p \leq 0.0001$; July: $p \leq 0.0001$; August: $p \leq 0.0001$). For irrigation systems, statistical differences were observed among the treatments in July and August (June: $p = 0.1740$; July: $p \leq 0.0001$; August: $p = 0.0244$). Interactions between the fertilizer rate and the irrigation system were observed in July only (June: $p = 0.2120$; July: $p = 0.0422$; August: $p = 0.2839$). Red maples fertilized using both low and high fertilizer rates, regardless of irrigation system, had greater NDVI compared to both controls. Controls in both irrigation systems had statistically similar and significantly lower NDVI compared to others in June and August, while in July, the control of overhead irrigation was significantly higher than that of drip irrigation (Figure 3).

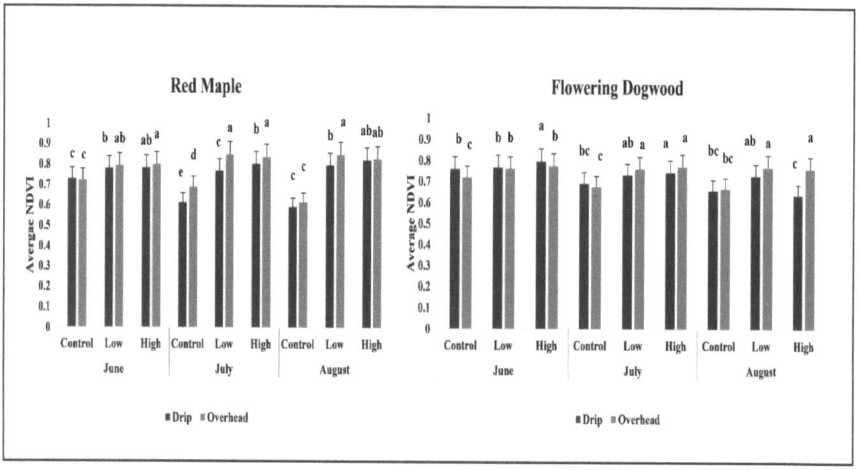

Figure 3. Normalized Difference Vegetative Index (NDVI) for containerized red maple (**left**) and flowering dogwoods (**right**) over three months using different fertilizer rates in overhead and drip irrigation systems. Different letters above the error bar (standard error) represent significant differences in NDVI (within each month) among the treatments ($\alpha = 0.05$, Fisher's LSD test).

Among flowering dogwoods, significant differences were observed for fertilizer rates throughout the study period (June: $p \leq 0.0001$; July: $p \leq 0.0001$; August: $p = 0.0099$); however, statistical differences were only observed in June and August for irrigation system (June: $p \leq 0.0001$; July: $p = 0.2699$; August: $p = 0.0120$). Interactions between fertilizer rates and irrigation systems occurred in June (June: $p = 0.0346$; July: $p = 0.2250$; August: $p = 0.1009$). In June, drip-irrigated flowering dogwoods with a high rate of fertilizer had the greatest NDVI (Figure 3). In July, both rates of fertilizer for overhead-irrigated plants and the high fertilizer rate for drip-irrigated plants led to greater NDVI compared to both the no-fertilizer controls. In August, both rates of fertilizer in overhead irrigation led to greater NDVI compared to all other treatments except for the low fertilizer rate for drip irrigation.

A positive correlation was observed between the UAV-mounted NDVI sensor and the handheld NDVI sensor. In red maples, 77%, 73%, and 83% correlations were observed in June, July, and August, respectively (June: $r = 0.77$, $p \leq 0.0001$; July: $r = 0.73$, $p \leq 0.0001$; August: $r = 0.83$, $p \leq 0.0001$; Figure 4). In flowering dogwood, 79%, 88%, and 83% correlations were observed in June, July, and August (June: $r = 0.79$, $p \leq 0.0001$; July: $r = 0.88$, $p \leq 0.0001$; August: $r = 0.83$, $p \leq 0.0001$; Figure 4). An NDVI mosaic of red maples and flowering dogwoods treated with different rates of fertilizers is shown in Figure 5.

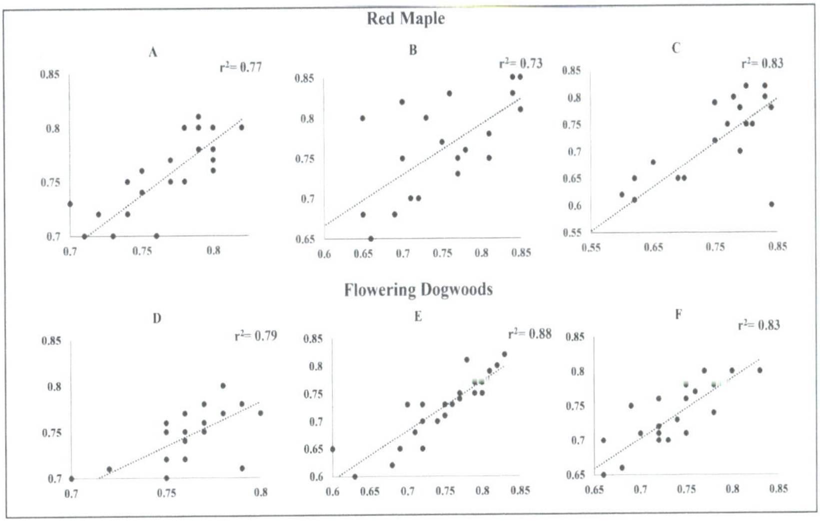

Figure 4. Pearson's correlation of the mean NDVI values between a handheld NDVI meter and a Sentera NDVI sensor mounted in a UAV for containerized red maple [June (**A**), July (**B**), August (**C**)] and flowering dogwoods [June (**D**), July (**E**), August (**F**)] treated with different fertilizer rates in overhead and drip irrigation systems.

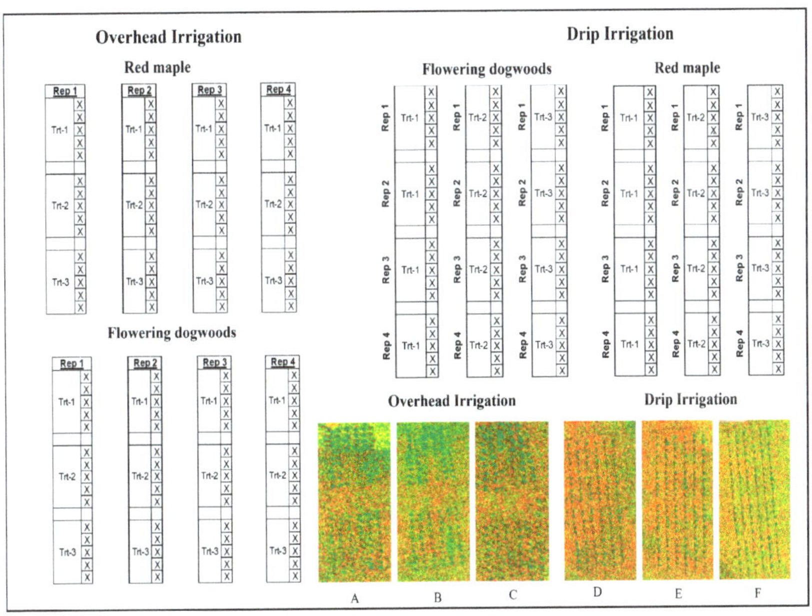

Figure 5. Layout of this study (**top**) [Trt 1: control; Trt 2: low rate of fertilizer; Trt 3: high rate of fertilizer] and Normalized Difference Vegetative Index (NDVI) mosaics combined (**bottom right**) for containerized red maple and flowering dogwoods treated with different fertilizer rates over three months for overhead [June (**A**), July (**B**), August (**C**)] and drip [June (**D**), July (**E**), August (**F**)] irrigation systems.

3.4. Relative Chlorophyll Content

In red maple, for relative chlorophyll content, significant differences occurred during July and August but not in June for fertilizer rates and the interaction between the fertilizer rate and the irrigation system (fertilizer rate: June: $p = 0.3010$; July: $p \leq 0.0001$; August: $p \leq 0.0001$; interaction: June: $p = 0.1424$; July: $p = 0.0085$; August: $p = 0.0192$). For the irrigation system, significant differences were only observed in August (June: $p = 0.4277$; July: $p = 0.9410$; August: $p = 0.0408$). In June, the high fertilizer rate in drip irrigation led to greater relative chlorophyll content compared to the low fertilizer rate in overhead irrigation and the no-fertilizer drip irrigation. In July and August, both fertilizer rates, regardless of irrigation system, led to greater relative chlorophyll content compared to both no-fertilizer controls (Figure 6).

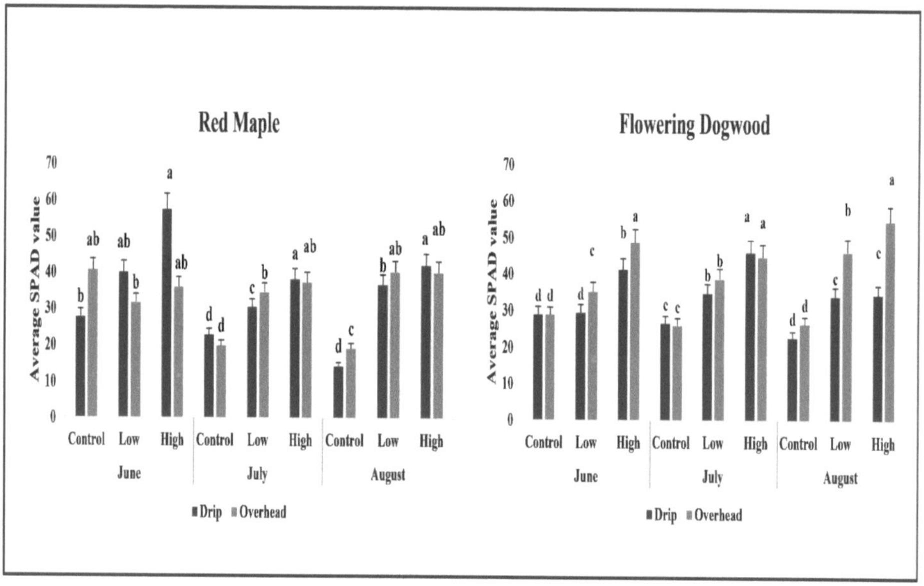

Figure 6. Average SPAD value for containerized red maple (**left**) and flowering dogwoods (**right**) over three months using different fertilizer rates in overhead and drip irrigation systems. Different letters above the error bar (standard error) represent significant differences in leaf greenness (within each month) among the treatments ($\alpha = 0.05$, Fisher's LSD test).

In flowering dogwood, significant differences were observed for fertilizer rates in June, July, and August (June: $p \leq 0.0001$; July: $p \leq 0.0001$; August: $p \leq 0.0001$). For irrigation systems (June: $p \leq 0.0001$; July: $p = 0.5668$; August: $p \leq 0.0001$) and interactions between fertilizer rates and irrigation systems (June: $p = 0.0031$; July: $p = 0.2067$; August: $p \leq 0.0106$), significant differences were observed in June and August only. In June, the high fertilizer rates in both irrigation systems and the low fertilizer rate in the overhead irrigation led to greater relative chlorophyll content compared to both no-fertilizer controls (Figure 6). In July and August, both fertilizer rates in both irrigation systems led to greater relative chlorophyll content compared to both no-fertilizer controls.

3.5. Visual Observation

Controls in both irrigation systems and both tree species showed significantly higher chlorotic symptoms compared to the other treatments. In red maple, significant differences were observed in the fertilizer rates throughout the study period (June: $p \leq 0.0001$; July: $p \leq 0.0001$; August: $p \leq 0.0001$). For the irrigation system, significant differences were observed in July and August (June: $p = 0.1640$; July: $p = 0.0066$; August: $p = 0.0011$). The interaction between fertilizer rates and irrigation systems was only observed in August (June: $p = 0.1452$; July: $p = 0.8452$; August: $p = 0.0003$; Figure 7). Throughout the study period, the no-fertilizer controls in both irrigation systems had the greatest chlorotic symptoms among all other treatments. Within the no-fertilizer controls, drip irrigation led to greater chlorotic symptoms than overhead irrigation in July and August.

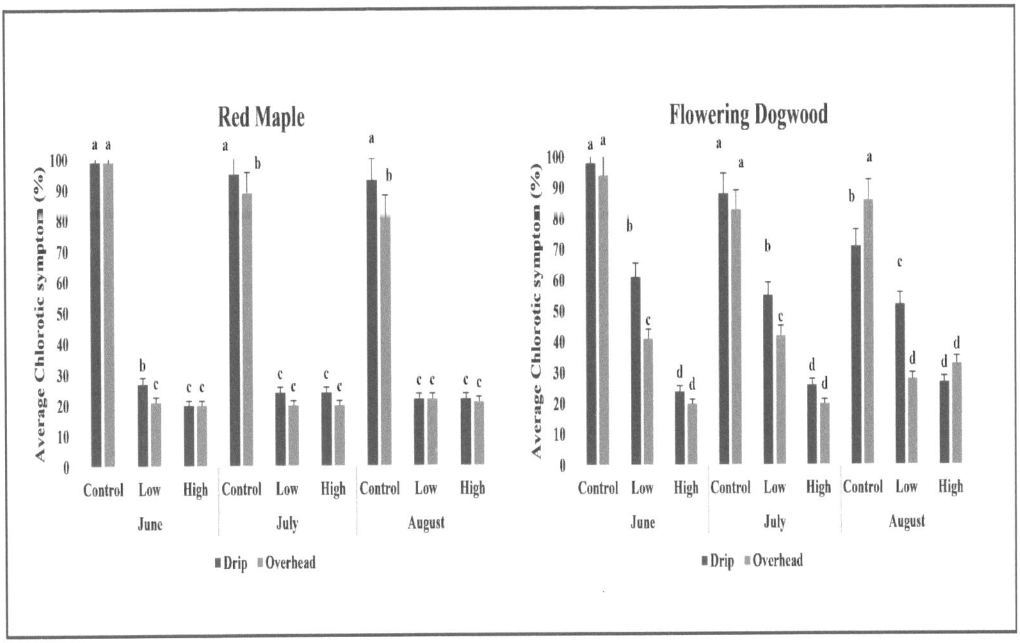

Figure 7. Average percentage of foliage showing chlorotic symptoms (visual observation) for containerized red maple (**left**) and flowering dogwoods (**right**) over three months using different fertilizer rates in overhead and drip irrigation systems. Different letters above the error bar (standard error) represent significant differences in chlorotic symptoms (within each month) among the treatments ($\alpha = 0.05$, Fisher's LSD test).

In flowering dogwood, significant differences were observed in fertilizer rate throughout the study period (June: $p \leq 0.0001$; July: $p \leq 0.0001$; August: $p \leq 0.0001$). For the irrigation system (June: $p = 0.0027$; July: $p = 0.0119$; August: $p = 0.8077$) and the interaction between fertilizer rates and irrigation systems (June: $p = 0.0434$; July: $p = 0.4740$; August: $p \leq 0.0001$; Figure 7), significant differences were observed in June and July and June and August, respectively. Throughout this study, both no-fertilizer controls had greater chlorotic symptoms compared to all other treatments.

3.6. Substrate pH and EC

In red maple, significant differences were observed for fertilizer rates throughout the study period (June: $p = 0.0361$; July: $p = 0.0173$; August: $p = 0.0015$), whereas no significant differences were observed for irrigation systems (June: $p = 0.5749$; July: $p = 0.8637$; August: $p = 0.8311$) or the interaction between fertilizer rates and irrigation systems (June: $p = 0.7895$; July: $p = 0.7817$; August: $p = 0.3177$). In June, substrate pH was greater for low fertilizer rate (both irrigation systems) and the no-fertilizer drip irrigation compared to the high fertilizer rate in overhead irrigation (Figure 8). In July and August, substrate pH decreased as the fertilizer rate increased regardless of the irrigation system. For substrate EC, significant differences were observed for fertilizer rates throughout the study period (June: $p \leq 0.0001$; July: $p = 0.0052$; August: $p = 0.0079$). For irrigation systems (June: $p \leq 0.0001$; July:

$p = 0.5049$; August: $p = 0.0159$) and the interaction between fertilizer rates and irrigation systems (June: $p \leq 0.0001$; July: $p = 0.2978$; August: $p = 0.0358$), significant differences were observed in June and August. Overall, substrate EC increased with fertilizer rate, and substrate EC tended to be numerically greater for the drip-irrigated plants. In June, substrate EC was greater for drip irrigation involving low and high fertilizer rates compared to overhead irrigation (Figure 9). In August, substrate EC was greatest for the high fertilizer rate in drip irrigation compared to all other treatments.

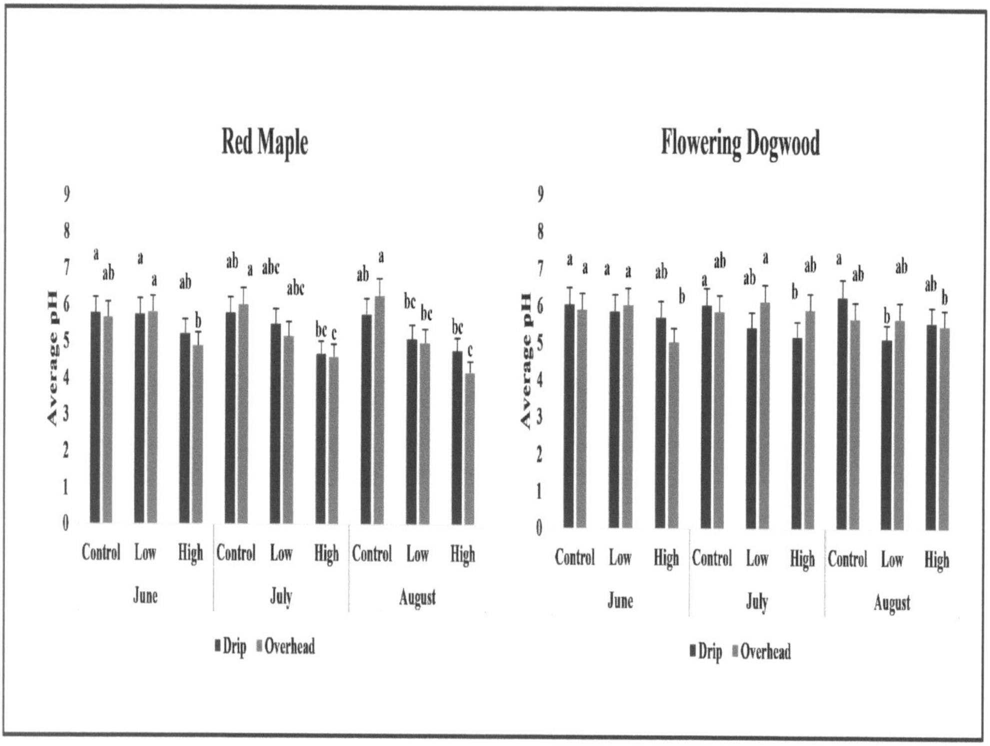

Figure 8. Average substrate pH for containerized red maple (**left**) and flowering dogwoods (**right**) over three months using different fertilizer rates in overhead and drip irrigation systems. Different letters above the error bar (standard error) represent significant differences in average pH (within each month) among the treatments ($\alpha = 0.05$, Fisher's LSD test).

In flowering dogwood, no significant differences in substrate pH were observed for fertilizer rates (June: $p = 0.0526$; July: $p = 0.3315$; August: $p = 0.0851$), irrigation systems (June: $p = 0.3199$; July: $p = 0.0695$; August: $p = 0.8459$), or interactions between fertilizer rates and irrigation systems (June: $p = 0.2909$; July: $p = 0.1861$; August: $p = 0.1003$). For substrate EC, significant differences were observed for fertilizer rates in June, July, and August (June: $p \leq 0.0001$; July: $p \leq 0.0001$; August: $p \leq 0.0001$). Significant differences were only observed for irrigation systems (June: $p = 0.2649$; July: $p = 0.0061$; August: $p = 0.9789$) and interactions between fertilizer rates and irrigation systems in July (June: $p = 0.2758$; July: $p = 0.0076$; August: $p = 0.9783$). Throughout this study, substrate EC was greater for the high fertilizer rate (both irrigation systems) compared to all other treatments (Figure 9). The irrigation system only influenced substrate EC in July, when EC was the greatest for the high-fertilizer drip irrigation.

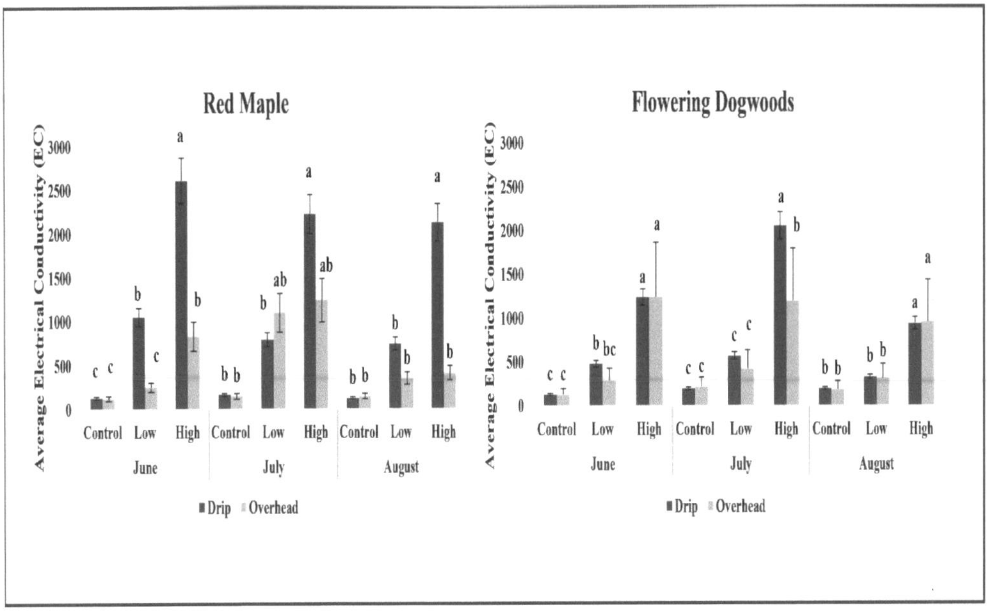

Figure 9. Average substrate electrical conductivity (EC) for containerized red maple (**left**) and flowering dogwoods (**right**) over three months using different fertilizer rates in overhead and drip irrigation systems. Different letters above the error bar (standard error) represent significant differences in average EC (within each month) among the treatments ($\alpha = 0.05$, Fisher's LSD test).

3.7. Nutrient Content

Leaf samples were analyzed to determine the leaf nutrient content at the end of this study. In red maple, significant differences in leaf N content were observed for the fertilizer rate but not for irrigation systems or the interaction between fertilizer rates and irrigation systems (fertilizer rate: $p = 0.0003$; irrigation system: $p = 0.4465$; interaction: $p = 0.4018$). For P content, significant differences were observed for fertilizer rates and irrigation systems but not the interaction (fertilizer rate: $p = 0.0084$; irrigation system: $p = 0.0014$; interaction: $p = 0.0576$). Similar to P content, K content had significant differences for fertilizer rates and irrigation systems but not for their interaction (fertilizer rate: $p = 0.0170$; irrigation system: $p \leq 0.0001$; interaction: $p = 0.0850$). For N content, the high rate of fertilizer in both irrigation systems and the low rate of fertilizer in drip irrigation led to greater N content compared to both no-fertilizer controls (drip and overhead irrigation) (Figure 10). Leaf P content was greater in drip irrigation (low and high fertilizer rates) compared to the no-fertilizer controls of both irrigation systems. Additionally, no differences were observed for P content among the overhead irrigation treatments. All treatments involving drip irrigation led to greater K content compared to those involving overhead irrigation. However, no differences were observed between the no-fertilizer controls and each fertilizer rate within each irrigation system (Figure 10).

In flowering dogwood, significant differences were observed in leaf N content among the treatments for fertilizer rates ($p \leq 0.0001$), irrigation systems ($p = 0.0004$), and their interactions ($p = 0.0242$). All treatments in both irrigation systems had higher leaf N content compared to the controls of both irrigation systems. For P content, significant differences were observed for fertilizer rates ($p \leq 0.0001$) and irrigation systems ($p = 0.0045$) but not for their interactions ($p = 0.0810$). Both rates of fertilizer in the overhead irrigation system and the high rate of fertilizer in the drip irrigation system led to significantly higher P content compared to both controls. For K content, significant differences were observed among the treatments for the fertilizer rate ($p = 0.0026$) and the interactions between fertilizer rates and irrigation systems ($p = 0.0463$) but not for irrigation systems ($p = 0.5003$). Only overhead irrigation led to significantly higher potassium content compared to other treatments (Figure 10).

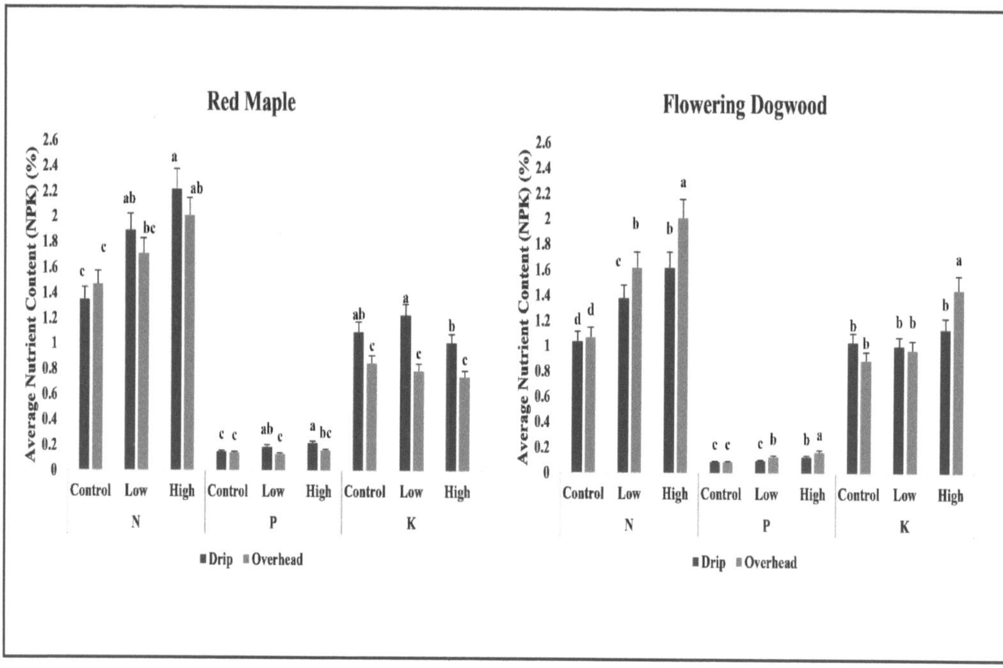

Figure 10. Average nutrient content [nitrogen (N), phosphorus (P), and potassium (K)] expressed in percentage of the nutrient content in the leaf tissue at the end of the study period for containerized red maple (**left**) and flowering dogwoods (**right**) using different fertilizer rates in overhead and drip irrigation systems. Different letters above the error bar (standard error) represent significant differences in average N, P, and K content among the treatments ($\alpha = 0.05$, Fisher's LSD test).

4. Discussion

In this study, we observed that plant growth varied depending on plant species and irrigation systems. Red maples showed an increase in plant height, whereas no significant plant height increase was observed for flowering dogwoods. Similarly, stem diameter was also greater in fertilized red maples, irrespective of irrigation systems. Unlike plant height, flowering dogwoods had a greater increase in stem diameter in fertilized plants compared to non-fertilized plants with drip irrigation for June but not in July. A study conducted by Worrall et al. [37] found that container-grown woody ornamentals responded variedly to controlled-release fertilizers; however, their response to the fertilizer was significant, positive, and linear. Similarly, a positive response of fertilizer on plant health was observed by Clark and Zheng [38] when they compared the high and low rates of controlled-release fertilizer on containerized nursery crops. However, different research has found that flowering dogwoods have struggled to show significant growth in the first year of the growing season. A study conducted by Witte et al. [39] found flowering dogwoods did not respond to fertilizer in the first growing season. Similarly, a study conducted by Warren [40] found that nitrogen had no effect on the growth of flowering dogwood until 56 days, but they had increased stem diameter. Thus, Warren [40] suggested applying limited nitrogen for flowering dogwoods in the first year of transplantation as it impacts root development.

In our study, no-fertilizer controls had lower NDVI compared to the fertilizer-applied plants in both irrigation systems. Similar results were also recorded by Guan et al. [41]

while assessing the NDVI of rice and wheat crops at different rates of fertilizer application. Edalat et al. [42] recorded higher NDVI for the high rate of fertilizer application while comparing conventional and reduced till systems for nitrogen management. An increased rate of fertilizer yielding higher NDVI was also found in Geraniums (*Pelargonium* sp. L.) by Dunn et al. [43] and Wang et al. [44,45].

However, our study found some inconsistencies in the NDVI values depending on the irrigation system. In flowering dogwood, in the drip irrigation system, a high rate of fertilizer application led to greater NDVI compared to the other treatments, but in overhead irrigation, both rates of fertilizer application were similar in June. In the drip irrigation system, the control and low rate of fertilizer had similar NDVI in July, whereas the control and high rate of fertilizer were statistically similar in August. This could be because of the effectiveness of drip irrigation during the initial phase of plant establishment compared to overhead irrigation. In a similar context, Fox [46] explained that drip irrigation was found effective in young transplants as they were applied to root zones. In addition, Kjelgren and Cleveland [47] and Gilman et al. [48] explained the effectiveness of drip irrigation in the establishment of young transplants. The researcher explained that young transplants have smaller and younger root systems and, thus, require lower water amounts but in greater frequency. The other reasons could be due to the constant rehumidification of the substrate by drip irrigation. The major benefit drip irrigation provides is maintaining precise control over the substrate moisture at the necessary interphase period [49]. Repetition of a similar volume of water throughout the day to constantly moisturize the substrate was found to ensure plant growth [50]. Also, repeated application of smaller volumes of water also reduces the leaching of mineral nutrients and surface runoff [51,52]. Interestingly, it was more prominent in flowering dogwoods and not in red maples. Red maples were able to grow taller and with more new roots in frequent irrigation compared to periodic irrigation [53]. These findings could be used to explain the variation in NDVI between red maples and flowering dogwoods in different irrigation systems. A strong and positive correlation was found between handheld NDVI and UAV-mounted NDVI sensors, similar to the findings of Neupane et al. [54].

We observed that relative chlorophyll content was significantly higher in both rates of fertilizer application compared to no-fertilizer controls in both tree species irrespective of irrigation system, which was expected since pine bark substrate does not contain mineral nutrients in concentrations that will support substantial plant growth. Chang and Robison [55] found that leaf chlorophyll content was positively correlated with nitrogen content in woody species. Higher chlorophyll content was also measured during foliar analysis by Demotes-Mainard et al. [56] and Cabrera [10] in *Lagestroemia indica* L. Similar to our findings, Freidenreich et al. [57] also observed no significant differences between the two rates of fertilizer in total chlorophyll content. The chlorophyll content observed from SPAD-502 was similar to those observed in the visual observation, as fertilizer-treated trees were healthy-looking and had significantly fewer chlorotic symptoms.

In our study, it was observed that overhead-irrigated maples had lower substrate pH after July, whereas flowering dogwoods with high fertilizer rates in overhead irrigation had lower pH compared to the no-fertilizer controls in June. This could be because of the excessive leaching due to overhead irrigation. Gamble and Daniels [58] and Daniels et al. [59] explained that intensive leaching caused the loss of bases, thus lowering the pH. Similarly, in our study, the high rate of fertilizer led to greater EC. Similar results were observed by Richards and Reed [60] while analyzing the growth response of New Guinea Impatiens using controlled-release fertilizers. Bayer et al. [61] also observed comparable results regarding the pH and EC of the leachates. They found that a high rate of fertilizer application and regulated irrigation led to higher EC.

While analyzing leaf tissue, we found that N and P content was greater in higher rates of fertilizer application for both red maples and flowering dogwoods. Potassium content was greater in drip-irrigated red maples yet contrastingly greater in overhead-irrigated flowering dogwood. This inconsistency in K content in different irrigation systems is explained by Neilsen et al. [62], who evaluated the effects of different irrigation systems in 'Gala' apples. Similarly, ShalekBriski et al. [63] also observed that irrigation systems impacted the foliar K content while using different irrigation systems.

There were a few unclear observations recorded in our study, such as variations in foliar concentration of nutrients in distinct species in the same irrigation system and variations in pH in different species. A study conducted by Wu et al. [64] explained that deficit irrigation results in decreased nitrogen uptake in alfalfa grass, while the other study conducted by Amankwaa-Yeboah et al. [65] suggested that deficit irrigation can be utilized for increasing nutrient uptake and crop water efficiency. This suggests that more detailed and in-depth studies are needed in the future to provide clear evidence of nutrient uptake in different irrigation systems. Similarly, inconsistencies in plant growth and NDVI of flowering dogwoods in different irrigation systems could not be addressed in our study. The results showed that not only NDVI but SPAD values of the flowering dogwoods have lowered with drip irrigation in the later month, highly suggesting that some other factors like water distribution and/or physiology of the flowering dogwoods could be in play for flowering dogwoods. These areas could be identified as a prospect for further research, which will help us better understand the effect of fertilizers and nutrients on different tree species in different irrigation systems.

5. Conclusions

As a low rate of fertilizer did not have significant differences among different variables compared to a high rate of fertilizers, we can conclude that the application of a low rate of fertilizer for ornamental tree production, such as red maples and flowering dogwoods, will be beneficial. Freidenreich et al. [57] also recommended the application of 20 g of Florikan while comparing 0, 10, 20, 30, 40, and 50 g for *Justicia brandegeana* (Brandegee) in the container production system. Irrigation systems could be used based on tree species and stage of stress, as drip irrigation was found effective in earlier days of plant establishment; however, overhead irrigation was more effective later in the summer. Similarly, studies conducted by Novair et al. [15] and Samadi et al. [16] also suggested that the addition of organic amendment and biofertilizers can be helpful for increasing nutrient uptake and overall plant health in low irrigation conditions. In this study, we also found a high correlation between handheld NDVI and UAV-mounted NDVI. The application of aerial imagery and sensor technology provides higher efficiency and economic benefits for ornamental growers [54]. The findings of this study provide beneficial insights into ornamental tree growers to optimize nutrient management, irrigation strategy, and the application of technology in a container-based production system.

Author Contributions: Conceptualization, F.B.-G. and A.W.; methodology, F.B.-G., A.W. and K.N.; software, K.N.; validation, F.B.-G., A.W. and K.N.; formal analysis, A.W. and K.N.; writing—original draft preparation, K.N.; writing—review and editing, K.N., A.W. and F.B.-G.; visualization, K.N.; supervision, F.B.-G. and A.W.; project administration, F.B.-G.; funding acquisition, F.B.-G. and A.W. All authors have read and agreed to the published version of the manuscript.

Funding: This work was supported by the National Institute of Food and Agriculture, United States Department of Agriculture Capacity Building grant, under award number 2019-38821-29062.

Data Availability Statement: The original contributions presented in the study are included in the article, further inquiries can be directed to the corresponding author.

Acknowledgments: We acknowledge Grayson Lane Delay for his contribution to data collection for this project.

Conflicts of Interest: The authors declare no conflicts of interest.

References

1. Dobermann, A.; Bruulsema, T.; Cakmak, I.; Gerard, B.; Majumdar, K.; McLaughlin, M.; Reidsma, P.; Vanlauwe, B.; Wollenberg, L.; Zhang, F. Responsible plant nutrition: A new paradigm to support food system transformation. *Glob. Food Secur.* **2022**, *33*, 100636. [CrossRef]
2. Mee, C.; Siva, K.B.; Ahmad, H.M.H. Detecting and monitoring plant nutrient stress using remote sensing approaches: A review. *Asian J. Plant Sci.* **2017**, *16*, 1–8.
3. Kirkby, E.A. Introduction, definition, and classification of nutrients. In *Marschner's Mineral Nutrition of Plants*; Elsevier: Amsterdam, The Netherlands, 2023; pp. 3–9.
4. Yang, K.; Li, Y. Effects of water stress and fertilizer stress on maize growth and spectral identification of different stresses. *Spectrochim. Acta Part A Mol. Biomol. Spectrosc.* **2023**, *297*, 122703. [CrossRef] [PubMed]
5. Gianquinto, G.; Orsini, F.; Fecondini, M.; Mezzetti, M.; Sambo, P.; Bona, S. A methodological approach for defining spectral indices for assessing tomato nitrogen status and yield. *Eur. J. Agron.* **2011**, *35*, 135–143. [CrossRef]
6. Hawkesford, M.; Horst, W.; Kichey, T.; Lambers, H.; Schjoerring, J.; Møller, I.S.; White, P. Functions of macronutrients. In *Marschner's Mineral Nutrition of Higher Plants*; Elsevier: Amsterdam, The Netherlands, 2012; pp. 135–189.
7. Fageria, N.K.; Baligar, V.C. Enhancing nitrogen use efficiency in crop plants. *Adv. Agron.* **2005**, *88*, 97–185.
8. Li, S.-X.; Wang, Z.-H.; Malhi, S.; Li, S.-Q.; Gao, Y.-J.; Tian, X.-H. Nutrient and water management effects on crop production, and nutrient and water use efficiency in dryland areas of China. *Adv. Agron.* **2009**, *102*, 223–265.
9. Rose, M.A.; Rose, M.; Wang, H. Fertilizer concentration and moisture tension affect growth and foliar N, P, and K contents of two woody ornamentals. *HortScience* **1999**, *34*, 246–250. [CrossRef]
10. Cabrera, R.I. Nitrogen balance for two container-grown woody ornamental plants. *Sci. Hortic.* **2003**, *97*, 297–308. [CrossRef]
11. Tisdale, S.L.; Nelson, W.L. Soil fertility and fertilizers. *Soil Sci.* **1966**, *101*, 346. [CrossRef]
12. Hawkesford, M.J.; Cakmak, I.; Coskun, D.; De Kok, L.J.; Lambers, H.; Schjoerring, J.K.; White, P.J. Functions of macronutrients. In *Marschner's Mineral Nutrition of Plants*; Elsevier: Amsterdam, The Netherlands, 2023; pp. 201–281.
13. Kant, S.; Kafkafi, U.; Pasricha, N.; Bansal, S. Potassium and abiotic stresses in plants. In *Potassium for Sustainable Crop Production*; Potash Institute of India: Gurgaon, India, 2002; Volume 233, p. 251.
14. Zahoor, R.; Dong, H.; Abid, M.; Zhao, W.; Wang, Y.; Zhou, Z. Potassium fertilizer improves drought stress alleviation potential in cotton by enhancing photosynthesis and carbohydrate metabolism. *Environ. Exp. Bot.* **2017**, *137*, 73–83. [CrossRef]
15. Novair, S.B.; Hosseini, H.M.; Etesami, H.; Pirmoradian, N.; Asgari Lajayer, B.; Price, G.W. Straw application improved soil biological properties and the growth of rice plant under low water irrigation. *Environ. Res.* **2024**, *255*, 119138. [CrossRef] [PubMed]
16. Samadi, A.; Derafshi, M.; Hassani, A.; Gholamhoseini, M.; Asgari Lajayer, B.; Astatkie, T.; Price, G.W. Effects of biofertilizers and Potassium Sulfate on nutrients uptake and physiological characteristics of Maize (*Zea mays* L.) under drought stress. *J. Crop Health* **2023**, *76*, 209–218. [CrossRef]
17. Klock-Moore, K.A.; Broschat, T.K. Irrigation systems and fertilizer affect petunia growth. *HortTechnology* **2001**, *11*, 416–418. [CrossRef]
18. Ncama, K.; Magwaza, L.; Shezi, S. Integrated water management and its significance prior to, during, and after harvesting of fresh horticultural produce. *Postharvest Manag. Fresh Prod.* **2023**, 195–216. [CrossRef]
19. Gent, M.P.; McAvoy, R.J. Water and nutrient uptake and use efficiency with partial saturation ebb and flow watering. *HortScience* **2011**, *46*, 791–798. [CrossRef]
20. Yang, P.; Wu, L.; Cheng, M.; Fan, J.; Li, S.; Wang, H.; Qian, L. Review on drip irrigation: Impact on crop yield, quality, and water productivity in China. *Water* **2023**, *15*, 1733. [CrossRef]
21. Rowe, D.B.; Kolp, M.R.; Greer, S.E.; Getter, K.L. Comparison of irrigation efficiency and plant health of overhead, drip, and sub-irrigation for extensive green roofs. *Ecol. Eng.* **2014**, *64*, 306–313. [CrossRef]
22. Davies, M.J.; Harrison-Murray, R.; Atkinson, C.J.; Grant, O.M. Application of deficit irrigation to container-grown hardy ornamental nursery stock via overhead irrigation, compared to drip irrigation. *Agric. Water Manag.* **2016**, *163*, 244–254. [CrossRef]
23. Chowhan, P.; Chakraborty, A.P. Remote Sensing Technology—A new dimension in detection, quantification and tracking of abiotic and biotic stresses. In *Plant Stress: Challenges and Management in the New Decade*; Springer: Berlin/Heidelberg, Germany, 2022; pp. 445–457.
24. Cadet, É.; Samson, G. Detection and discrimination of nutrient deficiencies in sunflower by blue-green and chlorophyll-a fluorescence imaging. *J. Plant Nutr.* **2011**, *34*, 2114–2126. [CrossRef]
25. Gulzar, S.; Hassan, A.; Nawchoo, I.A. A review of nutrient stress modifications in plants, alleviation strategies, and monitoring through remote sensing. In *Plant Micronutrients: Deficiency and Toxicity Management*; Springer: Cham, Switzerland, 2020; pp. 331–343.
26. Jiang, H.; Ali, M.A.; Jiao, Y.; Yang, B.; Dong, L. In-situ, real-time monitoring of nutrient uptake on plant chip integrated with nutrient sensor. In Proceedings of the 2017 19th International Conference on Solid-State Sensors, Actuators and Microsystems (TRANSDUCERS), Kaohsiung, Taiwan, 18–22 June 2017; pp. 289–292.
27. Mahajan, G.; Sahoo, R.; Pandey, R.; Gupta, V.; Kumar, D. Using hyperspectral remote sensing techniques to monitor nitrogen, phosphorus, sulphur and potassium in wheat (*Triticum aestivum* L.). *Precis. Agric.* **2014**, *15*, 499–522. [CrossRef]

28. Conrad, C.; Usman, M.; Morper-Busch, L.; Schönbrodt-Stitt, S. Remote sensing-based assessments of land use, soil and vegetation status, crop production and water use in irrigation systems of the Aral Sea Basin. A review. *Water Secur.* **2020**, *11*, 100078. [CrossRef]
29. Zwack, J.A.; Graves, W.R.; Townsend, A.M. Variation among red and Freeman maples in response to drought and flooding. *HortScience* **1999**, *34*, 664–668. [CrossRef]
30. Day, S.; Seiler, J.; Persaud, N. A comparison of root growth dynamics of silver maple and flowering dogwood in compacted soil at differing soil water contents. *Tree Physiol.* **2000**, *20*, 257–263. [CrossRef] [PubMed]
31. Grant, O.M.; Davies, M.J.; Longbottom, H.; Atkinson, C.J. Irrigation scheduling and irrigation systems: Optimising irrigation efficiency for container ornamental shrubs. *Irrig. Sci.* **2009**, *27*, 139–153. [CrossRef]
32. Beeson, R.; Knox, G. Analysis of efficiency of overhead irrigation in container production. *HortScience* **1991**, *26*, 848–850. [CrossRef]
33. Witcher, A.L.; Baysal-Gurel, F.; Blythe, E.K.; Fare, D.C. Container size and shade duration affect growth of flowering dogwood. *HortTechnology* **2019**, *29*, 842–853. [CrossRef]
34. Süß, A.; Danner, M.; Obster, C.; Locherer, M.; Hank, T.; Richter, K.; Consortium, E. *Measuring Leaf Chlorophyll Content with the Konica Minolta SPAD-502Plus*; EnMAP Consortium: Potsdam, Germany, 2015.
35. LeBude, A.; Bilderback, T. *The Pour-Through Extraction Procedure: A Nutrient Management Tool for Nursery Crops*; Bull. AG-717-W; North Carolina Cooperative Extension Service: Gatesville, NC, USA, 2009.
36. Messiga, A.J.; Dyck, K.; Ronda, K.; van Baar, K.; Haak, D.; Yu, S.; Dorais, M. Nutrients leaching in response to long-term fertigation and broadcast nitrogen in blueberry production. *Plants* **2020**, *9*, 1530. [CrossRef] [PubMed]
37. Worrall, R.; Lamont, G.; O'Connell, M.; Nicholls, P. The growth response of container-grown woody ornamentals to controlled-release fertilizers. *Sci. Hortic.* **1987**, *32*, 275–286. [CrossRef]
38. Clark, M.J.; Zheng, Y. Use of species-specific controlled-release fertilizer rates to manage growth and quality of container nursery crops. *HortTechnology* **2015**, *25*, 370–379. [CrossRef]
39. Witte, W.T.; Windham, M.T.; Windham, A.S.; Hale, F.A.; Clatterbuck, W.K. *Dogwoods for American Gardens*; University of Tennessee, Agricultural Extension Service: Knoxville, TN, USA, 2000.
40. Warren, S.L. Growth and nutrient concentration in flowering dogwood after nitrogen fertilization and dormant root pruning. *J. Arboric.* **1993**, *19*, 57–63. [CrossRef]
41. Guan, S.; Fukami, K.; Matsunaka, H.; Okami, M.; Tanaka, R.; Nakano, H.; Sakai, T.; Nakano, K.; Ohdan, H.; Takahashi, K. Assessing correlation of high-resolution NDVI with fertilizer application level and yield of rice and wheat crops using small UAVs. *Remote Sens.* **2019**, *11*, 112. [CrossRef]
42. Edalat, M.; Naderi, R.; Egan, T.P. Corn nitrogen management using NDVI and SPAD sensor-based data under conventional vs. reduced tillage systems. *J. Plant Nutr.* **2019**, *42*, 2310–2322. [CrossRef]
43. Dunn, B.L.; Shrestha, A.; Goad, C. Determining nitrogen fertility status using optical sensors in geranium with controlled release fertilizer. *J. Appl. Hortic.* **2015**, *17*, 7–11. [CrossRef]
44. Wang, Y.-W.; Dunn, B.L.; Arnall, D.B. Assessing nitrogen status in potted geranium through discriminant analysis of ground-based spectral reflectance data. *HortScience* **2012**, *47*, 343–348. [CrossRef]
45. Wang, Y.-W.; Dunn, B.L.; Arnall, D.B.; Mao, P.-S. Use of an active canopy sensor and SPAD chlorophyll meter to quantify geranium nitrogen status. *HortScience* **2012**, *47*, 45–50. [CrossRef]
46. Fox, L.C. Effects of Transplanting and Irrigation Regime on Growth and Gas Exchange of Select Tree Species in a Semi-Arid Climate. Master's Thesis, Texas Tech University, Lubbock, TX, USA, 2004.
47. Kjelgren, R.; Cleveland, B. Growth and water relations of Kentucky coffee tree and silver maple following transplanting. *J. Environ. Hortic.* **1994**, *12*, 96–99. [CrossRef]
48. Gilman, E.F.; Black, R.J.; Dehgan, B. Irrigation volume and frequency and tree size affect establishment rate. *Arboric. Urban For.* **1998**, *24*, 1–9. [CrossRef]
49. Tyshchenko, A.; Tyshchenko, O.; Piliarska, O.; Biliaieva, I.; Kuts, H.; Lykhovyd, P.; Halchenko, N. Seed productivity of alfalfa varieties depending on the conditions of humidification and growth regulators in the southern steppe of Ukraine. *Sci. Pap. Ser. Manag. Econ. Eng. Agric. Rural. Dev.* **2020**, *20*, 551.
50. Warren, S.L.; Bilderback, T.E. Timing of low pressure irrigation affects plant growth and water utilization efficiency. *J. Environ. Hortic.* **2002**, *20*, 184–188. [CrossRef]
51. Criscione, K.S.; Fields, J.S.; Owen, J.S.; Fultz, L.; Bush, E. Evaluating stratified substrates effect on containerized crop growth under varied irrigation strategies. *HortScience* **2022**, *57*, 400–413. [CrossRef]
52. Witcher, A.L. Evaluation of Fertilizer and Irrigation Production Systems for Large Nursery Containers. Master's Thesis, Louisiana State University and Agricultural & Mechanical College, Baton Rouge, LA, USA, 2003.
53. Marshall, M.D.; Gilman, E.F. Effects of nursery container type on root growth and landscape establishment of *Acer rubrum* L. *J. Environ. Hortic.* **1998**, *16*, 55–59. [CrossRef]
54. Neupane, K.; Witcher, A.; Baysal-Gurel, F. Evaluation of physiological changes in flowering dogwood under drought conditions in a container production system. *HortScience* **2023**, *58*, 1077–1084. [CrossRef]
55. Chang, S.X.; Robison, D.J. Nondestructive and rapid estimation of hardwood foliar nitrogen status using the SPAD-502 chlorophyll meter. *For. Ecol. Manag.* **2003**, *181*, 331–338. [CrossRef]

56. Demotes-Mainard, S.; Boumaza, R.; Meyer, S.; Cerovic, Z.G. Indicators of nitrogen status for ornamental woody plants based on optical measurements of leaf epidermal polyphenol and chlorophyll contents. *Sci. Hortic.* **2008**, *115*, 377–385. [CrossRef]
57. Freidenreich, A.; Barraza, G.; Jayachandran, K.; Khoddamzadeh, A.A. Precision agriculture application for sustainable nitrogen management of Justicia brandegeana using optical sensor technology. *Agriculture* **2019**, *9*, 98. [CrossRef]
58. Gamble, E.; Daniels, R. Parent Material of Upper-and Middle-Coastal-Plain Soils in North Carolina. *Soil Sci. Soc. Am. J.* **1974**, *38*, 633–637. [CrossRef]
59. Daniels, R.; Gamble, E.; Wheeler, W. Age of soil landscapes in the Coastal Plain of North Carolina. *Soil Sci. Soc. Am. J.* **1978**, *42*, 98–105. [CrossRef]
60. Richards, D.L.; Reed, D.W. New Guinea impatiens growth response and nutrient release from controlled-release fertilizer in a recirculating subirrigation and top-watering system. *HortScience* **2004**, *39*, 280–286. [CrossRef]
61. Bayer, A.; Ruter, J.; van Iersel, M.W. Optimizing irrigation and fertilization of *Gardenia jasminoides* for good growth and minimal leaching. *HortScience* **2015**, *50*, 994–1001. [CrossRef]
62. Neilsen, G.; Parchomchuk, P.; Neilsen, D.; Berard, R.; Hague, E. Leaf nutrition and soil nutrients are affected by irrigation frequency and method for NP-fertigated Gala'Apple. *J. Am. Soc. Hortic. Sci.* **1995**, *120*, 971–976. [CrossRef]
63. ShalekBriski, A.; Brorsen, B.W.; Biermacher, J.T.; Rohla, C.T.; Chaney, W. Effect of irrigation method on tree growth, foliar nutrient levels, and nut characteristics of young pecan trees in the southern Great Plains. *HortTechnology* **2019**, *29*, 109–113. [CrossRef]
64. Wu, W.; Liu, M.; Wu, X.; Wang, Z.; Yang, H. Effects of deficit irrigation on nitrogen uptake and soil mineral nitrogen in alfalfa grasslands of the inland arid area of China. *Agric. Water Manag.* **2022**, *269*, 107724. [CrossRef]
65. Amankwaa-Yeboah, P.; Aruna Akoriko, F.; Amponsah, W.; Yeboah, S.; Brempong, M.B.; Keteku, A.K. Combining deficit irrigation and nutrient amendment enhances the water productivity of tomato (*Solanum lycopersicum* L.) in the tropics. *Front. Sustain. Food Syst.* **2023**, *7*, 1199386. [CrossRef]

horticulturae

Article

Comprehensive Analysis of Genes Associated with the Reactive Oxygen Species Metabolism in *Citrus sinensis* during Pathogen Infection

Guiyan Huang [1,2], Fuxuan Li [1], Yanan Hu [1], Zhigang Ouyang [1,2,*] and Ruimin Li [1,2,*]

1 College of Life Sciences, Gannan Normal University, Ganzhou 341000, China; huangguiyan@gnnu.edu.cn (G.H.); lifuxuan@gnnu.edu.cn (F.L.); huyanan@gnnu.edu.cn (Y.H.)
2 National Navel Orange Engineering Research Center, Ganzhou 341000, China
* Correspondence: oyfry@163.com (Z.O.); ruimin_li_ry@aliyun.com (R.L.)

Abstract: Reactive oxygen species (ROS) are pivotal in signal transduction processes in plant–pathogen interactions. The ROS signaling pathways involved in *Candidatus* Liberibacter asiaticus (CLas) and *Xanthomonas citri* subspecies citri (Xcc) infections in *Citrus sinensis* (sweet orange) are unclear. In this study, we comprehensively identified ROS metabolism-associated genes, including 9 NADPH oxidase (*RBOH*), 14 superoxide dismutase (*SOD*), 1 catalase (*CAT*), 9 peroxiredoxin (*PrxR*), 5 ascorbate peroxidase (*APX*), 4 glutathione peroxidase (*GPX*), 3 monodehydroascorbate reductase (*MDAR*), 2 dehydroascorbate reductase (*DHAR*), 2 glutathione reductase (*GR*), 24 thioredoxin (*Trx*), and 18 glutaredoxin (*GLR*) genes in *C. sinensis*. An analysis revealed variable gene structures but conserved motifs and domains in ROS subfamilies. A comparative synteny analysis with *Arabidopsis thaliana* and *Vitis vinifera* indicated evolutionary conservation of most ROS metabolism-associated genes, with some originating from gene duplication events post-species divergence in *C. sinensis*. Expression profiling revealed five up-regulated genes and four down-regulated genes during both CLas and Xcc infections. Promoter analysis revealed numerous stress-responsive elements in the promoter of ROS metabolism-associated genes. Protein–protein interaction network analysis highlighted the involvement of ROS metabolism in various biological processes. A comparison of ROS metabolism-associated genes between *C. sinensis* and *Poncirus trifoliata* indicated multiple gene gain and loss events within ROS subfamilies of *C. sinensis*. This study enhances our understanding of ROS metabolism in *C. sinensis* and sheds light on citrus–pathogen interactions.

Keywords: *Citrus sinensis*; reactive oxygen species; huanglongbing; citrus bacterial canker; sweet orange

Citation: Huang, G.; Li, F.; Hu, Y.; Ouyang, Z.; Li, R. Comprehensive Analysis of Genes Associated with the Reactive Oxygen Species Metabolism in *Citrus sinensis* during Pathogen Infection. *Horticulturae* **2024**, *10*, 625. https://doi.org/10.3390/horticulturae10060625

Academic Editor: Harald Scherm

Received: 7 April 2024
Revised: 8 June 2024
Accepted: 9 June 2024
Published: 11 June 2024

1. Introduction

Citrus huanglongbing (HLB) and citrus bacterial canker are two major bacterial diseases in citrus production that cause severe losses. HLB is mainly caused by the phloem-limited bacterium *Candidatus* Liberibacter asiaticus (CLas). CLas infection strongly affects various biological processes and metabolic pathways in the leaves and fruits of citrus plants [1–5]. RNA-seq analysis of fruit peel tissues from *Citrus sinensis* cv. Valencia revealed an up-regulation of genes associated with the light reactions of photosynthesis, ATP synthesis, various WRKY transcription factor genes, and genes involved in the signaling pathways of ethylene, salicylic acid, and jasmonic acid. Conversely, genes associated with sucrose and starch metabolism, in addition to those involved in cytokinin and gibberellin pathways, were inhibited [4]. However, in HLB-tolerant *C. hystrix* (kaffir lime), the genes related to starch synthesis and the photosynthetic process remained undisturbed, while an induction of peroxidases, Cu/Zn-SOD, and POD4 genes was observed [2]. The HLB-tolerant 'LB8-9' mandarin (*C. reticulata*) demonstrated superior reactive oxygen species (ROS) scavenging activity compared with the HLB-susceptible 'Hamlin' *C. sinensis* [6]. In HLB-tolerant citrus varieties, the stimulation of ROS-scavenging genes, such as catalases and ascorbate peroxidases, may contribute to the mitigation of ROS in plants, especially in phloem tissues [7]. Prior research has revealed that HLB is an immunity-triggered disease prompted by pathogens, characterized by significant ROS accumulation in plants infected with CLas [8]. ROS accumulation initiates programmed cell death, a process that can be curtailed through the use of antioxidants [8]. However, the identification of genes related to ROS metabolism, including NADPH oxidase (*RBOH*), superoxide dismutase (*SOD*), catalase (*CAT*), peroxiredoxin (*PrxR*), thioredoxin (*Trx*), ascorbate peroxidase (*APX*), glutathione peroxidase (*GPX*), monodehydroascorbate reductase (*MDAR*), dehydroascorbate reductase (*DHAR*), glutaredoxin (*GLR*), and glutathione reductase (*GR*), within the genome of *C. sinensis* and their roles in the process of citrus–CLas interactions remain unclear. Citrus bacterial canker is another bacterial disease in citrus production caused by *Xanthomonas citri* subspecies citri (Xcc) [9]. Since Xcc can be cultured in vitro (unlike CLas), its pathogenesis is fairly clear. However, the correlation between ROS metabolism-associated genes and Xcc infection remains inadequately elucidated.

Comprehending the mechanisms of transcriptional regulation of plants in response to pathogen attack is integral to elucidating the interactions between plants and pathogens [10]. ROS serve as primary signaling entities in plant physiology, orchestrating a plethora of processes encompassing growth, development, responses to abiotic stress, and particularly, resistance against pathogens [11]. The regulation of ROS levels is critical in plants, as an imbalance can lead to oxidative stress, causing damage to cellular structures [12]. The RBOHs are primary producers of ROS in plants. They generate superoxide by transferring an electron from NADPH to molecular oxygen [13]. SOD, CAT, and PrxR are key enzymes in ROS detoxification. SOD catalyzes the dismutation of superoxide into oxygen and hydrogen peroxide, which is then detoxified by CAT and PrxR [14]. Furthermore, APX and GPX, two key enzymes in the ascorbate-glutathione cycle, transform hydrogen peroxide into water, playing a critical role in maintaining ROS homeostasis [12,15]. Moreover, MDAR, DHAR, and GR are crucial for the regeneration of antioxidants in the ascorbate-glutathione cycle, thereby maintaining the balance of ROS [16]. Additionally, Trx and GLR, two small redox-active proteins, modulate the redox state of target proteins and scavenge ROS, contributing to the redox homeostasis in plant cells [17]. ROS emerges as a vital element in the preliminary signal transduction pathways incited by pathogen detection [18,19]. The genes integral to ROS metabolism govern important roles in ROS signaling. ROS engendered by RBOH proteins not only function as a direct defense against invading pathogens but also perform as secondary messengers, instigating consequent defense reactions [20]. Enzymes such as SOD, CAT, PrxR, APX, GPX, MDAR, DHAR, GR, Trx and GLR, via their ROS detoxification activities, facilitate positive ROS signaling regulation, harmonizing the balance between ROS production and scavenging, thereby assuring optimal ROS signaling [11,14,21].

In this study, we aimed to identify and elucidate the crucial genes involved in ROS metabolism, which are responsive to pathogenic stress in *C. sinensis*. Furthermore, we sought to elucidate the functional roles of ROS metabolism-related genes during the infection stages of CLas and Xcc. We systematically identified and analyzed the expression profiles of the ROS metabolism-associated genes including *RBOH, SOD, CAT, PrxR, APX, GPX, MDAR, DHAR, GR, Trx,* and *GLR* in *C. sinensis*. Especially, differentially expressed ROS metabolism-associated genes' responses to HLB and Xcc were achieved. Moreover, we analyzed the synteny events between *C. sinensis* and *Arabidopsis thaliana* and *Vitis vinifera*. We also characterized the *cis*-acting elements in the promoter regions of all identified ROS metabolism-associated genes. Moreover, the protein–protein interaction (PPI) network analysis of the core differential expressed ROS metabolism-associated genes was constructed. Additionally, the potential gain and loss events of ROS metabolism-associated genes between HLB-susceptible *C. sinensis* and HLB-tolerant *P. trifoliata* were analyzed. Our study lays the foundation for further in-depth research into the functions of ROS metabolism-associated genes and provides new insights into the immune response mechanisms of citrus to bacterial diseases.

2. Materials and Methods

2.1. Identification of Genes Related to Metabolism of ROS

The gene family members of *RBOH, SOD, CAT, PrxR, APX, GPX, MDAR, DHAR, GR, Trx,* and *GLR* in the genome of *C. sinensis* (Citrus Pan-enome to Breeding Database, CPBD, version 3.0) were identified through blastp utilizing the sequences of their *A. thaliana* homologs, applying an E-value threshold of less than 1×10^{-5} [22,23]. A subsequent confirmation of the identified gene candidates was accomplished through sequence analysis utilizing SMART and blastp in the National Center for Biotechnology Information (NCBI) database [24,25]. In addition, the potential candidates underwent further validation via InterPro, which facilitated the identification of conserved domains within the protein sequences [26]. Eventually, the authenticated sequences were classified into different gene families. The chromosomal locations of *RBOH, SOD, CAT, PrxR, APX, GPX, MDAR, DHAR, GR, Trx,* and *GLR* genes were illustrated utilizing the 'Gene Location Visualize from GTF/GFF' plugin within the TBtools software suite version 2.096 [27].

2.2. Gene Structure Analysis and Motif and Conserved Domain Identification

The exon–intron gene structure of *RBOH, SOD, CAT, PrxR, APX, GPX, MDAR, DHAR, GR, Trx,* and *GLR* genes in *C. sinensis* was delineated using the annotation files from the CPBD [23]. Additionally, the motifs present in the protein sequences of these genes were identified employing the MEME suite [28]. Furthermore, the conserved domains within these proteins were discerned utilizing the Batch CD-Search tool available on NCBI [29]. The visualization for the phylogenetic tree, gene structures, motifs, and conserved domains was conducted using the 'Gene Structure View (Advanced)' plugin within the TBtools software suite [27].

2.3. Synteny Analysis

Genomic sequences and Generic Feature Format Version 3 (GFF3) files of *A. thaliana, C. sinensis,* and *V. vinifera* were obtained for synteny analysis. Specifically, the TAIR10 chromosome files and GFF3 file of *A. thaliana* were acquired from the TAIR database [30], while the genome sequences and GFF3 file of *C. sinensis* (version 3.0) were sourced from CPBD [23]. Moreover, the genomic sequences and GFF3 file of *V. vinifera* (release 52) were retrieved from EnsemblPlants for analysis [31]. Synteny analysis was executed with MCScanX software version 1.1 utilizing the downloaded genomic sequences and GFF3 files [32]. Visualization of the identified synteny events containing genes related to the metabolism of ROS within *C. sinensis* was accomplished using the 'Dual Synteny Plot' plugin available in the TBtools software suite [27].

2.4. Expression Profile Analysis

The expression profiles of *RBOH, SOD, CAT, PrxR, APX, GPX, MDAR, DHAR, GR, Trx,* and *GLR* genes in *C. sinensis* during CLas and Xcc infection were analyzed. In the investigation of CLas infection, the RNA-seq analysis involved the use of leaves from both CLas-free and CLas-infected *C. sinensis* cv. Newhall, each with three biological replicates [5]. Similarly, in the study of Xcc infection, RNA-seq analysis was conducted using leaves from Xcc-free and Xcc-infected *C. sinensis* cv. Newhall at 24 and 48 h post-infection, with three biological replicates for each condition. The RNA-seq data have been deposited in the NCBI Sequence Read Archive (SRA) database under the accession numbers PRJNA953196 [5] and PRJNA1088720. The raw RNA-seq data were first trimmed using Trimmomatic version 0.40 with default parameters [33]. Next, the gene expression profiles in *C. sinensis* were determined by utilizing Kallisto with a k-mer length of 31, as described by Bray et al. [34]. The expression values for each gene in different samples were normalized using the Z score transformation method [35]. Heatmaps were generated using the 'HeatMap' plugin within the TBtools software suite [27].

2.5. RT-qPCR Validation

The extraction of total RNA from samples for RT-qPCR was carried out using the *EasyPure*® Plant RNA Kit (Transgen Biotech, Beijing, China). The RNA was reverse-transcribed to cDNA using *EasyScript*® One-Step gDNA Removal and cDNA Synthesis SuperMix (Transgen Biotech). Then, the RT-qPCR procedure was conducted on a StepOnePlus real-time fluorescence quantitative PCR instrument (ABI, South San Francisco, CA, USA) with *TransStart*® Top Green qPCR SuperMix (Transgen Biotech). The specific primers used for RT-qPCR analysis are listed in Table S1 with the *GAPDH* gene of *C. sinensis* as the internal control.

2.6. Hydrogen Peroxide Content, CAT, and SOD Enzyme Activities

The supernatant from the ground leaves was extracted using a cold 50 mM phosphate buffer (pH 7.8) for subsequent measurements of hydrogen peroxide levels as well as CAT and SOD enzyme activities. The hydrogen peroxide content was determined using the potassium iodide method [8]. For CAT enzyme activity measurement, the supernatant was preheated in a constant temperature water bath at 25 °C, and then 50 µL was taken and mixed with 1.95 mL of 0.01 mol/L hydrogen peroxide. After adding hydrogen peroxide, the mixture was immediately placed into a UV spectrophotometer and the absorbance changes were measured at a wavelength of 240 nm, with absorbance readings taken every 1 min for 4 min. An enzyme activity unit (U) was defined as a decrease of 0.1 in absorbance per min. For SOD enzyme activity measurement, a reaction mixture (2 mL 14.5 mmol/L methionine, 0.8 mL 5 mmol/L nitro blue tetrazolium solution, 0.8 mL 5 mmol/L riboflavin solution, 0.8 mL EDTA) was added to the test tubes as a blank control. In the experimental group, 50 µL of supernatant extract was added to the control group. The absorbance was measured at 560 nm using a spectrophotometer, and an enzyme unit (U) was defined as the amount of enzyme needed to inhibit the reaction by 50%.

2.7. Identification of Cis-Acting Elements in the Promoters

The 2000 bp sequences upstream the coding sequence of *RBOH*, *SOD*, *CAT*, *PrxR*, *APX*, *GPX*, *MDAR*, *DHAR*, *GR*, *Trx*, and *GLR* genes in *C. sinensis* were obtained from the genome and subsequently queried against the PlantCARE database [36]. The abiotic and biotic cis-acting elements were displayed utilizing the 'Simple BioSequence Viewer' plugin in the TBtools software suite [27].

2.8. PPI Network Analysis In Silico

The PPI networks analysis of the ROS metabolism genes with similar expression patterns under CLas and Xcc infection was conducted using the STRING online database [37]. The 'Single/Multiple Proteins by Sequence' search module was utilized to construct PPI networks by querying against *C. sinensis*.

2.9. Gene Gain and Loss Events

The Orthologous Gene Search feature of the CPBD database was employed to identify gene gain and loss events in genes associated with ROS metabolism in *C. sinensis*, in comparison with *Poncirus trifoliata* (trifoliate orange) [23]. Orthologous genes corresponding to each ROS metabolism-associated gene in *C. sinensis* from *P. trifoliata* were gathered. Subsequently, an analysis of gene gain and loss was performed utilizing the 'Gene Lost & Gain Analysis (based on Notung)' plugin within the TBtools software suite [27]. Protein sequences of RBOH, SOD, CAT, PrxR, APX, GPX, MDAR, DHAR, GR, Trx, and GLR from *C. sinensis* were retrieved from CPBD [23]. Multiple sequence alignment was performed utilizing MAFFT [38], followed by alignment refinement with Gblocks [39], model selection using PartitionFinder 2 [40], and phylogenetic tree reconstruction via IQ-tree [41].

2.10. Statistical Analysis

The statistical analyses in this study were completed using SPSS software (version 25), which included *t*-tests and one-way ANOVA tests. Pairwise comparisons for the one-way ANOVA tests were conducted using the Student–Newman–Keuls method.

3. Results

3.1. Identification of ROS Metabolism-Associated Genes in C. sinensis

We utilized methods such as BLAST and conserved domain analysis to identify *RBOH, SOD, CAT, PrxR, APX, GPX, MDAR, DHAR, GR, Trx,* and *GLR* genes in *C. sinensis*. Altogether, 91 ROS metabolism-associated genes, including 9 *RBOH*, 14 *SOD*, 1 *CAT*, 9 *PrxR*, 5 *APX*, 4 *GPX*, 3 *MDAR*, 2 *DHAR*, 2 *GR*, 24 *Trx* and 18 *GLR* genes, were identified in *C. sinensis* (Figure 1A, Table S2). The distribution of these genes was non-uniform, with clusters observed on the chromosomes of *C. sinensis*, suggesting an intriguing pattern (Figure 1B). For example, the 7 *SOD* genes, from *CsSOD1* to *CsSOD7*, were closely positioned within a confined segment on chromosome 3 (Figure 1B).

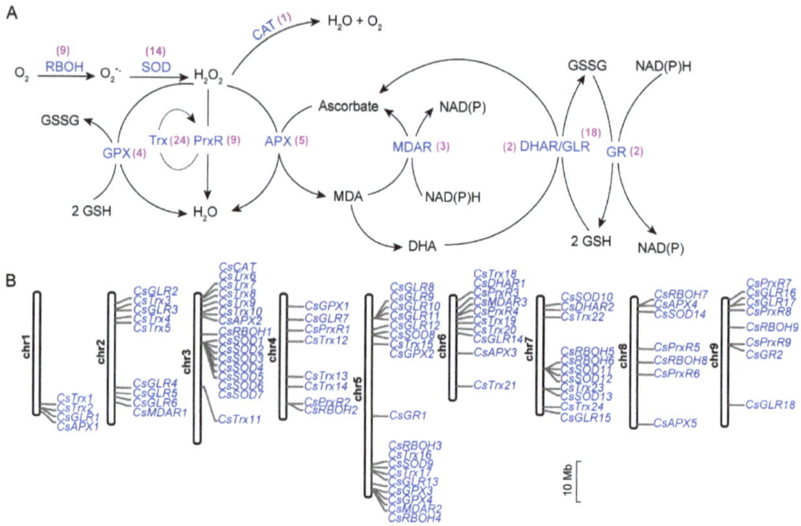

Figure 1. Identification of reactive oxygen species (ROS) metabolism-associated genes in *Citrus sinensis*. (**A**) The ROS metabolism pathway. Enzymes catalyzing corresponding metabolic reactions are highlighted in blue, while the number of genes encoding these enzymes in *C. sinensis* is represented in purple. (**B**) Chromosomal localization of ROS metabolism-associated genes. The scale represents 10 million base pairs (Mb). RBOH: NADPH oxidase. SOD: superoxide dismutase. CAT: catalase. PrxR: peroxiredoxin. Trx: thioredoxin. APX: ascorbate peroxidase. GPX: glutathione peroxidase. MDAR: monodehydroascorbate reductase. DHAR: dehydroascorbate reductase. GLR: glutaredoxin. GR: glutathione reductase.

3.2. Characterization of ROS Metabolism-Associated Genes in C. sinensis

To elucidate the characteristics of ROS metabolism-associated genes, we analyzed the gene structures, conserved motifs, and conserved domains of these genes in *C. sinensis*. The exon counts of ROS metabolism-associated genes varied from 1 to 16 (*CsMDAR2*), while 11 out of 18 *GLR* genes comprised a single exon devoid of introns (Figure 2A). An analysis of the conserved motifs of ROS metabolism-associated proteins indicated that members of the same gene family have similar or identical conserved motifs (Figure 2B). For example, in some proteins such as CsMDAR1, no conserved motifs were detected, while multiple conserved motifs were identified in members of the RBOH family (Figure 2B). There are also differences in the presence of conserved motifs within the same gene family. For instance, CsRBOH8 contains 1 conserved motif, while CsRBOH1 contains 10 conserved motifs (Figure 2B). Most of the ROS metabolism-associated proteins contained a conserved domain (Figure 2C). However, CsCAT contained two conserved Catalase domains, and CsSOD1 contains two conserved Sod_Cu domains. Conversely, no conserved domains were predicted in CsDHAR1, CsRBOH9, and CsTrx17 (Figure 2C).

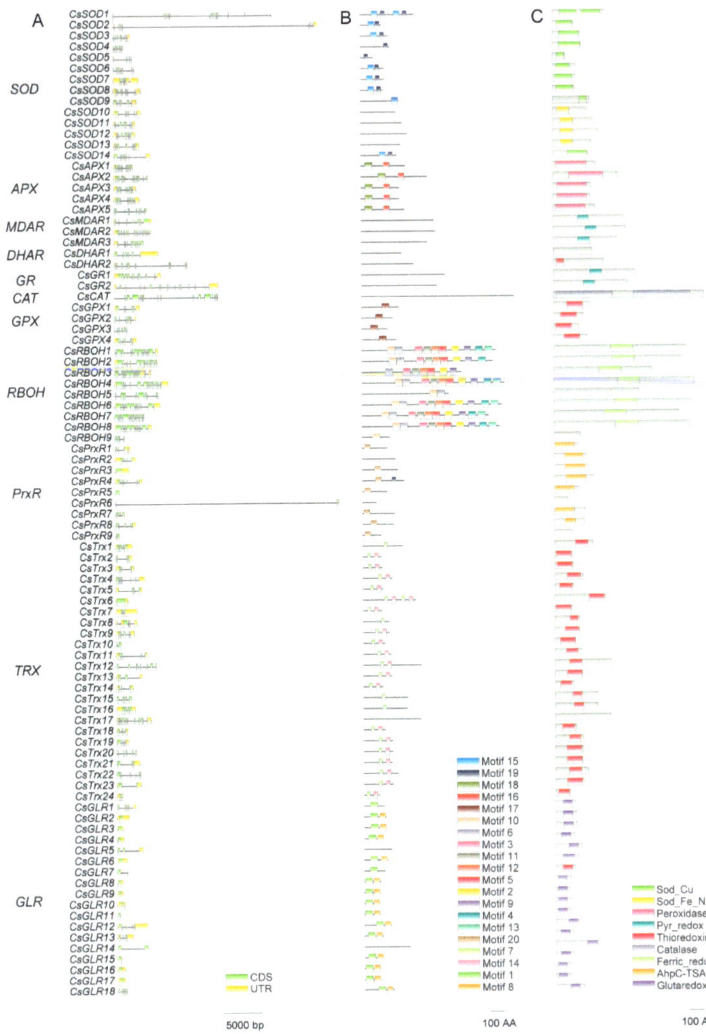

Figure 2. Characterization of ROS metabolism-associated genes in *C. sinensis*. (**A**) Gene structure. Line represents introns, green rectangles represent coding sequences (CDs), and yellow rectangles represent untranslated regions (UTRs). Scale represents 5000 base pairs (bp). (**B**) Conserved motifs. Rectangles of different colors represent 20 different motifs. Scale represents 100 amino acids (AA). (**C**) Conserved domains. Rectangles of different colors represent nine different domains. Scale represents 100 (AA).

3.3. Identification of Synteny Events between C. sinensis and Other Species

Synteny analysis is instrumental in elucidating gene evolution [32]. To comprehend the functionality of ROS metabolism-associated genes through the lens of their evolutionary relationships, synteny analysis between *C. sinensis* and *A. thaliana* and *V. vinifera* was performed. A total of 73 Synteny events were detected between *C. sinensis* and *A. thaliana*, and 70 Synteny events were identified between *C. sinensis* and *V. vinifera* (Figure 3). These findings indicate that a significant proportion of genes associated with ROS metabolism in *C. sinensis* have evolutionary origins shared with *A. thaliana* and *V. vinifera* (Figure 3). This information is valuable for elucidating the timing of gene occurrence relative to the divergence from a common ancestor.

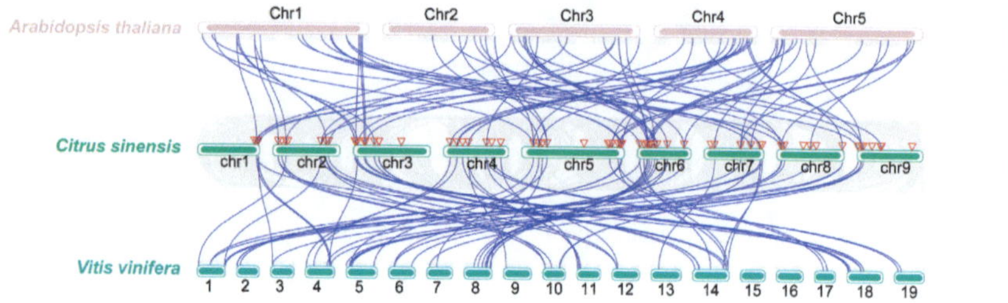

Figure 3. Identification of synteny events for ROS metabolism-associated genes between *C. sinensis* and *Arabidopsis thaliana* and *Vitis vinifera*. Thick rosy-brown lines represent chromosomes of *A. thaliana*, thick light-sea-green lines represent chromosomes of *C. sinensis*, and thick cyan lines represent chromosomes of *V. vinifera*. Blue lines represent synteny events for ROS metabolism-associated genes. The triangles represent ROS metabolism-associated genes in *C. sinensis*.

3.4. Dynamic Expression Profiles of ROS Metabolism-Associated Genes in C. sinensis under CLas Infection

CLas infection significantly induces ROS accumulation in *C. sinensis* [8]. To analyze the expression changes of ROS metabolism-associated genes in response to CLas infection, we utilized transcriptome data to examine the expression patterns of ROS metabolism-associated genes in both CLas-free and CLas-infected samples. Eleven ROS metabolism-associated genes including four *SOD* genes, three *GLR* genes, two *RBOH* genes, and two *PrxR* genes showed no expression in either CLas-free or CLas-infected samples (Figure 4A). The remaining 80 genes exhibited varied expression patterns, with some showing a higher expression in CLas-free samples and others showing a higher expression in CLas-infected samples (Figure 4A). To further elucidate the differential genes induced by CLas infection, we identified 30 differentially expressed ROS-related genes, including 16 up-regulated and 14 down-regulated genes. These differentially expressed genes span various gene families, indicating diverse response patterns of ROS metabolism-associated genes to CLas infection (Figure 4B).

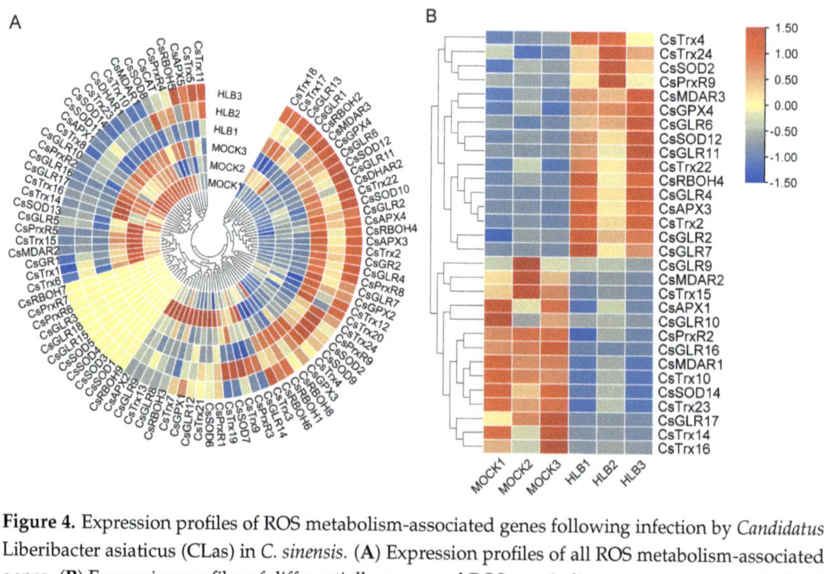

Figure 4. Expression profiles of ROS metabolism-associated genes following infection by *Candidatus* Liberibacter asiaticus (CLas) in *C. sinensis*. (**A**) Expression profiles of all ROS metabolism-associated genes. (**B**) Expression profiles of differentially expressed ROS metabolism-associated genes. 'MOCK' represents CLas-free samples, 'HLB' represents CLas-infected samples. Numbers represent different biological replicates. Bar represents gradient colors to correspond with normalized expression values.

3.5. Dynamic Expression Profiles of ROS Metabolism-Associated Genes in *C. sinensis* under Xcc Infection

To clarify the expression changes of ROS metabolism-associated genes during Xcc infection, we also conducted transcriptome analysis to examine the expression patterns of ROS metabolism-associated genes at different time points post-inoculation with Xcc. Eight genes were found to be non-expressed in both the control and Xcc-inoculated samples (Figure 5A). These 8 genes, including *CsSOD1/3/4/5*, *CsGLR3/15*, *CsPrxR6*, and *CsRBOH6*, were part of the 11 genes identified earlier that were not expressed in both CLas-free and CLas-infection samples (Figure 4A). A subsequent analysis of differentially expressed genes revealed a total of 24 genes with altered expression levels 48 h post-inoculation, consisting of 12 up-regulated and 12 down-regulated genes (Figure 5B). It is worth noting the expression pattern of *CsAPX3*, which exhibited up-regulation at 24 h post-inoculation but down-regulation at 48 h post-inoculation when compared with the control samples (Figure 5B).

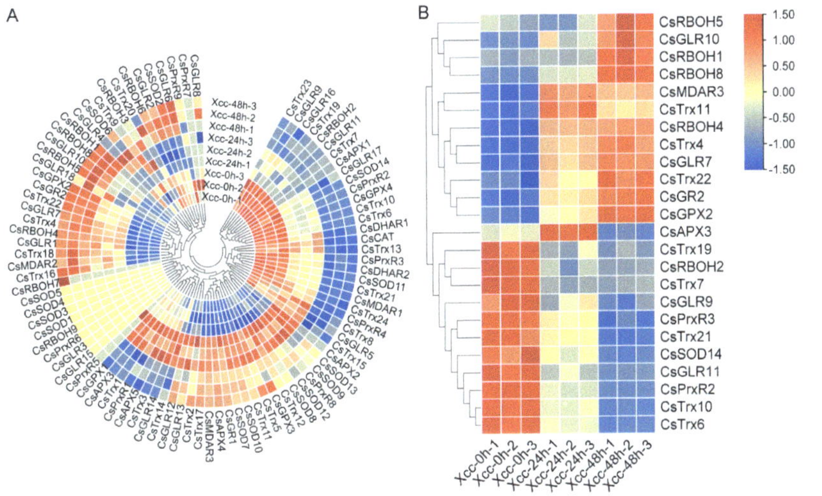

Figure 5. Expression profiles of ROS metabolism-associated genes following incubation by *Xanthomonas citri* subspecies citri (Xcc) in *C. sinensis*. (**A**) Expression profiles of all ROS metabolism-associated genes. (**B**) Expression profiles of differentially expressed ROS metabolism-associated genes. Times '0 h', '24 h', and '48 h' represent hours post incubation. Numbers represent different biological replicates. Bar represents gradient colors to correspond with normalized expression values.

3.6. Identification of ROS Metabolism-Associated Genes with Similar/Opposite Expression Patterns under CLas and Xcc Infection

To explore important genes involved in the response of *C. sinensis* to CLas and Xcc infection, we performed an analysis to pinpoint genes associated with ROS metabolism displaying comparable or contrasting expression profiles following CLas and Xcc infections. It was found that there were five common genes up-regulated in *C. sinensis* after CLas and Xcc infection, namely *CsMDAR3*, *CsRBOH4*, *CsTrx4*, *CsGLR7*, and *CsTrx22* (Figure 6A). Additionally, there were two genes that were up-regulated in *C. sinensis* after CLas infection but down-regulated after Xcc infection, which were *CsAPX3* and *CsGLR11* (Figure 6B). Furthermore, four genes were identified to be down-regulated in *C. sinensis* after CLas and Xcc infections, namely *CsGLR9*, *CsPrxR2*, *CsTrx10*, and *CsSOD14* (Figure 6C). Finally, there was one gene that was down-regulated after CLas infection but up-regulated after Xcc infection in *C. sinensis*, which was *CsGLR10* (Figure 6D). The differentially expressed genes with similar expression patterns during CLas and Xcc infections may participate in common signal transduction pathways triggered by bacterial infections.

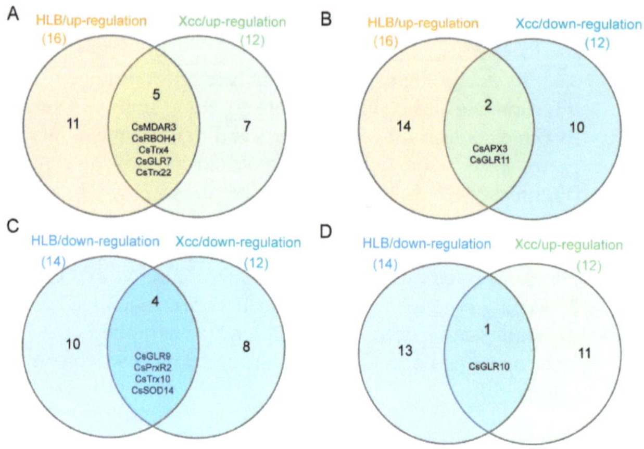

Figure 6. Identification of ROS metabolism-associated genes with similar/opposite expression patterns under *Candidatus* Liberibacter asiaticus (CLas) and *Xanthomonas citri* subspecies citri (Xcc) infection in *C. sinensis*. (**A**) Common up-regulated genes under CLas and Xcc infection. (**B**) Common genes of up-regulated genes during CLas infection and down-regulated genes during Xcc infection. (**C**) Common down-regulated genes under CLas and Xcc infection. (**D**) Common genes of down-regulated genes during CLas infection and up-regulated genes during Xcc infection.

3.7. RT-qPCR Analysis

To validate the reliability of the RNA-seq analysis, RT-qPCR analysis was conducted to assess the expression profiles of four selected differentially expressed genes, including two up-regulated (*CsMDAR3* and *CsRBOH4*) and two down-regulated (*CsPrxR2* and *CsTrx10*) genes in response to CLas and Xcc infection. The RT-qPCR results demonstrated significantly higher expression levels of *CsMDAR3* (Figure 7A) and *CsRBOH4* (Figure 7B) in CLas-infected samples, along with significantly lower expression levels of *CsPrxR2* (Figure 7C) and *CsTrx10* (Figure 7D). Similarly, in Xcc-infected samples, *CsMDAR3* (Figure 7E) and *CsRBOH4* (Figure 7F) exhibited significantly elevated expression levels, while *CsPrxR2* (Figure 7G) and *CsTrx10* (Figure 7H) showed significantly reduced expression levels.

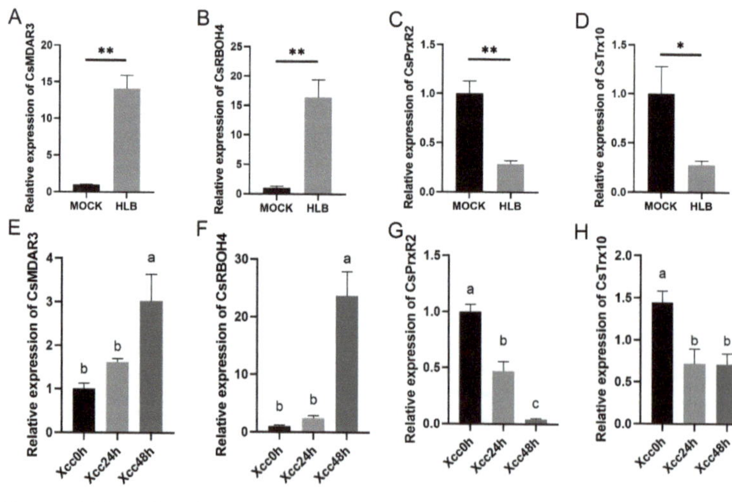

Figure 7. RT-qPCR analysis of ROS metabolism-associated genes in *C. sinensis* under *Candidatus* Liberibacter asiaticus (CLas) and *Xanthomonas citri* subspecies citri (Xcc) infection in *C. sinensis*.

Expression profiles of (**A**) *CsMDAR3*, (**B**) *CsRBOH4*, (**C**) *CsPrxR2*, and (**D**) *CsTrx10* under CLas infection. Expression profiles of (**E**) *CsMDAR3*, (**F**) *CsRBOH4*, (**G**) *CsPrxR2*, and (**H**) *CsTrx10* under Xcc infection. '*' indicates significant difference with a p-value < 0.05. '**' indicates significant difference with a p-value < 0.01. Lowercase letters annotated on bars indicate significant differences with a p-value < 0.05. 'MOCK' represents healthy plants, while 'HLB' represents plants infected with HLB. 'Xcc0h' represents control for citrus canker inoculation, and 'Xcc24h' and 'Xcc48h' represent samples collected at 24 h and 48 h post citrus canker inoculation, respectively. Each measurement was performed with three biological replicates.

3.8. Hydrogen Peroxide Content, CAT, and SOD Enzyme Activities during CLas and Xcc Infection

Pathogen invasion significantly alters the redox homeostasis within plant cells [21]. After CLas infection, we noted a significant accumulation of hydrogen peroxide (Figure 8A), and the activities of CAT (Figure 8B) and SOD (Figure 8C) were significantly increased in CLas infected leaves of *C. sinensis*. Similarly, we found that Xcc infection also caused a significant increase in hydrogen peroxide levels (Figure 8D), as well as a significant elevation in CAT (Figure 8E) and SOD (Figure 8F) activities. However, we observed that the activities of CAT and SOD continued to significantly increase (Figure 8E,F), while the level of hydrogen peroxide did not show a significant change between samples of 24 h and 48 h post Xcc infection (Figure 8D).

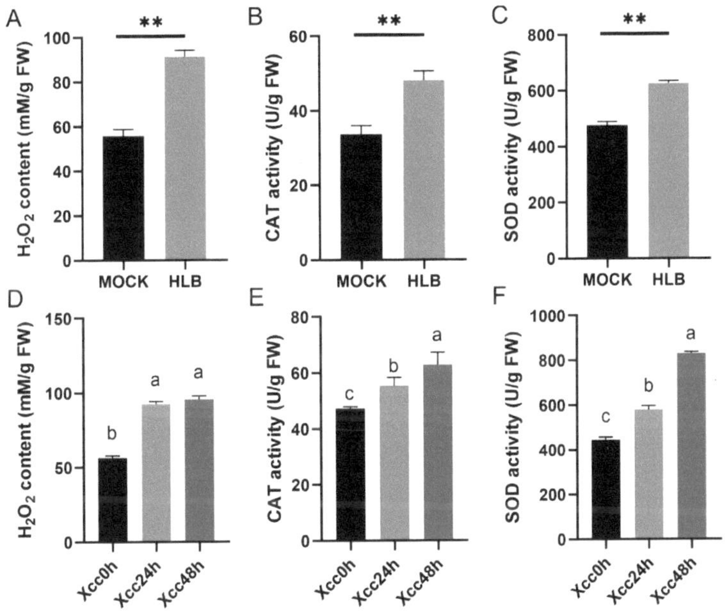

Figure 8. Determination of hydrogen peroxide content, CAT, and SOD enzyme activities in *C. sinensis* under *Candidatus* Liberibacter asiaticus (CLas) and *Xanthomonas citri* subspecies citri (Xcc) infection in *C. sinensis*. (**A**) Hydrogen peroxide content, (**B**) CAT activities, and (**C**) SOD activities under CLas infection. (**D**) Hydrogen peroxide content, (**E**) CAT activities, and (**F**) SOD activities under Xcc infection. '**' indicates significant difference with a p-value < 0.01. Lowercase letters annotated on bars indicate significant differences with a p-value < 0.05. 'MOCK' represents healthy plants, while 'HLB' represents plants infected with HLB. 'Xcc0h' represents control for citrus canker inoculation, and 'Xcc24h' and 'Xcc48h' represent samples collected at 24 h and 48 h post citrus canker inoculation, respectively. Each measurement was performed with three biological replicates.

3.9. PPI Network Analysis of Key Candidate ROS Metabolism-Associated Genes In Silico

In order to identify more cooperation factors involved in ROS metabolism, we performed a PPI network analysis of the common up-regulated and down-regulated genes during CLas and Xcc infections in silico (Figure 9). CsMDAR3 exhibited interactions with DAO domain-containing proteins, stress-response A/B barrel domain-containing proteins, and Rieske domain-containing proteins (Figure 9A). CsRBOH4 interacted with calcium-dependent protein kinases (Figure 9B). CsTrx4 and CsTrx22 primarily interacted with Thioredoxin domain-containing proteins and Peroxiredoxins (Figure 9C,D). CsGLR7 exhibited variable interaction partners such as Ribonucleoside-diphosphate reductases, Glutathione reductases, and Expansin (Figure 9E). CsGLR9 interacted with Ribonucleoside-diphosphate reductases and Glutathione reductases (Figure 9F). The interaction partners of CsPrxR2 included Tryptophan synthases and Peptide-methionine (R)-S-oxide reductases (Figure 9G). CsTrx10 interacted with Thioredoxin domain-containing proteins and Peroxiredoxins (Figure 9H). The interaction partners of CsSOD14 included DJ-1_PfpI domain-containing proteins and catalases (Figure 9I). Interacting partners play a crucial role in the catalytic activity of these enzymes, but their specific biological functions require further investigation.

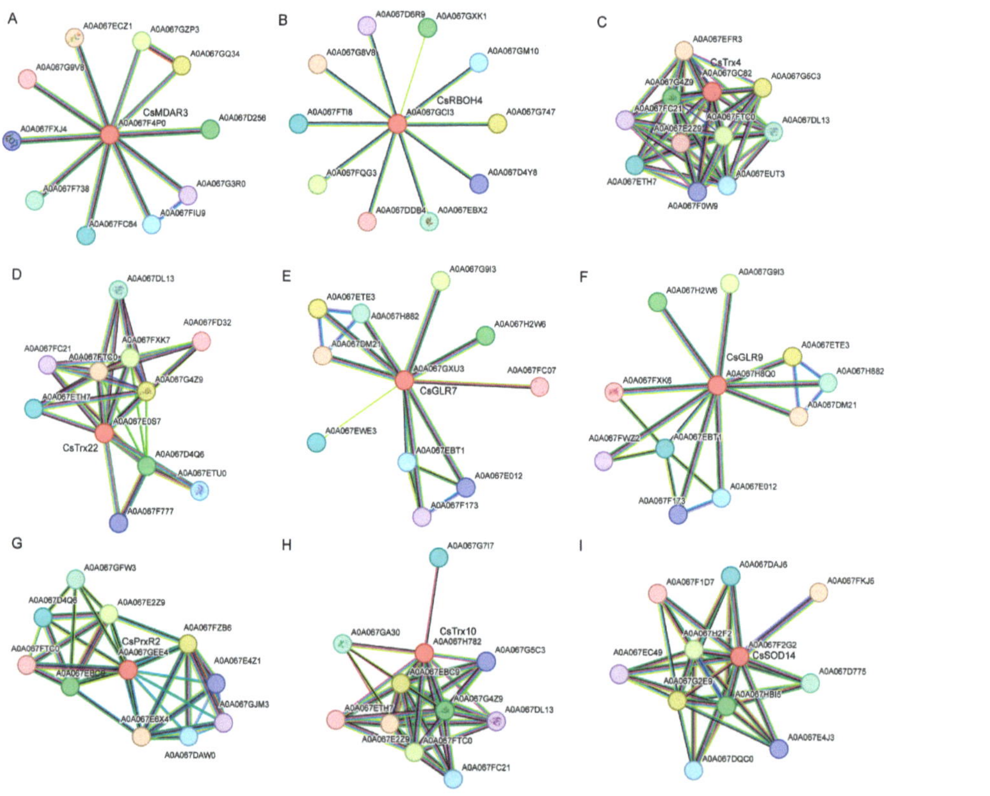

Figure 9. Protein–protein interaction (PPI) network analysis of key ROS metabolism-associated genes under *Candidatus* Liberibacter asiaticus and *Xanthomonas citri* subspecies citri infection in *C. sinensis* in silico. PPI network analysis of (**A**) CsMADR3, (**B**) CsROBH4, (**C**) CsTrx4, (**D**) CsTrx22, (**E**) CsGLR7, (**F**) CsGLR9, (**G**) CsPrxR2, (**H**) CsTrx10, and (**I**) CsSOD14. Raw networks were downloaded from STRING database and modified using Cytoscape.

3.10. Cis-Acting Elements in the Promoter of ROS Metabolism-Associated Genes in *C. Sinensis*

To elucidate the potential expression regulatory mechanisms of ROS metabolism-associated genes, we analyzed the promoter sequences of these genes, focusing on cis-acting elements involved in the hormone signaling response and abiotic stress response. In the promoter region of *CsSOD11*, as many as 31 cis-acting elements are involved in hormone signaling and abiotic stress responses, while no such cis-acting elements were found in the promoter sequences of *CsTrx1*, *CsTrx14*, and *CsTrx22* (Figure S1). Furthermore, the distribution of cis-acting elements is irregular in the promoter region of ROS metabolism-associated genes in *C. sinensis*, even within the same gene family, with completely different distributions of cis-acting elements (Figure S1). It is worth noting that although no cis-acting elements responding to hormone signals or abiotic stress were found in the promoter region of *CsTrx22*, *CsTrx22* was up-regulated during CLas and Xcc infection (Figure 6A). This suggests that there are still deficiencies in predicting cis-acting elements using the current method, and how CsTrx22 responds to pathogen infection may involve new regulatory mechanisms.

3.11. Gene Gain and Loss Events in ROS Metabolism-Associated Genes between C. sinensis and P. trifoliata

Previous studies suggested that *P. trifoliata* is one of the germplasms resistant to HLB, while *C. sinensis* is more susceptible to the disease, although there are differences between different varieties [42–44]. Given the ROS accumulation observed in CLas-infected tissues, we systematically identified the occurrence of gene gain and loss events within ROS metabolism-associated genes between *C. sinensis* and *P. trifoliata* by analyzing their orthologous relationships (Table 1). Gene gain and loss events were found in the *RBOH*, *SOD*, *PrxR*, *APX*, and *Trx* gene families (Table 1, Figures 8 and S2–S11). Specifically, one gene loss event was identified in the *RBOH* gene family, along with three gene gain events and five gene loss events in the *SOD* gene family. Furthermore, three gene gain events and three gene loss events were documented in the PrxR gene family, while one gene gain event was observed in the *APX* gene family. Lastly, one gene gain event and one gene loss event were found in the *Trx* gene family (Table 1).

Table 1. Potential gain and loss events of ROS metabolism-associated genes between HLB-susceptible *C. sinensis* and HLB-tolerant *P. trifoliata*.

Gene Family	*Citrus sinensis*	
	Gain	Lost
RBOH	0	1
SOD	3	5
CAT	0	0
PrxR	3	3
APX	1	0
GPX	0	0
MDAR	0	0
DHAR	0	0
GR	0	0
Trx	1	1
GLR	0	0

Taking the *SOD* gene family as an example, a sequence alignment and a phylogenetic analysis were performed on all *SOD* gene members in *C. sinensis and P. trifoliata*. The examination of individual evolutionary branches facilitated the identification of gene gain and loss events (Figure 10).

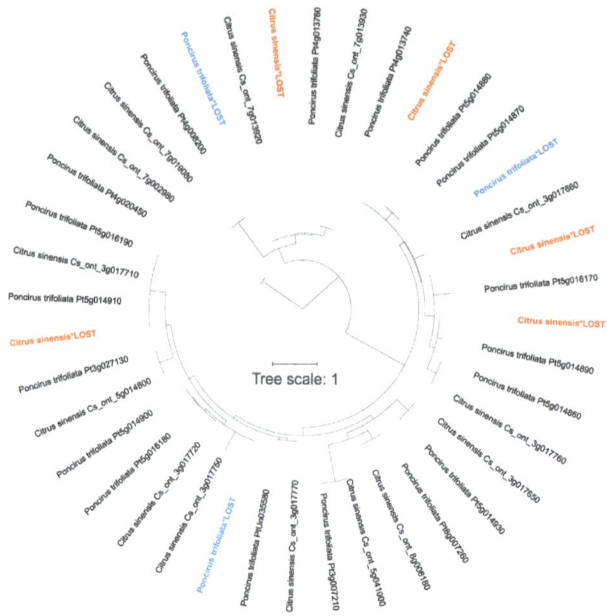

Figure 10. Gene gain and loss events in *SOD* gene family between *C. sinensis* and *P. trifoliata*. Gene loss events in *C. sinensis* are highlighted in bold orange color, while gene loss events in *P. trifoliata* are highlighted in bold blue color.

4. Discussion

ROS are continuously produced as by-products of aerobic metabolism in plant cells, with their levels intricately regulated through a complex network of metabolic pathways [45]. These pathways, regulated by ROS metabolism-associated genes, are responsible for scavenging and detoxifying ROS, maintaining a balance between ROS production and elimination [46]. The balance between ROS production and scavenging is vital for the maintenance of cellular homeostasis [47]. The coordinated expression of ROS metabolism-associated genes ensures this balance, maintains cellular redox homeostasis, and allows for ROS to act as crucial signaling molecules in plant physiology [48].

There are significant differences in the number of ROS metabolism-associated genes among different species. Taking SOD as an example, there are 8 *SOD* genes in *A. thaliana* [22], 14 in *C. sinensis*, 29 in *Brassica juncea* [49], 18 in *B. rapa* [49], 25 in *Musa acuminata* [50], 18 in *Gossypium raimondii* and *G. arboreum* [51], and 8 in *Salvia miltiorrhiza* [52]. More surprising is that there are 18 *SOD* genes identified in *P. trifoliata*, a closely related species of *C. sinensis*. Since SOD functions as the initial step in catalyzing the conversion of O^{2-} to hydrogen peroxide, the knockout mutant of *AtSOD2* caused SOD activity to decline significantly in chloroplast and cytosol [53]. Nevertheless, functional overlap among *SOD* genes in *A. thaliana* has been noted, as some of them originated from gene duplication events [54]. Moreover, in *C. sinensis*, there is a single *CAT* gene responsible for eliminating hydrogen peroxide, whereas in *A. thaliana*, there are three gene loci encoding CAT [55]. Additionally, three *CAT* genes were identified in *Oryza sativa* [56], *Zea mays* [57] and *Nicotiana plumbaginifolia* [58]. Furthermore, two *CAT* genes were identified within *P. trifoliata* in this study. To determine the number of *CAT* genes in *C. sinensis* and potential evolutionary events, we conducted a homology search in genome versions 1 and 2 of *C. sinensis*, revealing two *CAT* genes located at adjacent genetic loci. Considering that the sequence of *CAT* in the genome of *C. sinensis* (version 3) is notably long and contains two conserved catalase domains, we have reason to suspect an annotation error of this gene in the genome of *C. sinensis* (version 3). Therefore, it is reasonable to conclude that *C. sinensis* contains two *CAT* genes.

The *APX* gene family in *C. sinensis* comprises a total of 5 members, while *A. thaliana* has 9 members [22], and there are 11 *APX* genes in the genome of *Populus trichocarpa* [59]. In addition, while the genome of *A. thaliana* contained 27 *GLR* genes and 32 *Trx* genes [22], 18 *GLR* genes and 24 *Trx* genes were found in the genome of *C. sinensis*. This indicates that the quantity of ROS metabolism-associated genes in *C. sinensis* is generally lower than that in *A. thaliana*, with the exception of the SOD gene family.

Given that CLas infection significantly induces the accumulation of ROS in the phloem tissues, thereby triggering an exaggerated immune response in plants, our focus lies on determining the altered expression patterns of ROS metabolism-associated genes following CLas infection. Thirty differentially expressed genes, including 16 up-regulated genes and 14 down-regulated genes were observed in *C. sinensis* during CLas infection. Among them, 1 *RBOH* gene *CsRBOH4* was up-regulated. A previous study indicated that *RBOH* genes are activated in the process of CLas infection, thus leading to ROS accumulation [1]. The homologous gene of *CsRBOH4* in *A. thaliana* is *AtRBOHF*. AtRBOHF is a crucial factor involved not only in cell death associated with the hypersensitive response but also in modulating plant defense responses to pathogens [60]. Significantly, CLas infection leads to a notable increase in the accumulation of hydrogen peroxide and facilitates cell death in citrus plants [8]. Therefore, CsRBOH4 may serve as a key player in the interaction between CLas and citrus plants. Interestingly, the differentially expressed genes did not include genes from the *CAT* and *GR* gene families. One reason is that the number of genes in the *CAT* and *GR* gene families is small, and another reason is that these two genes are widely and importantly involved in biological processes within cells. Xcc infection also induced a differential expression of multiple genes, including 12 up-regulated and 12 down-regulated genes. Four *RBOH* genes were up-regulated, including *CsRBOH4*, while one *RBOH* gene, namely *CsRBOH2*, was down-regulated. The homologous gene of *CsRBOH2* in *A. thaliana* is *AtRBOHE*. Research has demonstrated the involvement of AtRBOHE in the development of various tissues such as roots and pollen, as well as its participation in plant–*Trichoderma* interactions [61,62]. The negatively regulated expression of *CsRBOH2* in *C. sinensis* during Xcc infection suggests an antagonistic relationship between plant development and stress responses. The identification of differentially expressed genes revealed that multiple genes exhibit similar or contrasting expression patterns in response to both CLas and Xcc infections. It is worthy to note that *CsRBOH4* was up-regulated during both CLas and Xcc infections, suggesting that CsRBOH4 may play a central role in the *C. sinensis* response to pathogen invasion. Nevertheless, *CsAPX3* and *CsGLR11* were observed to be up-regulated during CLas infection but down-regulated during Xcc infection, highlighting distinct responses of *C. sinensis* to the infections caused by the two pathogens. Upon *Pseudomonas syringae* infection, *APX* was significantly up-regulated in *Actinidia chinensis* [63]. In contrast, upon *Erwinia amylovora* infection, the *APX* gene demonstrated down-regulation in *Pyrus communis* [64]. Therefore, this contrasting regulatory pattern of genes in response to different pathogens could provide novel clues for further elucidating pathogenic mechanisms.

In plant–pathogen interactions, ROS metabolism-associated genes play a dual role [12,21,45]. These genes play a role in triggering the oxidative burst, a quick and temporary increase in ROS levels that acts as an initial defense mechanism against pathogens [65]. However, some pathogens have developed mechanisms to control the generation of ROS and alter the functions of ROS metabolism enzymes, ultimately impacting the plant's defense against these pathogens [66,67]. In this study, results showed a notable rise in hydrogen peroxide levels after CLas and Xcc infection, suggesting that *C. sinensis* relies on increased hydrogen peroxide to combat pathogen attacks post-infection. SOD facilitates the conversion of superoxide anions into hydrogen peroxide, while CAT plays a pivotal role in eliminating intracellular hydrogen peroxide [14]. Interestingly, enzymatic activity assessments in our study indicated a marked elevation in SOD and CAT activities of *C. sinensis* subsequent to being infected by CLas and Xcc. Nevertheless, RNA-seq analysis revealed no substantial alteration in *CAT* gene expression and a noteworthy decrease in

CsSOD14 expression post CLas and Xcc infection. The findings imply inconsistencies between transcriptional and protein levels, emphasizing the complexity of the ROS metabolic pathway. Within this pathway, these genes cooperate to sustain the intricate balance of ROS.

5. Conclusions

In summary, this study systematically identified ROS metabolism-associated genes in the genome of *C. sinensis* using bioinformatics methods, including gene families such as *RBOH, SOD, CAT, PrxR, APX, GPX, MDAR, DHAR, GR, Trx*, and *GLR*. Through gene structure analysis, conserved motifs, and conserved domain prediction, sequence attributes of genes related to ROS metabolism were analyzed. A synteny analysis was used to explore the collinearity relationships of ROS metabolism-associated genes in sweet orange with those in Arabidopsis and grape. The changes in the expression patterns of genes related to ROS metabolism in sweet orange during CLas infection and Xcc infection processes were investigated, revealing multiple differentially expressed genes. Furthermore, five genes were found to be up-regulated in both CLas infection and Xcc infection processes, while four genes were down-regulated. A PPI network analysis was conducted to explore the protein interaction partners of these genes. An analysis of the promoter cis-elements of genes related to sweet orange ROS metabolism provided clues to their expression regulation patterns. Lastly, gene gain and loss analysis clarified events of gene gain and loss related to ROS metabolism genes in the sweet orange and trifoliate orange genomes. This study provides data support for investigating the reactive oxygen species burst during the sweet orange response to CLas infection and offers clues for further understanding the citrus–CLas interaction.

Supplementary Materials: The following supporting information can be downloaded at: https://www.mdpi.com/article/10.3390/horticulturae10060625/s1, Figure S1. Identification of *cis*-acting elements in the promoter regions of ROS metabolism-associated genes in *Citrus sinensis;* Figure S2. Gene gain and loss events in the *APX* gene family between *Citrus sinensis* and *Poncirus trifoliata;* Figure S3. Gene gain and loss events in the *CAT* gene family between *Citrus sinensis* and *Poncirus trifoliata;* Figure S4. Gene gain and loss events in the *DHAR* gene family between *Citrus sinensis* and *Poncirus trifoliata;* Figure S5. Gene gain and loss events in the *GLR* gene family between *Citrus sinensis* and *Poncirus trifoliata;* Figure S6. Gene gain and loss events in the *GPX* gene family between *Citrus sinensis* and *Poncirus trifoliata;* Figure S7. Gene gain and loss events in the *GR* gene family between *Citrus sinensis* and *Poncirus trifoliata;* Figure S8. Gene gain and loss events in the *MDAR* gene family between *Citrus sinensis* and *Poncirus trifoliata;* Figure S9. Gene gain and loss events in the *PrxR* gene family between *Citrus sinensis* and *Poncirus trifoliata;* Figure S10. Gene gain and loss events in the *RBOH* gene family between *Citrus sinensis* and *Poncirus trifoliata;* Figure S11. Gene gain and loss events in the *Trx* gene family between *Citrus sinensis* and *Poncirus trifoliata;* Table S1. Primers used in this study; Table S2. Details of ROS metabolism-associated genes in *Citrus sinensis;* Table S3. Expression profiles of ROS metabolism-associated genes in *Citrus sinensis* during *Candidatus* Liberibacter asiaticus infection; Table S4. Expression profiles of ROS metabolism-associated genes in *Citrus sinensis* during *Xanthomonas citri* subspecies citri infection; Table S5. Annotation of proteins interacting with up-regulated ROS-associated members in *Citrus sinensis;* Table S6. Annotation of proteins interacting with down-regulated ROS-associated members in *Citrus sinensis;* Table S7. Gene gain and loss events of ROS metabolism-associated genes between *Citrus sinensis* and *Poncirus trifoliata.*

Author Contributions: Conceptualization, R.L. and Z.O.; methodology, G.H., F.L., Y.H. and R.L.; software, G.H., F.L., Y.H. and R.L.; validation, R.L.; formal analysis, R.L.; investigation, G.H., F.L., Y.H. and R.L.; resources, R.L.; data curation, G.H., F.L., Y.H. and R.L.; writing—original draft preparation, G.H., F.L., Y.H. and R.L.; writing—review and editing, G.H., Z.O. and R.L.; visualization, R.L.; supervision, R.L.; project administration, R.L.; funding acquisition, R.L, G.H. and F.L. All authors have read and agreed to the published version of the manuscript.

Funding: This work was supported by the National Natural Science Foundation of China (32160621, 32260659), the Innovation Program of the short-term Leader of Jiangxi Province (jxsq2020102129), and the Innovation Fund Designated for Graduate Students of Jiangxi Province (YC2023-S850).

Data Availability Statement: All data generated and analyzed during this study are included in this published article. The original RNA-seq data has been uploaded to the NCBI_SRA database (PRJNA953196 and PRJNA1088720).

Acknowledgments: The authors are grateful for the helpful discussions from the laboratory members.

Conflicts of Interest: The authors declare no conflicts of interest.

References

1. Ribeiro, C.; Xu, J.; Hendrich, C.; Pandey, S.S.; Yu, Q.; Gmitter, F.G.; Wang, N. Seasonal Transcriptome Profiling of Susceptible and Tolerant Citrus Cultivars to Citrus Huanglongbing. *Phytopathology* **2022**, *113*, 286–298. [CrossRef]
2. Hu, Y.; Zhong, X.; Liu, X.; Lou, B.; Zhou, C.; Wang, X. Comparative transcriptome analysis unveils the tolerance mechanisms of *Citrus hystrix* in response to '*Candidatus* Liberibacter asiaticus' infection. *PLoS ONE* **2017**, *12*, e0189229. [CrossRef] [PubMed]
3. Fu, S.; Shao, J.; Zhou, C.; Hartung, J.S. Transcriptome analysis of sweet orange trees infected with '*Candidatus* Liberibacter asiaticus' and two strains of Citrus Tristeza Virus. *BMC Genom.* **2016**, *17*, 349. [CrossRef]
4. Martinelli, F.; Uratsu, S.L.; Albrecht, U.; Reagan, R.L.; Phu, M.L.; Britton, M.; Buffalo, V.; Fass, J.; Leicht, E.; Zhao, W. Transcriptome profiling of citrus fruit response to huanglongbing disease. *PLoS ONE* **2012**, *7*, e38039. [CrossRef]
5. Liu, C.; Chang, X.; Li, F.; Yan, Y.; Zuo, X.; Huang, G.; Li, R. Transcriptome analysis of Citrus sinensis reveals potential responsive events triggered by Candidatus Liberibacter asiaticus. *Protoplasma* **2024**, *261*, 499–512. [CrossRef]
6. Tang, L.; Vashisth, T. New insight in Huanglongbing-associated mature fruit drop in citrus and its link to oxidative stress. *Sci. Hortic.* **2020**, *265*, 109246. [CrossRef]
7. Gao, C.; Li, C.; Li, Z.; Liu, Y.; Li, J.; Guo, J.; Mao, J.; Fang, F.; Wang, C.; Deng, X.; et al. Comparative transcriptome profiling of susceptible and tolerant citrus species at early and late stage of infection by "*Candidatus* Liberibacter asiaticus". *Front. Plant Sci.* **2023**, *14*, 1191029. [CrossRef]
8. Ma, W.; Pang, Z.; Huang, X.; Xu, J.; Pandey, S.S.; Li, J.; Achor, D.S.; Vasconcelos, F.N.C.; Hendrich, C.; Huang, Y.; et al. Citrus Huanglongbing is a pathogen-triggered immune disease that can be mitigated with antioxidants and gibberellin. *Nat. Commun.* **2022**, *13*, 529. [CrossRef] [PubMed]
9. Hu, Y.; Zhang, J.; Jia, H.; Sosso, D.; Li, T.; Frommer, W.B.; Yang, B.; White, F.F.; Wang, N.; Jones, J.B. Lateral organ boundaries 1 is a disease susceptibility gene for citrus bacterial canker disease. *Proc. Natl. Acad. Sci. USA* **2014**, *111*, E521–E529. [CrossRef] [PubMed]
10. Moore, J.W.; Loake, G.J.; Spoel, S.H. Transcription Dynamics in Plant Immunity. *Plant Cell* **2011**, *23*, 2809–2820. [CrossRef]
11. Baxter, A.; Mittler, R.; Suzuki, N. ROS as key players in plant stress signalling. *J. Exp. Bot.* **2014**, *65*, 1229–1240. [CrossRef] [PubMed]
12. Das, K.; Roychoudhury, A. Reactive oxygen species (ROS) and response of antioxidants as ROS-scavengers during environmental stress in plants. *Front. Environ. Sci.* **2014**, *2*, 53. [CrossRef]
13. Marino, D.; Dunand, C.; Puppo, A.; Pauly, N. A burst of plant NADPH oxidases. *Trends Plant Sci.* **2012**, *17*, 9–15. [CrossRef] [PubMed]
14. Jiménez, A.; Sevilla, F.; Martí, M.C. Reactive oxygen species homeostasis and circadian rhythms in plants. *J. Exp. Bot.* **2021**, *72*, 5825–5840. [CrossRef] [PubMed]
15. Choudhury, S.; Panda, P.; Sahoo, L.; Panda, S.K. Reactive oxygen species signaling in plants under abiotic stress. *Plant Signal. Behav.* **2013**, *8*, e23681. [CrossRef] [PubMed]
16. Hasanuzzaman, M.; Bhuyan, M.H.M.B.; Anee, T.I.; Parvin, K.; Nahar, K.; Mahmud, J.A.; Fujita, M. Regulation of Ascorbate-Glutathione Pathway in Mitigating Oxidative Damage in Plants under Abiotic Stress. *Antioxidants* **2019**, *8*, 384. [CrossRef] [PubMed]
17. Zhang, T.; Xiao, J.; Zhao, Y.; Zhang, Y.; Jie, Y.; Shen, D.; Yue, C.; Huang, J.; Hua, Y.; Zhou, T. Comparative physiological and transcriptomic analyses reveal ascorbate and glutathione coregulation of cadmium toxicity resistance in wheat genotypes. *BMC Plant Biol.* **2021**, *21*, 459. [CrossRef] [PubMed]
18. Paiva, C.N.; Bozza, M.T. Are Reactive Oxygen Species Always Detrimental to Pathogens? *Antioxid. Redox Signal.* **2013**, *20*, 1000–1037. [CrossRef] [PubMed]
19. Vernon, P.J.; Tang, D. Eat-Me: Autophagy, Phagocytosis, and Reactive Oxygen Species Signaling. *Antioxid. Redox Signal.* **2012**, *18*, 677–691. [CrossRef]
20. Lodde, V.; Morandini, P.; Costa, A.; Murgia, I.; Ezquer, I. cROStalk for Life: Uncovering ROS Signaling in Plants and Animal Systems, from Gametogenesis to Early Embryonic Development. *Genes* **2021**, *12*, 525. [CrossRef]
21. Camejo, D.; Guzmán-Cedeño, Á.; Moreno, A. Reactive oxygen species, essential molecules, during plant–pathogen interactions. *Plant Physiol. Biochem.* **2016**, *103*, 10–23. [CrossRef] [PubMed]
22. Inupakutika, M.A.; Sengupta, S.; Devireddy, A.R.; Azad, R.K.; Mittler, R. The evolution of reactive oxygen species metabolism. *J. Exp. Bot.* **2016**, *67*, 5933–5943. [CrossRef] [PubMed]
23. Liu, H.; Wang, X.; Liu, S.; Huang, Y.; Guo, Y.-X.; Xie, W.-Z.; Liu, H.; ul Qamar, M.T.; Xu, Q.; Chen, L.-L. Citrus Pan-Genome to Breeding Database (CPBD): A comprehensive genome database for citrus breeding. *Mol. Plant* **2022**, *15*, 1503–1505. [CrossRef] [PubMed]

24. Letunic, I.; Khedkar, S.; Bork, P. SMART: Recent updates, new developments and status in 2020. *Nucleic Acids Res.* **2021**, *49*, 458–460. [CrossRef] [PubMed]
25. Johnson, M.; Zaretskaya, I.; Raytselis, Y.; Merezhuk, Y.; McGinnis, S.; Madden, T.L. NCBI BLAST: A better web interface. *Nucleic Acids Res.* **2008**, *36*, W5–W9. [CrossRef] [PubMed]
26. Paysan-Lafosse, T.; Blum, M.; Chuguransky, S.; Grego, T.; Pinto, B.L.; Salazar, G.A.; Bileschi, M.L.; Bork, P.; Bridge, A.; Colwell, L.; et al. InterPro in 2022. *Nucleic Acids Res.* **2023**, *51*, D418–D427. [CrossRef] [PubMed]
27. Chen, C.; Wu, Y.; Li, J.; Wang, X.; Zeng, Z.; Xu, J.; Liu, Y.; Feng, J.; Chen, H.; He, Y.; et al. TBtools-II: A "one for all, all for one" bioinformatics platform for biological big-data mining. *Mol Plant* **2023**, *16*, 1733–1742. [CrossRef] [PubMed]
28. Bailey, T.L.; Johnson, J.; Grant, C.E.; Noble, W.S. The MEME suite. *Nucleic Acids Res.* **2015**, *43*, 39–49. [CrossRef] [PubMed]
29. Lu, S.; Wang, J.; Chitsaz, F.; Derbyshire, M.K.; Geer, R.C.; Gonzales, N.R.; Gwadz, M.; Hurwitz, D.I.; Marchler, G.H.; Song, J.S.; et al. CDD/SPARCLE: The conserved domain database in 2020. *Nucleic Acids Res.* **2020**, *48*, D265–D268. [CrossRef] [PubMed]
30. Lamesch, P.; Berardini, T.Z.; Li, D.; Swarbreck, D.; Wilks, C.; Sasidharan, R.; Muller, R.; Dreher, K.; Alexander, D.L.; Garcia-Hernandez, M.; et al. The Arabidopsis Information Resource (TAIR): Improved gene annotation and new tools. *Nucleic Acids Res.* **2012**, *40*, D1202–D1210. [CrossRef]
31. Bolser, D.M.; Staines, D.M.; Perry, E.; Kersey, P.J. Ensembl Plants: Integrating Tools for Visualizing, Mining, and Analyzing Plant Genomic Data. *Methods Mol. Biol.* **2017**, *1533*, 1–31. [CrossRef] [PubMed]
32. Wang, Y.; Tang, H.; Debarry, J.D.; Tan, X.; Li, J.; Wang, X.; Lee, T.H.; Jin, H.; Marler, B.; Guo, H.; et al. MCScanX: A toolkit for detection and evolutionary analysis of gene synteny and collinearity. *Nucleic Acids Res.* **2012**, *40*, e49. [CrossRef] [PubMed]
33. Bolger, A.M.; Lohse, M.; Usadel, B. Trimmomatic: A flexible trimmer for Illumina sequence data. *Bioinformatics* **2014**, *30*, 2114–2120. [CrossRef] [PubMed]
34. Bray, N.L.; Pimentel, H.; Melsted, P.; Pachter, L. Near-optimal probabilistic RNA-seq quantification. *Nat. Biotechnol.* **2016**, *34*, 525–527. [CrossRef] [PubMed]
35. Cheadle, C.; Vawter, M.P.; Freed, W.J.; Becker, K.G. Analysis of Microarray Data Using Z Score Transformation. *J. Mol. Diagn.* **2003**, *5*, 73–81. [CrossRef] [PubMed]
36. Lescot, M.; Déhais, P.; Thijs, G.; Marchal, K.; Moreau, Y.; Van de Peer, Y.; Rouzé, P.; Rombauts, S. PlantCARE, a database of plant cis-acting regulatory elements and a portal to tools for in silico analysis of promoter sequences. *Nucleic Acids Res.* **2002**, *30*, 325–327. [CrossRef] [PubMed]
37. Szklarczyk, D.; Kirsch, R.; Koutrouli, M.; Nastou, K.; Mehryary, F.; Hachilif, R.; Gable, A.L.; Fang, T.; Doncheva, N.T.; Pyysalo, S.; et al. The STRING database in 2023: Protein-protein association networks and functional enrichment analyses for any sequenced genome of interest. *Nucleic Acids Res.* **2023**, *51*, 638–646. [CrossRef] [PubMed]
38. Katoh, K.; Rozewicki, J.; Yamada, K.D. MAFFT online service: Multiple sequence alignment, interactive sequence choice and visualization. *Brief. Bioinform.* **2019**, *20*, 1160–1166. [CrossRef] [PubMed]
39. Castresana, J. Selection of Conserved Blocks from Multiple Alignments for Their Use in Phylogenetic Analysis. *Mol. Biol. Evol.* **2000**, *17*, 540–552. [CrossRef]
40. Lanfear, R.; Frandsen, P.B.; Wright, A.M.; Senfeld, T.; Calcott, B. PartitionFinder 2: New Methods for Selecting Partitioned Models of Evolution for Molecular and Morphological Phylogenetic Analyses. *Mol. Biol. Evol.* **2017**, *34*, 772–773. [CrossRef]
41. Minh, B.Q.; Schmidt, H.A.; Chernomor, O.; Schrempf, D.; Woodhams, M.D.; von Haeseler, A.; Lanfear, R. IQ-TREE 2: New Models and Efficient Methods for Phylogenetic Inference in the Genomic Era. *Mol. Biol. Evol.* **2020**, *37*, 1530–1534. [CrossRef] [PubMed]
42. Peng, Z.; Bredeson, J.V.; Wu, G.A.; Shu, S.; Rawat, N.; Du, D.; Parajuli, S.; Yu, Q.; You, Q.; Rokhsar, D.S.; et al. A chromosome-scale reference genome of trifoliate orange (*Poncirus trifoliata*) provides insights into disease resistance, cold tolerance and genome evolution in Citrus. *Plant J.* **2020**, *104*, 1215–1232. [CrossRef] [PubMed]
43. Rawat, N.; Kumar, B.; Albrecht, U.; Du, D.; Huang, M.; Yu, Q.; Zhang, Y.; Duan, Y.-P.; Bowman, K.D.; Gmitter, F.G., Jr.; et al. Genome resequencing and transcriptome profiling reveal structural diversity and expression patterns of constitutive disease resistance genes in Huanglongbing-tolerant *Poncirus trifoliata* and its hybrids. *Hortic. Res.* **2017**, *4*, 17064. [CrossRef] [PubMed]
44. Huang, M.; Roose, M.L.; Yu, Q.; Du, D.; Yu, Y.; Zhang, Y.; Deng, Z.; Stover, E.; Gmitter, F.G., Jr. Construction of High-Density Genetic Maps and Detection of QTLs Associated with Huanglongbing Tolerance in Citrus. *Front. Plant Sci.* **2018**, *9*, 1694. [CrossRef] [PubMed]
45. Mansoor, S.; Ali Wani, O.; Lone, J.K.; Manhas, S.; Kour, N.; Alam, P.; Ahmad, A.; Ahmad, P. Reactive Oxygen Species in Plants: From Source to Sink. *Antioxidants* **2022**, *11*, 225. [CrossRef] [PubMed]
46. Janků, M.; Luhová, L.; Petřivalský, M. On the Origin and Fate of Reactive Oxygen Species in Plant Cell Compartments. *Antioxidants* **2019**, *8*, 105. [CrossRef] [PubMed]
47. Decros, G.; Baldet, P.; Beauvoit, B.; Stevens, R.; Flandin, A.; Colombie, S.; Gibon, Y.; Pétriacq, P. Get the Balance Right: ROS Homeostasis and Redox Signalling in Fruit. *Front. Plant Sci.* **2019**, *10*, 1091. [CrossRef] [PubMed]
48. Hasanuzzaman, M.; Parvin, K.; Bardhan, K.; Nahar, K.; Anee, T.I.; Masud, A.A.; Fotopoulos, V. Biostimulants for the Regulation of Reactive Oxygen Species Metabolism in Plants under Abiotic Stress. *Cells* **2021**, *10*, 2537. [CrossRef] [PubMed]
49. Verma, D.; Lakhanpal, N.; Singh, K. Genome-wide identification and characterization of abiotic-stress responsive SOD (superoxide dismutase) gene family in *Brassica juncea* and *B. rapa*. *BMC Genom.* **2019**, *20*, 227. [CrossRef]
50. Feng, X.; Lai, Z.; Lin, Y.; Lai, G.; Lian, C. Genome-wide identification and characterization of the superoxide dismutase gene family in *Musa acuminata* cv. Tianbaojiao (AAA group). *BMC Genom.* **2015**, *16*, 823. [CrossRef]

51. Wang, W.; Xia, M.; Chen, J.; Deng, F.; Yuan, R.; Zhang, X.; Shen, F. Genome-wide analysis of superoxide dismutase gene family in *Gossypium raimondii* and *G. arboreum*. *Plant Gene* **2016**, *6*, 18–29. [CrossRef]
52. Han, L.-M.; Hua, W.-P.; Cao, X.-Y.; Yan, J.-A.; Chen, C.; Wang, Z.-Z. Genome-wide identification and expression analysis of the superoxide dismutase (SOD) gene family in *Salvia miltiorrhiza*. *Gene* **2020**, *742*, 144603. [CrossRef] [PubMed]
53. Cohu, C.M.; Abdel-Ghany, S.E.; Gogolin Reynolds, K.A.; Onofrio, A.M.; Bodecker, J.R.; Kimbrel, J.A.; Niyogi, K.K.; Pilon, M. Copper Delivery by the Copper Chaperone for Chloroplast and Cytosolic Copper/Zinc-Superoxide Dismutases: Regulation and Unexpected Phenotypes in an Arabidopsis Mutant. *Mol. Plant* **2009**, *2*, 1336–1350. [CrossRef] [PubMed]
54. Shams, M.; Pokora, W.; Khadivi, A.; Aksmann, A. Superoxide dismutase in Arabidopsis and Chlamydomonas: Diversity, localization, regulation, and role. *Plant Soil* **2024**. [CrossRef]
55. Su, T.; Wang, P.; Li, H.; Zhao, Y.; Lu, Y.; Dai, P.; Ren, T.; Wang, X.; Li, X.; Shao, Q.; et al. The Arabidopsis catalase triple mutant reveals important roles of catalases and peroxisome-derived signaling in plant development. *J. Integr. Plant Biol.* **2018**, *60*, 591–607. [CrossRef] [PubMed]
56. Schmidt, R.; Mieulet, D.; Hubberten, H.-M.; Obata, T.; Hoefgen, R.; Fernie, A.R.; Fisahn, J.; San Segundo, B.; Guiderdoni, E.; Schippers, J.H.M.; et al. SALT-RESPONSIVE ERF1 Regulates Reactive Oxygen Species–Dependent Signaling during the Initial Response to Salt Stress in Rice. *Plant Cell* **2013**, *25*, 2115–2131. [CrossRef] [PubMed]
57. Ge, S.; Wei, Y.; Zhang, J.; Shao, W.; Li, J.; Ding, D.; Liu, M.; Xu, J. Bioinformatic analysis of catalase gene family of Arabidopsis and Maize. *Mol. Plant Breed.* **2021**, *12*, 34. [CrossRef]
58. Willekens, H.; Villarroel, R.; Van Montagu, M.; Inzé, D.; Van Camp, W. Molecular identification of catalases from *Nicotiana plumbaginifolia* (L.). *FEBS Lett.* **1994**, *352*, 79–83. [CrossRef]
59. Leng, X.; Wang, H.; Zhang, S.; Qu, C.; Yang, C.; Xu, Z.; Liu, G. Identification and Characterization of the APX Gene Family and Its Expression Pattern under Phytohormone Treatment and Abiotic Stress in *Populus trichocarpa*. *Genes* **2021**, *12*, 334. [CrossRef]
60. Chaouch, S.; Queval, G.; Noctor, G. AtRbohF is a crucial modulator of defence-associated metabolism and a key actor in the interplay between intracellular oxidative stress and pathogenesis responses in Arabidopsis. *Plant J.* **2012**, *69*, 613–627. [CrossRef]
61. Esparza-Reynoso, S.; Ávalos-Rangel, A.; Pelagio-Flores, R.; López-Bucio, J. Reactive oxygen species and NADPH oxidase-encoding genes underly the plant growth and developmental responses to Trichoderma. *Protoplasma* **2023**, *260*, 1257–1269. [CrossRef]
62. Truskina, J.; Boeuf, S.; Renard, J.; Andersen, T.G.; Geldner, N.; Ingram, G. Anther development in Arabidopsis thaliana involves symplastic isolation and apoplastic gating of the tapetum-middle layer interface. *Development* **2022**, *149*, dev200596. [CrossRef] [PubMed]
63. Nunes da Silva, M.; Vasconcelos, M.W.; Gaspar, M.; Balestra, G.M.; Mazzaglia, A.; Carvalho, S.M.P. Early Pathogen Recognition and Antioxidant System Activation Contributes to *Actinidia arguta* Tolerance against *Pseudomonas syringae Pathovars actinidiae* and *actinidifoliorum*. *Front. Plant Sci.* **2020**, *11*, 1022. [CrossRef] [PubMed]
64. Azarabadi, S.; Abdollahi, H.; Torabi, M.; Salehi, Z.; Nasiri, J. ROS generation, oxidative burst and dynamic expression profiles of ROS-scavenging enzymes of superoxide dismutase (SOD), catalase (CAT) and ascorbate peroxidase (APX) in response to Erwinia amylovora in pear (*Pyrus communis* L.). *Eur. J. Plant Pathol.* **2017**, *147*, 279–294. [CrossRef]
65. Singh, Y.; Nair, A.M.; Verma, P.K. Surviving the odds: From perception to survival of fungal phytopathogens under host-generated oxidative burst. *Plant Commun.* **2021**, *2*, 100142. [CrossRef] [PubMed]
66. Zhang, M.; Li, Q.; Liu, T.; Liu, L.; Shen, D.; Zhu, Y.; Liu, P.; Zhou, J.-M.; Dou, D. Two Cytoplasmic Effectors of Phytophthora sojae Regulate Plant Cell Death via Interactions with Plant Catalases. *Plant Physiol.* **2015**, *167*, 164. [CrossRef]
67. Yuan, H.; Jin, C.; Pei, H.; Zhao, L.; Li, X.; Li, J.; Huang, W.; Fan, R.; Liu, W.; Shen, Q.-H. The Powdery Mildew Effector CSEP0027 Interacts with Barley Catalase to Regulate Host Immunity. *Front. Plant Sci.* **2021**, *12*, 733237. [CrossRef]

horticulturae

Article

Boosting Tomato Resilience in Tanzania: Grafting to Combat Bacterial Wilt and Abiotic Stress

Shem Elias Msabila [1,2,*], Thibault Nordey [1,3,4,*], Zablon Ernest [1], Nickson Mlowe [1], Ravishankar Manickam [5], Srinivasan Ramasamy [5] and Joël Huat [4,6]

1 World Vegetable Center, Eastern and Southern Africa, Duluti, Arusha P.O. Box 10, Tanzania; zablon.ernest@worldveg.org (Z.E.); nickson.mlowe@worldveg.org (N.M.)
2 Department of Plant and Soil Science, University of Delaware, 210 South College Ave., Newark, DE 19716, USA
3 CIRAD (French Agricultural Research Centre for International Development), UPR HortSys, F-34398 Montpellier, France
4 HortSys, University of Montpellier, CIRAD, F-34090 Montpellier, France; joel.huat@cirad.fr
5 World Vegetable Center, Shanhua, P.O. Box 42, Tainan 74199, Taiwan; ravi.manickam@worldveg.org (R.M.); srini.ramasamy@worldveg.org (S.R.)
6 CIRAD, UPR HortSys, F-97455 Saint-Pierre, Reunion, France
* Correspondence: selias@udel.edu (S.E.M.); thibault.nordey@orange.fr (T.N.)

Abstract: The grafting technique has successfully mitigated crop losses from diseases and stress in vegetable production; however, vegetable grafting in Tanzania is very limited. Field and greenhouse experiments conducted in Tanzania's mainland and islands compared the response of tomato determinate cv. 'Tanya' to production challenges when ungrafted and grafted onto five tomato rootstocks ('Hawaii 7796', 'Tengeru 1997', 'Tengeru 2010', 'R3034', and 'Shelter'), one eggplant variety ('EG 203'), and one wild *Solanum* species (*Solanum elaeagnifolium*). The visual symptoms of bacterial wilt varied significantly with location and season, ranging from 8 to 100%, attributed to varying bacterial wilt pressures and strains of *Ralstonia solanacearum* isolated (Phylotype I sequevars 17, 18, and 31). 'EG203' and 'Hawaii 7796' emerged as the most effective rootstocks, reducing wilting by 49.8 and 51.0% and improving yield by 57.2% and 27.7% on average across experiments conducted in three locations (Moshi, Pemba, and Unguja) over two seasons. Combining reduced water supply with grafting resulted in an average reduction in wilting of 76%, while also boosting yields by an average of 3.6 times in experiments conducted in Arusha over two seasons. Grafting onto 'Hawaii 7796' and 'Shelter' significantly improved 'Tanya' yields by 38.3% and 41.6% on average over two seasons, only under standard nutrient application rates. While certain rootstocks improved crop performance, yields across various sites and seasons were significantly hampered by pest pressure. These findings support grafting's potential to mitigate damage from common stresses, emphasizing the need for further research to identify suitable rootstocks for optimizing returns on investments in grafted plants in Tanzania.

Keywords: tomato grafting; abiotic stress; biotic stress; water deficit; *Ralstonia solanacearum*; Tanzania

Citation: Msabila, S.E.; Nordey, T.; Ernest, Z.; Mlowe, N.; Manickam, R.; Ramasamy, S.; Huat, J. Boosting Tomato Resilience in Tanzania: Grafting to Combat Bacterial Wilt and Abiotic Stress. *Horticulturae* **2024**, *10*, 338. https://doi.org/10.3390/horticulturae10040338

Academic Editors: Aušra Brazaitytė, Changxia Li and Yue Wu

Received: 12 February 2024
Revised: 16 March 2024
Accepted: 27 March 2024
Published: 29 March 2024

1. Introduction

Adopting a diverse and vibrant dietary regimen that includes a variety of vegetables represents a scientifically supported strategy, ensuring the intake of crucial nutrients, vitamins, minerals, and antioxidants, and thereby substantiating a proactive approach toward sustained health and well-being. Limited availability contributes to the ongoing deficiency in vegetable consumption within developing countries. Moreover, vegetable production presents an economic opportunity to alleviate unemployment and poverty in these nations [1]. However, this potential is hindered by financial risks associated with weather conditions, pest and disease damage, and fluctuating sales prices [2]. Among the diseases, bacterial wilt caused by the pathogen *Ralstonia solanacearum* and root-knot nematodes (*Meloidogyne* spp.) alone or in combination limits tomato production. *R. solanacearum* is a soilborne plant pathogen that attacks the vascular system of plants. This results in the gradual withering, yellowing, and, eventually, death of the plant. A recent study reported a high diversity of *R. solanacearum* in Tanzania [3], resulting in significant losses for tomato crops. This pathogen has several hosts, including tomato plants, and the root-knot nematodes increase its access to its host plants. Root-knot nematodes (*Meloidogyne* spp.) are tiny, parasitic roundworms that infect the roots of plants through their feeding mechanisms. This leads to the formation of knots or galls on the root. These nematodes are a significant agricultural pest, affecting many crops, including tomato plants.

To combat crop damage caused by pests and diseases, vegetable growers are turning to the use of pesticides [4], which can have negative impacts on both their health [5] and that of consumers [6,7]. To achieve higher yields while minimizing chemical inputs, 'ecological intensification' principles recommend utilizing biological regulations in agroecosystems [8]. While using improved crop varieties is an effective method of sustainable production, it can also be time-consuming and costly, which hinders the development of suitable options [9].

Alternatively, vegetable grafting is becoming a popular technique for effectively merging the shoots (known as scions) of high-yield potential with the rootstocks that can withstand soilborne diseases and other environmental stresses [10]. While grafting is widely used in fruit and vegetable crops across Europe, Asia, and North America, it remains predominantly limited to fruit crops in sub-Saharan Africa (SSA). Numerous studies have highlighted the advantages of vegetable grafting in combating soilborne diseases and improving tolerance to abiotic stresses [11–13].

After analyzing 159 studies on tomato grafting, the authors of [14] discovered a concerning lack of research in SSA, with only three studies dedicated to this region. Most investigations on tomato grafting have been carried out in Europe, the Middle East, Asia, and, to a lesser extent, North America. This trend is also reflected in the top tomato-producing nations of 2018, according to the FAOSTAT database [10]. China (62 million tons), India (19 million tons), the European Union (17 million tons, with 6 million tons from Italy and 5 million tons from Spain making significant contributions), the United States (13 million tons), and Turkey (13 million tons) were the leaders in tomato production. Despite Africa's noteworthy 2018 tomato output (21 million tons), there is limited information on the merits of tomato grafting.

A current review [10] has highlighted the potential of vegetable grafting in SSA, specifically for tomato farmers who are struggling with soilborne pathogens such as bacterial wilt and root-knot nematodes, as well as abiotic challenges including water deficit, flooding, and poor soil fertility.

Grafting using resistant rootstocks to increase root resilience and endurance is a promising alternative. However, the practical implementation of grafted plants in SSA faces obstacles, such as the absence of appropriate rootstocks and insufficient evidence about their profitability in various production systems, such as open-field versus greenhouse cultivation.

Previous research on tomato grafting in sub-Saharan Africa has yielded conflicting results, making it difficult for farmers to make informed decisions [15–18]. To address this uncertainty, the 'GrAfrica' project funded by GIZ from 2017 to 2020 aimed to introduce grafted plantlets to help increase yields and income for small-scale tomato producers in Tanzania. The project included comprehensive training in tomato grafting for nursery operators and producers and experiments conducted in collaboration with national partners across diverse locations in Tanzania. The objectives of the experiments were to evaluate the response of a popular determinate tomato variety ('Tanya') when grown either ungrafted or grafted (i.e., onto rootstocks of tomatoes, eggplants, and wild species) in different 1. fertilization regimes, 2. soils infected by bacterial wilt, and 3. water deficit levels in soils infected by bacterial wilt. The findings of these experiments, presented in this article, offer valuable and reliable information to guide future efforts toward successfully implementing tomato grafting practices in Tanzania.

2. Materials and Methods

2.1. Plant Material

Tomato seedlings were produced in a nursery covered with a plastic film as a roof and with insect-proof nets (60 mesh) on the sides. Seedling trays with a hole capacity of 66 were filled with a heat-sterilized mixture of forest soil, well-decomposed manure, and sand at 2:1:1 [17]. The seedlings were irrigated daily with a watering can and fertilized weekly with a solution comprising 0.4 $g \cdot L^{-1}$ each of nitrogen, phosphorus, and potassium. 'Tanya', an open-pollinated determinate tomato variety, was used both ungrafted as a control in all the experiments and grafted onto five tomato rootstocks ('Hawaii 7796', 'Tengeru 1997', 'Tengeru 2010', 'R3034', and 'Shelter'), one eggplant rootstock ('EG 203'), and one wild species (*Solanum elaeagnifolium*) rootstock (Table 1). Despite its high sensitivity to bacterial wilt, Tanya is widespread in Tanzania for its elongated fruits and long shelf-life. Three-week-old seedlings were used as scions at the two- to three-true leaf stages. Tomato, eggplant, and wild species seedlings used as rootstock were sown, respectively, two days, two weeks, and three weeks before the tomato scions to obtain a similar stem diameter for grafting [19].

Table 1. Tomato, eggplant, and a wild *Solanum* species were used in the experiments. N.B. Seeds of the wild *Solanum* species were collected from the fields of Arusha and characterized morphologically.

Varieties	Species	Type	Remarks	Use	Origin	Experiments
Tanya	*S. lycopersicum* L.	Open-pollinated, determinate	Extended shelf-life, highly sensitive to nematodes and bacterial wilt	Scion and ungrafted control	WorldVeg	1, 2, and 3
Tengeru 1997	*S. lycopersicum* L.	Open-pollinated, semi-indeterminate	Partial resistance to bacterial wilt	Rootstock	WorldVeg	2 and 3
Tengeru 2010	*S. lycopersicum* L.	Open-pollinated, semi-indeterminate	Partial resistance to bacterial wilt	Rootstock and ungrafted control	WorldVeg	2
Hawaii 7796	*S. lycopersicum* L.	Open-pollinated, determinate	Used as a rootstock in Asia to improve resistance to bacterial wilt	Rootstock	INRA	1 and 2
R3034	*S. melongena*	Open-pollinated, determinate	Used as a rootstock in Asia to improve resistance to bacterial wilt	Rootstock	WorldVeg	1
Shelter	*S. lycopersicum* L.	Hybrid	Commercial rootstock	Rootstock	Rijk Zwaan	1
EG 203	*S. melongena*	Open-pollinated	Used as a rootstock in Asia to improve resistance to bacterial wilt and flooding.	Rootstock	WorldVeg	1 and 2
Silverleaf nightshade	*S. elaeagnifolium*	Wild *Solanum* species widespread in Tanzania	Resistance to bacterial wilt is not known.	Rootstock	Wild species	1 and 2

The splicing technique described by [20] was used for grafting. Using sterilized blades, the stems of the scions and rootstocks were cut obliquely above the cotyledons at an angle of 30°. The surfaces of the cut scions and rootstocks were then gently aligned and held together using a transparent plastic silicon clip measuring 1.6 mm in diameter. Plants were grafted above the cotyledons to have a height advantage and prevent the scion from contacting the soil through adventitious roots, which would have forfeited the grafting efforts after transplanting. For recovery after grafting, the grafted plants were placed in a dark and shaded 'healing chamber' with high humidity (~90%) thanks to a shallow layer of water on the polyethylene floor (for a complete description, see [20]). In the Pemba and Unguja locations, underground 'healing chambers' were used to lower the temperature and raise the relative humidity to increase the success rate of grafting [19]. After three days within the 'healing chambers', the doors were partially opened, allowing seedlings to acclimate to ambient humidity for two days. The seedlings were then taken out of the 'healing chambers' and placed in a nursery under shade nets for a recovery period of seven days, during which they were gradually exposed to sunlight. Care was taken when transplanting the grafted plants to ensure the graft union was above the soil surface to avoid the development of adventitious roots.

2.2. Experiments

Three sets of experiments were performed in four different locations in Tanzania (Arusha, Moshi, Pemba, and Unguja; see Figure 1) in two seasons from 2018 to 2019 to compare the response of 'Tanya' tomatoes when ungrafted and grafted to five tomato rootstocks ('Hawaii 7796', 'Tengeru 1997', 'Tengeru 2010', 'R3034', and 'Shelter'), one eggplant rootstock ('EG 203'), and one wild species rootstock (*Solanum elaeagnifolium*) (Table 1). All the rootstocks tested, except the wild species, were reportedly resistant to bacterial wilt. 'Tanya' tomatoes were used as scions and ungrafted controls in all the experiments, but to reflect local practices, the rootstocks tested at the experimental sites differed, as did planting density.

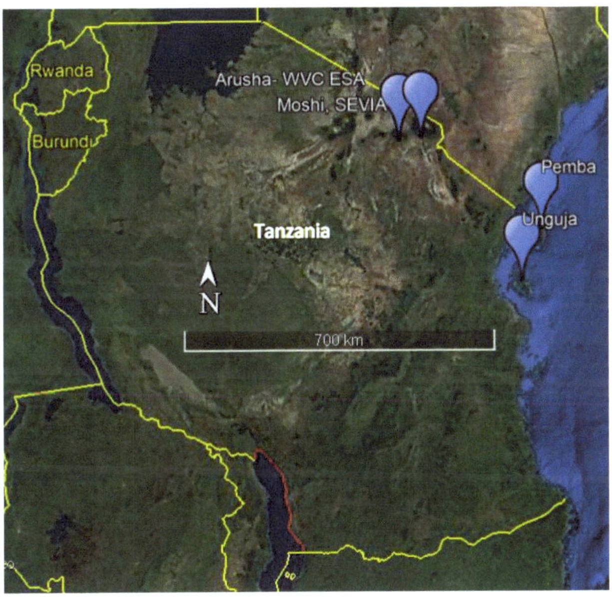

Figure 1. Location of experimental sites in Arusha, Moshi, Pemba, and Unguja, Tanzania.

We evaluated the effectiveness of grafting, as compared to an ungrafted control, on 'Tanya' tomatoes when paired with five different tomato rootstocks, namely 'Hawaii 7796', 'Tengeru 1997', 'Tengeru 2010', 'R3034', and 'Shelter'. One eggplant rootstock ('EG 203') and one wild species rootstock (*Solanum elaeagnifolium*) were also tested. It is worth mentioning that all rootstocks, except the wild species, were reportedly resistant to bacterial wilt. All experiments used 'Tanya' tomatoes as scions and as the ungrafted control. The rootstocks used varied across the experimental sites, and planting densities were adjusted to reflect local practices. Please refer to Figure 1 for a map of the experimental locations and Table 1 for additional information regarding the rootstocks tested.

Experimental sites with contrasted climate and bacterial wilt (*Ralstonia solanacearum*) pressures were selected to account for the different production challenges encountered in Tanzania. The northern sites, Arusha (latitude −3°22′22.82″, longitude 36°48′21.59″, altitude 1250 m asl) and Moshi (latitude −3°18′54.72″, longitude 37°15′8.28″, altitude 987 m asl), are located in the Tanzania highlands close to Mount Kilimanjaro and have an oceanic subtropical highland climate (Cwb), and a marine west coast (Cfb) climate, respectively, and loamy-clay soils. The sites in Pemba (latitude −5°4′3.72″, longitude 39°45′1.08″, altitude 13 m asl) and Unguja (latitude −6°6′24.48″, longitude 39°21′41.04″, altitude 46 m asl) are located on the eastern seaboard of Tanzania, more than 400 km away as the crow flies from the northern locations, and have a tropical monsoon (Am) and tropical savanna (Aw) climate, respectively, with clay to sandy-loam soils.

2.2.1. Experiment 1: Response of Ungrafted and Grafted Tomato Plants to Different Fertilization Regimes

In a field experiment conducted over two seasons at the World Vegetable Center for Eastern and Southern Africa near Arusha, the development and yields of ungrafted and grafted plants were assessed under different fertilization regimes. The experiment used a split-plot design with four replications and two fertilization rates (standard and low) on loamy to loamy-clay soils with good drainage and aeration. The standard fertilization rate consisted of 225 kg ha^{-1} of N, 90 kg ha^{-1} of P, and 250 kg ha^{-1} of K, designed to meet the NPK requirements of a tomato crop with a potential yield of 80 t ha^{-1} [21].

The soil was manually plowed to ensure optimal fertilization, and 5 t ha^{-1} of cow manure was incorporated during the last plowing. A week after transplanting, a side dressing of 17-17-17 NPK fertilizer (500 kg ha^{-1}) was added, followed by a side dressing of calcium ammonium nitrate (400 kg ha^{-1}) and muriate of potash (250 kg ha^{-1}) four weeks later for standard fertilization treatments. In comparison, plots with lower fertilization rates only received 5 t ha^{-1} of cow manure during plowing, equivalent to 15% of the NPK applied in the standard fertilization treatment. Adding 5 t ha^{-1} of cow manure during plowing is a good agricultural practice to slowly replenish nutrients in the soil. The subplot treatment included ungrafted 'Tanya' tomato plants (as controls) and tomato plants grafted onto 'Shelter', 'Hawaii 7796', and 'R3034' rootstocks. Each experimental unit (28 m^2) consisted of 66 plants (2.3 plants·m^{-2}) transplanted in bare soil across two raised beds (length 8 m, width 1 m, height 0.25 m) spaced 1.75 m apart, with 60 cm between plants in the same row. Irrigation with drip lines was scheduled every two days, with the quantity provided adjusted to meet plant requirements based on weather conditions and the stage of crop development. Chemical treatments were applied as needed to control insect pests and diseases using a knapsack sprayer (please refer to Table 2 for details).

One plant per experimental unit was randomly uprooted 10 weeks after transplanting (before the first harvest) to measure its leaf surface area and fresh weight (shoot and roots). The leaf surface area of each plant was assessed using a picture analysis method as described in previous studies [22,23]. Tomatoes were harvested thrice weekly at the mature green stage or more advanced ripening stages. They were then sorted based on their size (>7 cm length) and absence of defects (rot, discoloration, and cracking) to assess the marketable yields.

Air temperature, air relative humidity, and rainfall were measured at one-minute intervals and averaged every 30 min by a complete weather station (Vantage PRO2, Davis Instruments, Hayward, CA, USA) installed nearby.

Table 2. Pesticide treatments were applied during the field experiments on tomatoes in Tanzania. N.B. This information is not intended as a guideline or recommendation.

Experiments	Sites	Active Ingredients	Number of Applications in the First Season	Number of Applications in the Second Season	Targets
1	Arusha	Thiocyclam ($C_7H_{13}NO_4S_3$, Evisect, Arysta LifeScience, Noguères, France)	3	1	Whiteflies
		Lambda-cyhalothrin ($C_{23}H_{19}ClF_3NO_3$ Ninja, Posit. Inter. Limited, Daresalaam, Tanzania)	2	1	Whiteflies
		Emamectin benzoate ($C_{49}H_{75}NO_{13}$, Prove, Equatorial Africa Ltd., Daresalaam, Tanzania)	2	1	*Tuta absoluta*
		Metalaxyl and mancozeb (Ridomyl, Syngenta, Basel, Switzerland)	3	5	Early and late blight
		Copper (Mocrops, Daresalaam, Tanzania)	0	1	Bacterial spot
2	Pemba	Metalaxyl and mancozeb (Ridomyl, Syngenta, Basel, Switzerland)	4	4	Early and late blights
		Chlorothalonil (Daconil, Syngenta, Basel, Switzerland)	3	5	Powdery mildew
		Imidacloprid (Confidor, Bayer, Leverkusen, Germany)	4	4	Whiteflies
		Esfenvalerate (Sumi-alpha, Philagro, Pretoria, South Africa)	4	4	Whiteflies
		Azinphos-methyl (Gusathion, Bayer, Leverkusen, Germany)	1	2	Tomato fruit worms
		Methomyl (Lannate, Dupont, Tel Aviv, Israel)	1	2	Tomato fruit worms
		Emamectin benzoate ($C_{49}H_{75}NO_{13}$, Prove, Equatorial Africa, Daresalaam, Tanzania)	3	4	*Tuta absoluta*
	Unguja	Metalaxyl and mancozeb (Ridomyl, Syngenta, Basel, Switzerland)	5	4	Early and late blight
		Chlorothalonil (Daconil, Syngenta, Basel, Switzerland)	4	4	Powdery mildew
		Imidacloprid (Confidor, Bayer, Leverkusen, Germany)	4	2	Whiteflies
		Esfenvalerate (Sumi-alpha, Philagro, Tokyo, Japan)	4	2	Whiteflies
		Azinphos-methyl (Gusathion, Bayer, Leverkusen, Germany)	2	3	Tomato fruit worms
		Methomyl (Lannate, Dupont, Tel Aviv, Israel)	2	3	Tomato fruit worms
		Emamectin benzoate (Prove, Equatorial Africa, Daresalaam, Tanzania)	2	6	*Tuta absoluta*

Table 2. *Cont.*

Experiments	Sites	Active Ingredients	Number of Applications in the First Season	Number of Applications in the Second Season	Targets
	Moshi	Metalaxyl and mancozeb (Ridomyl, Syngenta, Basel, Switzerland)	3	7	Early and late blight
		Chlorothalonil (Daconil, Syngenta, Basel, Switzerland)	2	5	Powdery mildew
		Imidacloprid (Confidor, Bayer, Leverkusen, Germany)	5	3	Whiteflies
		Esfenvalerate (Sumi-alpha, Philagro, Tokyo, Japan)	5	3	Whiteflies
		Azinphos-methyl (Gusathion, Bayer, Leverkusen, Germany)	3	1	Tomato fruit worms
		Methomyl (Lannate, Dupont, Tel Aviv, Israel)	3	1	Tomato fruit worms
		Emamectin benzoate ($C_{49}H_{75}NO_{13}$, Prove, Equatorial Africa, Daresalaam, Tanzania)	8	2	*Tuta absoluta*
3	Arusha	Metalaxyl and mancozeb (Ridomyl, Syngenta, Basel, Switzerland)	1	2	Early and late blight
		Chlorothalonil (Daconil, Syngenta, Basel, Switzerland)	3	0	Powdery mildew
		Imidacloprid (Confidor, Bayer, Leverkusen, Germany)	3	4	Whiteflies
		Emamectin Benzoate ($C_{49}H_{75}NO_{13}$, Prove, Equatorial Africa, Daresalaam, Tanzania)	4	3	*Tuta absoluta*

2.2.2. Experiment 2: Response of Ungrafted and Grafted Tomato Plants to the Soil Infested with Bacterial Wilt Pathogen

After obtaining a positive streaming test, plots with repeated crop failures due to severe pressure from bacterial wilt were ideal for this study, which compared the symptoms and yields of grafted and ungrafted tomatoes. The trials were carried out over two seasons in Moshi, Pemba, and Unguja.

The Seeds of Expertise carried out the trials in Moshi for the Vegetable Sector of Africa (SEVIA) from May to September 2018 and from November 2018 to February 2019. A Latin square design with five treatments and five replications was used in both seasons. The treatments included ungrafted 'Tanya' as controls and 'Tanya' grafted onto 'Tanya' (self-grafted, to account for any grafting or healing effects), 'Hawaii 7796', 'EG 203', and onto a wild eggplant species (*S. elaeagnifolium*) rootstock. Each experimental unit (16 m^2) consisted of 20 plants (1.25 plants·m^{-2}) transplanted at the three- to four-leaf stage in two raised beds (length 400 cm, width 75 cm, height 25 cm) covered with a plastic mulch with 200 cm between-row spacing and 60 cm between plants in the same row.

The trials in Pemba and Unguja were conducted by the Tanzania Horticultural Association (TAHA) from November 2018 to January 2019 and June to November 2019. A complete randomized block design with three replications and five treatments was used in both seasons. The treatments included ungrafted 'Tanya' and 'Tengeru 2010' as controls and 'Tanya' grafted onto 'Tengeru 2010', 'Hawaii 7796', and 'Shelter' rootstocks. Each experimental unit (35 m^2) consisted of 50 plants transplanted (1.4 plants·m^{-2}) at the three-to four-leaf stage into two raised beds (length 10 m, width 1 m, height 25 cm) covered with plastic mulch with 1.5 m between-row spacing and 40 cm between plants in the same row.

In Moshi, Pemba, and Unguja, irrigation was provided by drip lines every two days to meet plant requirements, and the same standard fertilization as that used in Experiment 1 was applied. Chemical pesticide treatments were applied based on scouting results using a knapsack sprayer to control insect pests and diseases (see details in Table 2).

The bacterial wilt incidence was assessed every week throughout the growing season. Ooze-out tests (streaming tests) were performed on wilted plants to confirm infection by *R. solanacearum*, and five samples from each site were sent to ANSES (the French Agency for Food, Environmental, and Occupational Health and Safety Laboratory) on Reunion Island at the end of the second season for the molecular typing of bacterial wilt strains.

Marketable yields were measured in the same way as in Experiment 1. Data on daily average temperature, air moisture content, and total rainfall were collected from the weather stations at the nearest airport (www.ncei.noaa.gov, accessed on 13 March 2020).

2.2.3. Experiment 3: Response of Ungrafted and Grafted Tomato Plants to Water Deficit Soil Infested by Bacterial Wilt

Experiment 3 was conducted at the World Vegetable Center for Eastern and Southern Africa in a greenhouse with repeated crop failures caused by bacterial wilt (*R. solanacearum*). The experiment aimed to compare the wilting symptoms and yields of ungrafted and grafted tomato plants under two irrigation regimes (standard and with a 50% reduction). The greenhouse was 120 m^2 (15 m × 8 m) and had a double door covered with a plastic film as a roof and insect-proof nets (60 mesh) on the sides. The experiment was conducted in two seasons, from December 2018 to March 2019 and March 2019 to July 2019.

The experiment used a split-plot design with three replications (blocks) in both seasons. The whole-plot treatment consisted of two irrigation regimes, the subplot treatments included the ungrafted 'Tanya' as a control and 'Tanya' grafted onto 'Tengeru 1997'. Each experimental unit (7.5 m^2) consisted of 12 plants (1.6 plants$\cdot m^{-2}$) transplanted at the three- to four-true leaf stage to two raised beds (length 300 cm, width 75 cm, height 25 cm) covered with plastic mulch and spaced 1.25 m apart with 60 cm between plants in the same row.

The irrigation was carried out every two days with drip lines, and the quantity provided varied between treatments. Three tensiometers (Burdon, SDEC, 37310 Tauxigny, France) were used to monitor soil water potential and guide the amount of water added in plots to reach the target soil moisture level during irrigation. In plots with standard irrigation, the aim was to maintain a soil water potential between −0.03 and −0.02 MPa at 30 cm below ground to meet tomato plant water requirements [24]. In the plots with reduced irrigation, only half the amount of water was supplied compared to plots under a standard irrigation regime. This resulted in a soil water potential ranging from −0.08 to −0.02 MPa (please refer to Supplemental Figure S1 for more details). The fertilization process was the same as the standard fertilization treatment used in Experiment 1. Chemical treatments were applied using a knapsack sprayer, based on scouting, to control insect pests and diseases (please see Table 2 for more details). The bacterial wilt incidence was assessed as described in Experiment 2, and marketable yields were evaluated as described in Experiment 1. The temperature and air moisture content were recorded every 30 min using a data logger (HOBO Pro v2 U23-001, Onset Computer Corporation, Bourne, MA, USA). To avoid direct exposure to sunlight, the data loggers were suspended 1.8 m above the ground in the middle of the greenhouse under a perforated white cover with an open bottom section.

2.3. *Molecular Typing of Bacterial Wilt Strains*

At each experimental site, a 30 cm section above the plant root collar was gently cut from five wilted tomato plants at the end of the second season. After the samples were disinfected with 70% ethanol, they were packed in paper envelopes. They were sent to the ANSES (French Agency for Food, Environmental, and Occupational Health and

Safety) laboratory on Reunion Island for genetic identification. The isolation and genetic identification of bacterial wilt strains were performed following the method described by N'Guessan et al. [25]. Macerates of samples in distilled water were streaked on Kelman's triphenyl tetrazolium chloride (TZC) and modified Sequeira media, and incubated at 28 °C for two or three days. Bacterial colonies morphologically typical of *R. solanacearum* were purified and used for molecular typing. After applying multiplex-polymerase chain reactions (PCRs) to bacterial suspensions, the phylotype was assigned to each strain. The phylogeny of strains was characterized using a comparative analysis of the partial nucleotide sequences of all genes [26].

2.4. Statistical Analyses

The data from each experiment were analyzed separately. We checked all data for normality and homogeneity of variance using the Shapiro–Wilk and Bartlett tests, respectively. To identify significant differences between factors in the data concerning plant development in Experiment 1 (fresh weight and leaf surface area of plants) and yield in all three experiments, we used analysis of variance (ANOVA). Whenever we identified significant differences, we performed post hoc analyses using Fisher's least significant difference (LSD) test to compare treatments. However, in Experiments 2 and 3, the data did not follow a normal distribution. Therefore, we used the Kruskal–Wallis test to compare the percentage of wilted tomatoes. Furthermore, we performed post hoc analyses using Dunn's trial and the Holm method to adjust the p-value whenever we identified significant differences between treatments. Finally, we conducted all statistical analyses using R software version 4.0.0 with the agricolae package.

3. Results

3.1. The Response of Grafted Tomatoes to Different Fertilization Treatments

No wilt symptoms were observed in Experiment 1, which was conducted in Arusha to compare crop development and yields of ungrafted and grafted tomatoes with standard and low fertilization rates. The reduced plant vigor and yield during the second season resulted from the unsatisfactory control of early and late blight promoted by high rainfall (Table 3) despite an increase in fungicide treatments (Table 2). No significant differences were identified in the fresh mass of plants between fertilization treatments 10 weeks after transplanting in the first season (p-value = 0.204) and in the second season (p-value = 0.116) (Table 4). Although no significant differences were observed in the first season (p-value = 0.720), the leaf surface area measured in plants receiving the low fertilization treatment was smaller than with the standard treatment in the second season (p-value = 0.003). The reduction in fertilization significantly reduced yields in both seasons; the p-values were 0.019 and 0.018 in the first and second seasons, respectively. The yields of ungrafted 'Tanya' tomatoes decreased from 2.3 (standard fertilization) to 1.6 (low fertilization) $kg{\cdot}m^{-2}$ in the first season and from 1.2 to 0.6 $kg{\cdot}m^{-2}$ in the second season (Figure 2).

Ungrafted plants exhibited higher vigor levels, with greater fresh mass and leaf surface area than grafted plants, regardless of the fertilization treatment (Table 4). However, no significant differences were observed between the yields of ungrafted and grafted tomatoes when the fertilization rate was reduced, regardless of the rootstocks used and the season. Upon applying standard fertilization, higher yields were achieved from tomatoes grafted onto 'Hawaii 7796' (37% and 40% increases in the first and second seasons, respectively) and onto 'Shelter' (24% and 72% increases in the first and second seasons, respectively) in comparison to ungrafted tomatoes, but no differences were found in tomatoes grafted onto 'R3034'. Grafting onto 'Shelter' and 'Hawaii 7796' resulted in average yield increases of 38.3% and 24.4%, respectively, across both tested fertilization treatments.

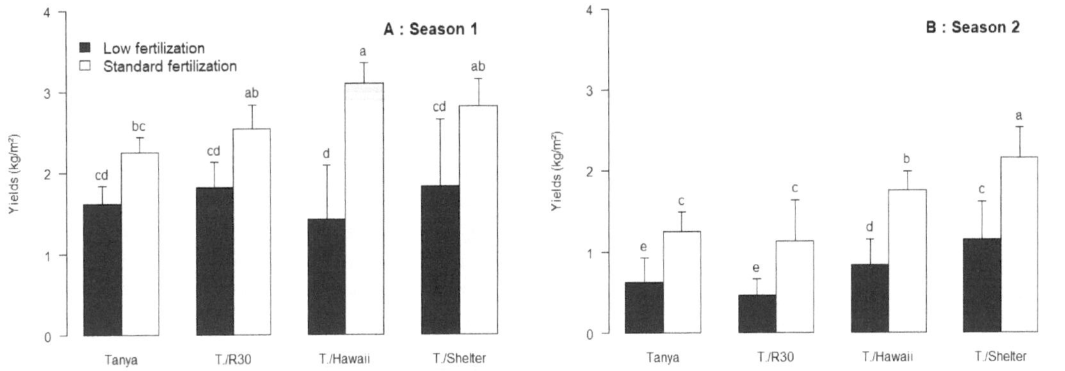

Figure 2. Comparison of marketable yields of grafted and ungrafted tomatoes under different fertilization treatments (standard or 15%) in two seasons. Data are the observed means of replications (±standard deviation, N = 4). Different lower-case letters mean significant differences based on Fisher's least significant difference (LSD) test (at p = 0.05).

Table 3. The average air temperature, rainfall, and average air relative humidity in trials conducted in the two seasons.

Experiment	Location	Season	Date of Transplanting	Date of the Last Harvest	Average Temperature (°C)	Rainfall (mm)	Average Relative Humidity (%)
1	Arusha (open field)	1	14 November 2018	15 February 2019	22.4	87.9	71.3
		2	8 March 2019	7 July 2019	20.1	287.4	82.3
2	Moshi (open field)	1	31 May 2018	11 September 2018	20.6	17.2	
		2	19 November 2018	19 February 2019	25.6	101.6	
2	Unguja (open field)	1	1 November 2018	31 January 2019	27.8	266.9	
		2	4 July 2019	15 October 2019	26.1	350.8	
2	Pemba (open field)	1	25 October 2018	25 January 2019	26.6	177.9	
		2	28 June 2019	15 Oct 2019	24.4	4.8	
3	Arusha (greenhouse)	1	3 December 2018	5 March 2019	28.1	0	64.5
		2	27 March 2019	18 June 2019	22.1	0	77.1

Table 4. Impacts of tomato grafting on plant vigor, i.e., plant fresh weight (aerial part and roots) and leaf surface area according to the fertilization treatment (standard or 15%), 10 weeks after transplanting. Data are the observed means of replications (N = 4), and values with the same letters are not significantly different based on Fisher's least significant difference (LSD) test (p = 0.05).

Season	Fertilization	Plants	Plant Fresh Weight (g)	Plant Leaf Surface (cm^2)
1	Low	Tanya	144.3 ± 18.6 ab	1108.2 ± 291.8 ab
		Tanya/Hawaii	101.1 ± 11.3 cd	993.3 ± 336.6 ab
		Tanya/R30	94.9 ± 11.1 d	1030.9 ± 355.3 ab
		Tanya/Shelter	124.6 ± 9.1 bc	848.8 ± 74.9 b
	Standard	Tanya	164.4 ± 32.0 a	1319.9 ± 57.5 a
		Tanya/Hawaii	100.0 ± 17.0 cd	972.3 ± 105.9 ab
		Tanya/R30	108.3 ± 14.0 cd	805.6 ± 82.5 b
		Tanya/Shelter	127.4 ± 22.9 bc	986.3 ± 317.2 ab

Table 4. *Cont.*

Season	Fertilization	Plants	Plant Fresh Weight (g)	Plant Leaf Surface (cm^2)
2	Low	Tanya	26.5 ± 9.5 bcd	1028.7 ± 23.7 ab
		Tanya/Hawaii	17.4 ± 1.1 e	861.5 ± 160.6 bcd
		Tanya/R30	18.6 ± 2.5 de	819.8 ± 73.0 cd
		Tanya/Shelter	21.9 ± 7.6 cde	1009.1 ± 36.6 abc
	Standard	Tanya	37.8 ± 7.8 a	1162.2 ± 179.3 a
		Tanya/Hawaii	28.9 ± 3.1 bc	940.1 ± 104.3 bcd
		Tanya/R30	24.3 ± 4.4 bcde	747.1 ± 112.7 d
		Tanya/Shelter	32.0 ± 8.3 ab	929.5 ± 86.9 bcd

3.2. Performances of Grafted Tomato in Soil Infested by Bacterial Wilt

The genetic analysis of wilted samples has confirmed that the wilting was caused by *R. solanacearum* in all locations. However, Phylotype I sequevar 31 strains were found in the experimental plots of Arusha and Moshi. In contrast, Phylotype I sequevar 17 strains were isolated in Pemba, and Phylotype I sequevar 18 strains were found in Unguja.

Rainfall and temperatures were higher in the coastal regions (Pemba and Unguja) compared to the northern highlands of Tanzania (Moshi and Arusha) (Table 3). In the coastal areas, the incidence of bacterial wilt was high throughout both seasons, and all ungrafted 'Tanya' and 'Tengeru 2010' tomatoes used as controls were wilted at the end of the experiments (Figure 3A–D), leading to a complete loss of yield (Table 4). Similarly, all plants grafted onto 'Tengeru 2010' were wilted at both sites by the end of the two seasons, resulting in almost zero yields (Table 4).

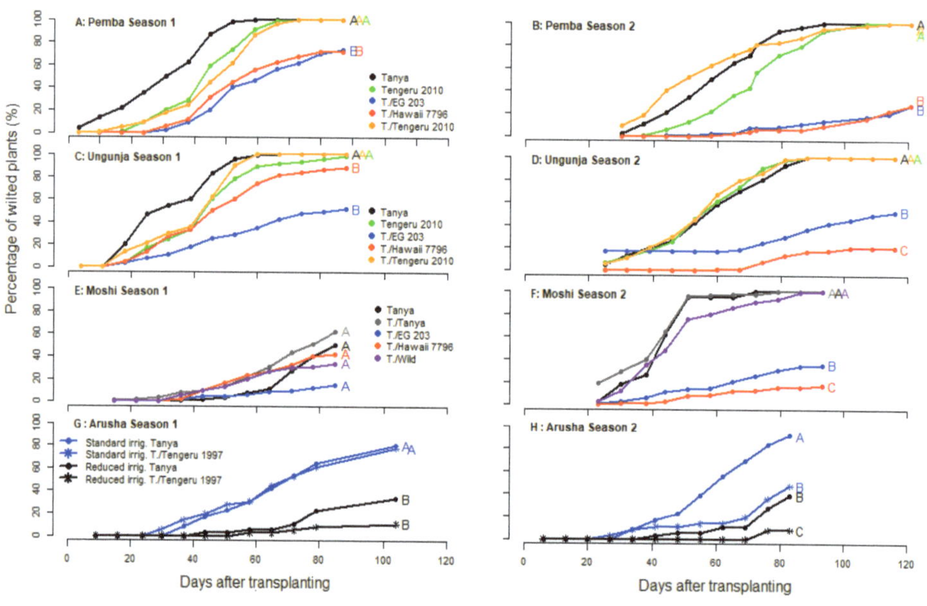

Figure 3. Bacterial wilt incidence (*R. solanacearum*) in the experiments conducted in two seasons in Pemba (**A**,**B**), Unguja (**C**,**D**), Moshi (**E**,**F**), and Arusha (**G**,**H**) with ungrafted 'Tanya' and 'Tengeru 2010' tomatoes and 'Tanya' grafted onto 'Tanya' (self-grafted), 'EG 203', 'Hawaii 7796', 'Tengeru 2010', 'Tengeru 1997', and a wild Solanaceous species (*S. elaeagnifolium*). Data are the observed means of replications (N = 5 in Moshi, N = 3 in Pemba, Unguja, and Arusha), and values with the same letters are not significantly different based on Dunn's test ($p = 0.05$).

Grafting tomatoes onto 'EG 203' slightly improved yields to 0.18 and 0.52 kg·m^{-2} in the first and second seasons by reducing plant wilting to 73% and 27% without significant differences. Grafting tomatoes onto 'Hawaii 7796' also reduced wilt to 71% and 28% in the first and second seasons, resulting in yields of 0.28 and 0.53 kg·m^{-2} (Table 5). Similar results were recorded in Unguja, with 51% and 52% of tomatoes grafted onto 'EG 203' exhibiting wilt symptoms at the end of the first and second seasons, respectively, resulting in yields of 0.28 and 0.53 kg·m^{-2}, as opposed to the 88% and 20% wilting of tomatoes grafted onto 'Hawaii 7996' with yields of 0.04 and 0.33 kg·m^{-2} (Table 5).

Table 5. Comparison of marketable yields of ungrafted 'Tanya' and 'Tengeru 2010' tomatoes and 'Tanya' tomatoes grafted onto tomato ('Tengeru 2010', 'Hawaii 7996') and eggplant ('EG 203') rootstocks at the different sites in Tanzania (Moshi, Pemba, Unguja, and Arusha) infested by bacterial wilt in two seasons. Data are the observed means of replications (N = 5 in Moshi and N = 3 in Pemba, Unguja, and Arusha), and values with the same letters are not significantly different based on the Fisher's least significant difference (LSD) test (p = 0.05).

Sites	Treatment	Plants	Marketable Yields (kg/m^2) Season 1	Marketable Yields (kg/m^2) Season 2
Pemba	Normal irrigation	Tanya	0.00 ± 0.00 c	0.00 ± 0.0 b
	Normal irrigation	Tengeru 2010	0.05 ± 0.06 abc	0.12 ± 0.11 b
	Normal irrigation	Tanya/EG 203	0.18 ± 0.04 a	0.52 ± 0.17 a
	Normal irrigation	Tanya/Hawaii 7796	0.10 ± 0.08 ab	0.55 ± 0.20 a
	Normal irrigation	Tanya/Tengeru 2010	0.01 ± 0.02 bc	0.04 ± 0.07 b
Unguja	Normal irrigation	Tanya	0.00 ± 0.00 b	0.01 ± 0.02 c
	Normal irrigation	Tengeru 2010	0.02 ± 0.04 b	0.07 ± 0.05 bc
	Normal irrigation	Tanya/EG 203	0.28 ± 0.02 a	0.53 ± 0.18 a
	Normal irrigation	Tanya/Hawaii 7796	0.04 ± 0.01 ab	0.33 ± 0.09 ab
	Normal irrigation	Tanya/Tengeru 2010	0.00 ± 0.00 b	0.01 ± 0.01 c
Moshi	Normal irrigation	Tanya	3.59 ± 1.85 a	0.00 ± 0.00 b
	Normal irrigation	Tanya/EG 203	3.59 ± 1.91 a	0.56 ± 0.11 a
	Normal irrigation	Tanya/Hawaii 7796	2.05 ± 1.13 a	0.85 ± 0.34 a
	Normal irrigation	Tanya/Tanya	2.46 ± 1.16 a	0.00 ± 0.00 b
	Normal irrigation	Tanya/Wild	2.32 ± 1.01 a	0.00 ± 0.01 b
Arusha	Normal irrigation	Tanya	0.49 ± 0.12 c	0.43 ± 0.05 c
	Normal irrigation	Tanya/Tengeru 1997	0.54 ± 0.07 c	0.55 ± 0.14 c
	Half-reduced irrigation	Tanya	1.02 ± 0.24 b	1.39 ± 0.36 b
	Half-reduced irrigation	Tanya/Tengeru 1997	1.22 ± 0.24 a	2.11 ± 0.47 a

In Moshi, the wilt incidence in tomatoes showed marked seasonal variations. At the end of the first season, 50% of ungrafted 'Tanya' tomatoes wilted, resulting in yields of 3.59 kg·m^{-2}. In contrast, by the end of the second season, 100% of ungrafted 'Tanya' tomatoes wilted, yielding zero yields (Figure 3F). Grafting tomatoes onto 'EG 203' reduced wilting to 14% and 35%, with 3.59 and 0.56 kg·m^{-2} yields in the first and second seasons, respectively. On the other hand, grafting tomatoes onto 'Hawaii 7996' resulted in 42% and 17% wilting and yields of 2.05 and 0.85 kg·m^{-2} in the first and second seasons, respectively. Self-grafting or using the wild eggplant (*S. elaeagnifolium*) rootstocks did not provide any advantage, as all plants wilted by the end of the second season with zero yields (Table 4). Taking into account results from all experimental sites and seasons, grafting onto 'EG203' and 'Hawaii 7796' increased yields by an average of 57.2% and 27.7%, respectively. It is noteworthy that these findings obscure significant variability between the sites, as the yields of grafted plants onto these rootstocks ranged from 0.1 to 3.59 kg·m^{-2}.

In Experiment 3, cutting the water supply by half reduced the wilting of ungrafted 'Tanya' tomatoes from 81% to 33% in the first season and from 92% to 39% in the second season (Figure 3G,H), increasing yields from 0.49 to 1.02 kg·m^{-2} in the first season and from 0.43 to 1.39 kg·m^{-2} in the second season. Although no difference was observed in the first season, grafting 'Tanya' onto 'Tengeru 1997' reduced wilting from 92% to 47%, with standard irrigation increasing yields from 0.43 to 0.55 kg·m^{-2}. The advantages of using 'Tengeru 1997' as rootstock were clearer under the 50% irrigation treatment since wilting was reduced from 33% to 11% and 39% to 8% compared to ungrafted plants in the first and second seasons. This increased yields from 1.02 to 1.22 kg·m^{-2} and 1.39 to 2.11 kg·m^{-2} (Table 5). The combination of reduced water supply and grafting led to an average reduction in wilting of 76% and an average yield improvement of 3.6 times in experiments conducted over two seasons in Arusha.

4. Discussion

The results of the present study show that grafting tomatoes systematically leads to higher plant survival depending on the rootstock used for grafting but does not systematically lead to higher yields. The advantages of using grafted plants vary with the rootstocks, cultivation practices, and the degree of pressure and diversity from soilborne pathogens. Grafting presumably aids pathogen avoidance when the rootstock is resistant to the pathogen. That is why scion adventitious roots confer susceptibility to the grafted plant when they come into contact with the soil [27].

Though grafting can be advantageous, the mechanism is still unclear. Sometimes, the different strains of *R. solanacearum* of Phylotype I (sequevars 17, 18, and 31) isolated from the experimental sites emphasize the genetic diversity of the pathogen in Tanzania [3], which was previously reported to be limited to strains of Phylotype III [28]. King and coauthors [27] concluded in their review that there is a likelihood that new races or pathotypes will evolve, weakening the existing resistant rootstocks due to selection pressure brought by vast rootstock usage.

The field trials showed that grafting tomatoes onto 'Hawaii 7796', 'Tengeru 1997', and 'EG 203' improved resistance to bacterial wilt to a certain extent, whereas grafting tomatoes onto 'Tengeru 2010' and onto the wild species (*S. elaeagnifolium*) did not. Laboratory experiments conducted by Lebeau [26] showed no wilting of 'Hawaii 7796' and 'EG 203' after inoculation with a strain of Phylotype I (sequevars 13 and 18), although latent infection tests revealed that some plants did carry the pathogen. Rootstock functioning can be related to pathogen pressure in the soil; Herman and Treves [29] pinpointed the existence of fusarium pathogens in grafted cucumber stems but at low levels as compared to the control when commercially available rootstocks were used, meaning under low pathogen pressure, it is unlikely for the pathogen to overpower the semi-resistant rootstocks completely. However, they can still enter the xylem and colonize it. The improvements in yield obtained by using grafted plants in soils infected by bacterial wilt measured in the present study are, thus, far from those previously reported in studies conducted in the USA [30] and in Asia [31]. The difference between the results of the experiments performed on other continents can be attributed to differences in *R. solanacearum* strains and the combination of scions and rootstocks. The yield potential of tomatoes in this region may not be similar due to other pressures like leaf diseases such as early and late blights, which limits the production potential. Less wilting and higher yields with 'Tanya' tomatoes grafted onto 'Hawaii 7996' grown in soil infested by bacterial wilt have also been previously reported at one of the experimental sites (Arusha) used in the present study [32]. The lower performances of the grafted plants in the present study may be explained by a higher incidence of bacterial wilt in our study conditions, like the marked differences in wilting and yields between seasons in Moshi on grafted plants (Figure 3E,F). The all-year-round high incidence of bacterial wilt in the coastal areas (Pemba and Unguja), where the climate is hot and humid, limited yields to less than 0.55 kg·m^{-2} in all three experiments, regardless of the rootstock used. The impact of the soil moisture on bacterial wilt incidence was further confirmed by the results of Experiment 3, which showed that reducing irrigation by half improved yields thanks to reduced wilting (Figure 3G,H). This is consistent with the results of a previous laboratory study showing that tomato wilt incidence was lower in dry soil [33], and improved yields were obtained from grafted tomatoes grown under a rain shelter [34]. The bacteria need a hot and wet environment for their multiplication and spread. Since the reduction in soil moisture reduces the spread of the pathogen, seedling mortality is reduced.

Rootstocks have the potential to alter yields negatively; Davis and coauthors [35] argue that there are optimal temperature and humidity ranges for the optimal performance of scion/rootstock combinations, and this may explain contrasting grafting benefits based on our geographical location situated with a different climatic condition from Asia or the US where these rootstocks had amazing results. In the present study, the maximum yields obtained from ungrafted 'Tanya' tomatoes in bacterial wilt-free soil with standard fertilization and irrigation did not exceed 2.2 $kg \cdot m^{-2}$. Although these yields are in agreement with those commonly observed for this variety when grown in the field [22,36], they are far below their potential yields (up to 5.0 $kg \cdot m^{-2}$) due to high pressure from insects and diseases during the experiments despite a large number of pesticide treatments (Table 2). This underlines the extent of crop losses caused by aerial pests, which limits the benefits of investing in grafted plants, and explains the reliance of Tanzanian tomato producers on pesticides [5].

Our results provide no evidence for significant improvement in yield to be achieved by grafting 'Tanya' tomatoes onto 'Hawaii 7796', 'R3034', or 'Shelter' with reduced fertilization. Rivard and Louws [37] advise using grafting only when there is disease pressure, meaning there can be a yield penalty. Previous studies reported improved plant nutrient uptake achieved by grafting tomatoes onto 'Maxifort' or 'Brigeor' rootstocks [38,39]. In agreement with these studies, higher yields were also reported for field-grown tomatoes (cv. 'Florida 47') grafted onto 'Multifort' or 'Beaufort' rootstocks with a nitrogen supply below the recommended rate [40]. The latter study confirmed higher yields from tomatoes (cv. 'Tribute') grafted onto 'Maxifort' with a low nitrogen supply. Still, contrasting results were obtained with the 'RST-106' rootstock, as improved yields were only obtained in one out of the two years of the experiment [41]. Plant responses to stress can be complex. Ullah and coauthors [42] summarized how plants could adopt different signaling pathways under environmental stress in conjunction with phytohormones like auxins. However, some response mechanisms were still unclear without a direct pattern due to the complexity of soil health, microbial diversity, and ecological functioning. In the present study, the absence of a significant difference between grafted and ungrafted tomatoes with reduced fertilization could be due to the choice of rootstocks used and/or to the high pressure of pests and diseases during the experiments that may have reduced the effect of grafting.

Overall, the results of the three experiments conducted in the present study provide no evidence for any advantage in using grafted plants even though the merits of some of the rootstocks tested ('EG 203' and 'Hawaii 7796') in increasing the tolerance of the plants to bacterial wilt and flooding have been widely demonstrated in Asia [20,31]. The lack of tangible results echoes observations of a meta-analysis indicating that among the 949 combinations of rootstock scions reviewed, only 37% produced higher yields than ungrafted tomatoes [14]. To benefit from tomato grafting onto suitable rootstocks, further efforts are required to identify rootstocks suitable for production in Tanzanian conditions, along with the appropriate management of other biotic and abiotic stresses that limit tomato production.

5. Conclusions

Several field experiments were conducted in various regions of Tanzania, including Arusha, Moshi, Pemba, and Unguja, to assess the response of the 'Tanya' determinate tomato variety. The experiments involved growing the tomatoes either ungrafted or grafted to rootstocks of tomato, eggplant, and wild species under different fertilization regimes in soils infected by bacterial wilt and water deficit levels in soils infected with bacterial wilt. The findings revealed that the rootstocks tested in the study, such as 'EG 203', 'Tengeru 1997', 'Tengeru 2010', 'Hawaii 7996', and *Solanum elaeagnifolium*, did not significantly improve tomato yields. However, some rootstocks were found to increase the plant's resistance to bacterial wilt to some extent. Moreover, the research showed that grafting tomatoes onto 'Shelter', 'Hawaii 7796', and 'R3034' did not enhance yields under the 15% low fertilization treatment. As a result, further research is necessary to identify rootstocks resistant to the various strains of bacterial wilt present in Tanzania. Such research could help boost the tolerance of tomatoes to common production limitations, such as water deficit and poor soil fertility.

Supplementary Materials: The following supporting information can be downloaded at https://www.mdpi.com/article/10.3390/horticulturae10040338/s1: Figure S1: Comparison of the soil water potential at a depth of 30 cm below the soil surface under standard (from −0.03 to −0.02 MPa, in black) and 50% irrigation (reduced from −0.08 to −0.02 MPa, in light grey) supply in Experiment 3 in Arusha. Data are averages (N = 3).

Author Contributions: Methodology, S.E.M.; software, T.N.; validation, R.M. and J.H.; formal analysis, T.N.; investigation, Z.E. and N.M.; resources, T.N.; data curation, S.E.M.; writing—original draft preparation, T.N. and S.E.M.; writing—review and editing, T.N., S.E.M. and S.R.; visualization, T.N.; supervision, J.H.; project administration, T.N.; funding acquisition, T.N. All authors have read and agreed to the published version of the manuscript.

Funding: This research was funded by GIZ under a grant from the Federal Ministry for Economic Cooperation and Development (BMZ, 16.7860.6-001-00) and core donors to the World Vegetable Center: Taiwan; the UK's Foreign, Commonwealth and Development Office (FCDO); the United States Agency for International Development (USAID); the Australian Centre for International Agricultural Research (ACIAR); Germany; Thailand; the Philippines; Korea; and Japan.

Data Availability Statement: The data presented in this study are available upon request from the corresponding author. They are not publicly available for a certain period and can later be accessed from https://worldveg.tind.io/ from 15 May 2024.

Acknowledgments: The authors gratefully acknowledge the staff of TAHA (the Tanzania Horticulture Association) and SEVIA (Seeds of Expertise for the Vegetable Sector in Africa) for their support in field experiments and Gilles Cellier (ANSES, Reunion Island) for his support in the molecular typing of bacterial wilt strains.

Conflicts of Interest: The authors declare no conflicts of interest.

References

1. Schreinemachers, P.; Simmons, E.B.; Wopereis, M.C.S. Tapping the Economic and Nutritional Power of Vegetables. *Glob. Food Secur.* **2018**, *16*, 36–45. [CrossRef]
2. Fakayode, S.; Rahji, M.; Adeniyi, S. Economic Analysis of Risks in Fruit and Vegetable Farming in Osun State, Nigeria. *Bangladesh J. Agric. Res* **2012**, *37*, 473–491. [CrossRef]
3. Cellier, G.; Nordey, T.; Cortada, L.; Gauche, M.; Rasoamanana, H.; Yahiaoui, N.; Rébert, E.; Prior, P.; Chéron, J.J.; Poussier, S.; et al. Molecular Epidemiology of *Ralstonia pseudosolanacearum* Phylotype I Strains in the Southwest Indian Ocean Region and Their Relatedness to African Strains. *Phytopathology®* **2023**, *113*, 423–435. [CrossRef] [PubMed]
4. De Bon, H.; Huat, J.; Parrot, L.; Sinzogan, A.; Martin, T.; Malézieux, E.; Vayssières, J.-F. Pesticide Risks from Fruit and Vegetable Pest Management by Small Farmers in Sub-Saharan Africa. A Review. *Agron. Sustain. Dev.* **2014**, *34*, 723–736. [CrossRef]
5. Ngowi, A.V.F.; Mbise, T.J.; Ijani, A.S.M.; London, L.; Ajayi, O.C. Smallholder Vegetable Farmers in Northern Tanzania: Pesticides Use Practices, Perceptions, Cost and Health Effects. *Crop Prot.* **2007**, *26*, 1617–1624. [CrossRef] [PubMed]
6. Lehmann, E.; Turrero, N.; Kolia, M.; Konaté, Y.; De Alencastro, L.F. Dietary Risk Assessment of Pesticides from Vegetables and Drinking Water in Gardening Areas in Burkina Faso. *Sci. Total Environ.* **2017**, *601–602*, 1208–1216. [CrossRef] [PubMed]
7. Mutengwe, M.T.; Aneck-Hahn, N.H.; Korsten, L.; Van Zijl, M.C.; De Jager, C. Pesticide Residues and Estrogenic Activity in Fruit and Vegetables Sampled from Major Fresh Produce Markets in South Africa. *Food Addit. Contam. Part A* **2015**, *33*, 95–104. [CrossRef]
8. Doré, T.; Makowski, D.; Malézieux, E.; Munier-Jolain, N.; Tchamitchian, M.; Tittonell, P. Facing up to the Paradigm of Ecological Intensification in Agronomy: Revisiting Methods, Concepts and Knowledge. *Eur. J. Agron.* **2011**, *34*, 197–210. [CrossRef]
9. Wezel, A.; Casagrande, M.; Celette, F.; Vian, J.-F.; Ferrer, A.; Peigné, J. Agroecological Practices for Sustainable Agriculture. A Review. *Agron. Sustain. Dev.* **2014**, *34*, 1–20. [CrossRef]
10. Nordey, T.; Schwarz, D.; Kenyon, L.; Manickam, R.; Huat, J. Tapping the Potential of Grafting to Improve the Performance of Vegetable Cropping Systems in Sub-Saharan Africa. A Review. *Agron. Sustain. Dev.* **2020**, *40*, 23. [CrossRef]

11. Louws, F.J.; Rivard, C.L.; Kubota, C. Grafting Fruiting Vegetables to Manage Soilborne Pathogens, Foliar Pathogens, Arthropods and Weeds. *Sci. Hortic.* **2010**, *127*, 127–146. [CrossRef]
12. Rouphael, Y.; Kyriacou, M.C.; Colla, G. Vegetable Grafting: A Toolbox for Securing Yield Stability under Multiple Stress Conditions. *Front. Plant Sci.* **2018**, *8*, 2255. [CrossRef] [PubMed]
13. Schwarz, D.; Rouphael, Y.; Colla, G.; Venema, J.H. Grafting as a Tool to Improve Tolerance of Vegetables to Abiotic Stresses: Thermal Stress, Water Stress and Organic Pollutants. *Sci. Hortic.* **2010**, *127*, 162–171. [CrossRef]
14. Grieneisen, M.L.; Aegerter, B.J.; Scott Stoddard, C.; Zhang, M. Yield and Fruit Quality of Grafted Tomatoes, and Their Potential for Soil Fumigant Use Reduction. A Meta-Analysis. *Agron. Sustain. Dev.* **2018**, *38*, 29. [CrossRef]
15. Ganiyu, S.; Popoola, R.; Enikuomehin, O.; Goke, B. Tube Grafting Reduces Incidence and Severity of Bacterial Wilt in Two Tomato Cultivars in Abeokuta, Nigeria. *J. Agric. Sci. Environ.* **2017**, *16*, 96–104. [CrossRef]
16. Ganiyu, S.A.; Popoola, A.R.; Enikuomehin, O.A.; Bodunde, J.G. Influence of Grafting on Growth and Yield Performance of Two Tomato Cultivars Grown in Open Field in Nigeria. *J. Plant Pathol.* **2018**, *100*, 43–50. [CrossRef]
17. Lin, L.J.; Luther, G.C.; Hanson, P. *Raising Healthy Tomato Seedlings*; AVRDC—The World Vegetable Center: Shanhua, Taiwan, 2015; pp. 15–795.
18. Mpinga, L.E. Tomato Grafting for Low-Resource Open-Field Tomato Production in Tanzania. Ph.D. Thesis, University of Florida, Gainesville, FL, USA, 2013.
19. Nordey, T.; Shem, E.; Huat, J. Impacts of Temperature and Rootstocks on Tomato Grafting Success Rates. *HortScience* **2020**, *55*, 136–140. [CrossRef]
20. Black, L.L.; Wu, D.L.; Wang, J.F.; Kalb, T.; Abbass, D.; Chen, J.H. *Grafting Tomatoes for Production in the Hot-Wet Season*; Asian Vegetable Research & Development Center (AVRDC Publication): Shanhua, Taiwan, 2003.
21. Hanson, P.; Chen, J.T.; Kuo, C.G.; Morris, R.; Opeña, R.T. Suggested Cultural Practices for Tomato. *Int. Coop. Guide AVRDC* **2000**, 1–8.
22. Nordey, T.; Deletre, E.; Mlowe, N.; Martin, T. Nethouses Protect Cucumber Plants from Insect Pests and Increase Yields in Eastern Africa. *J. Hortic. Sci. Biotechnol.* **2020**, *95*, 673–678. [CrossRef]
23. Nordey, T.; Deletre, E.; Mlowe, N.; Martin, T. Small Mesh Nets Protect Tomato Plants from Insect Pests and Increase Yields in Eastern Africa. *J. Hortic. Sci. Biotechnol.* **2020**, *95*, 222–228. [CrossRef]
24. Do Rosário, G.; Oliveira, M.; Calado, A.M.; Portas, C.A.M. Tomato Root Distribution under Drip Irrigation. *J. Am. Soc. Hortic. Sci.* **1996**, *121*, 644–648. [CrossRef]
25. N'Guessan, C.A.; Abo, K.; Fondio, L.; Chiroleu, F.; Lebeau, A.; Poussier, S.; Wicker, E.; Koné, D. So Near and Yet so Far: The Specific Case of *Ralstonia solanacearum* Populations from Côte d'Ivoire in Africa. *Phytopathology®* **2012**, *102*, 733–740. [CrossRef] [PubMed]
26. Lebeau, A.; Daunay, M.-C.; Frary, A.; Palloix, A.; Wang, J.-F.; Dintinger, J.; Chiroleu, F.; Wicker, E.; Prior, P. Bacterial Wilt Resistance in Tomato, Pepper, and Eggplant: Genetic Resources Respond to Diverse Strains in the *Ralstonia solanacearum* Species Complex. *Phytopathology®* **2011**, *101*, 154–165. [CrossRef] [PubMed]
27. King, S.R.; Davis, A.R.; Liu, W.; Levi, A. Grafting for Disease Resistance. *HortScience* **2008**, *43*, 1673–1676. [CrossRef]
28. Kanyagha, H. Characterization of *Ralstonia* spp. in Tanzania and Potential Intergrated Pest Management Strategies for Managing Bacterial Wilt in Tomatoes. Ph.D. Thesis, The Ohio State University, Columbus, OH, USA, 2021.
29. Herman, R.; Perl-Treves, R. Characterization and Inheritance of a New Source of Resistance to *Fusarium oxysporum* f. sp. *melonis* Race 1.2 in *Cucumis melo*. *Plant Dis.* **2007**, *91*, 1180–1186. [CrossRef] [PubMed]
30. Rivard, C.L.; O'Connell, S.; Peet, M.M.; Welker, R.M.; Louws, F.J. Grafting Tomato to Manage Bacterial Wilt Caused by *Ralstonia solanacearum* in the Southeastern United States. *Plant Dis.* **2012**, *96*, 973–978. [CrossRef] [PubMed]
31. Genova, C.; Schreinemachers, P.; Afari-Sefa, V. *An Impact Assessment of AVRDC's Tomato Grafting in Vietnam*; AVRDC—The World Vegetable Center: Shanhua, Taiwan, 2013.
32. Kitundu, J.A.; Dinssa, F.F.; Manickam, R.; Macharia, J.; Mbwambo, O.; Aloyce, A.; Ramasamy, S. Effect of Grafting on Yield of Commercial Tomato Cultivars Grown under Bacterial Wilt (*Ralstonia solanacearum*) Infested Soil. *Fruits* **2022**, *77*, 20220152857. [CrossRef] [PubMed]
33. Islam, T.M.; Toyota, K. Effect of Moisture Conditions and Pre-Incubation at Low Temperature on Bacterial Wilt of Tomato Caused by *Ralstonia solanacearum*. *Microb. Environ.* **2004**, *19*, 244–247. [CrossRef]
34. Biswas, S.; Akter, K.T.; Rakib, A.; Rhanom, M.S.R.; Rahman, M.M.; Ahmed, S. Performance of Tomato Varieties Grafted with Eggplant under Rain Shelter Condition. *Int. J. Nat. Soc. Sci.* **2019**, *6*, 63–70.
35. Davis, A.R.; Perkins-Veazie, P.; Hassell, R.; Levi, A.; King, S.R.; Zhang, X. Grafting Effects on Vegetable Quality. *HortScience* **2008**, *43*, 1670–1672. [CrossRef]
36. Meya, A.I.; Mamiro, D.P.; Kusolwa, P.M.; Maerere, A.P.; Sibuga, K.P.; Erbaugh, M.; Miller, S.A.; Mtui, H.D. Management of Tomato Late Blight Disease Using Reduced Fungicide Spray Regimes in Morogoro, Tanzania. *J. Agric. Sci.* **2014**, *13*, 2.
37. Rivard, C.L.; Louws, F.J. Grafting to Manage Soilborne Diseases in Heirloom Tomato Production. *HortScience* **2008**, *43*, 2104–2111. [CrossRef]
38. Savvas, D.; Öztekin, G.B.; Tepecik, M.; Ropokis, A.; Tüzel, Y.; Ntatsi, G.; Schwarz, D. Impact of Grafting and Rootstock on Nutrient to Water Uptake Ratios during the First Month after Planting of Hydroponically Grown Tomato. *J. Hortic. Sci. Biotechnol.* **2017**, *92*, 294–302. [CrossRef]

39. Schwarz, D.; Öztekin, G.B.; Tüzel, Y.; Brückner, B.; Krumbein, A. Rootstocks Can Enhance Tomato Growth and Quality Characteristics at Low Potassium Supply. *Sci. Hortic.* **2013**, *149*, 70–79. [CrossRef]
40. Djidonou, D.; Zhao, X.; Simonne, E.H.; Koch, K.E.; Erickson, J.E. Yield, Water-, and Nitrogen-use Efficiency in Field-grown, Grafted Tomatoes. *HortScience* **2013**, *48*, 485–492. [CrossRef]
41. Suchoff, D.H.; Gunter, C.C.; Schultheis, J.R.; Hassell, R.L.; Louws, F.J. The Effect of Grafting on Nitrogen Use in Determinate Field-Grown Tomatoes. *J. Hortic. Sci. Biotechnol.* **2019**, *94*, 102–109. [CrossRef]
42. Ullah, F.; Saqib, S.; Khan, W.; Ayaz, A.; Batool, A.; Wang, W.-Y.; Xiong, Y.-C. The Multifaceted Role of Sodium Nitroprusside in Plants: Crosstalk with Phytohormones under Normal and Stressful Conditions. *Plant Growth Regul.* **2024**, 1–18. [CrossRef]

horticulturae

Article

Joint Analysis of Small RNA and mRNA Sequencing Unveils miRNA-Mediated Regulatory Network in Response to Methyl Jasmonate in *Apocynum venetum* L.

Jinhua Tan [1,†], Xiaowei Han [2,3,†], Qi Liu [1], Tashi Dorjee [1], Yijun Zhou [1], Huigai Sun [2,3,*] and Fei Gao [1,*]

1 College of Life and Environmental Sciences, Minzu University of China, Beijing 100081, China; 18400218@muc.edu.cn (J.T.); 20400256@muc.edu.cn (Q.L.); 20400268@muc.edu.cn (T.D.); zhouyijun@muc.edu.cn (Y.Z.)

2 School of Pharmacy, Hebei University of Chinese Medicine, Shijiazhuang 050200, China; hanxiaowei@hebcm.edu.cn

3 Traditional Chinese Medicine Processing Technology Inheritance Base of the State Administration of Traditional Chinese Medicine (Hebei), Shijiazhuang 050200, China

* Correspondence: sunhuigai@hebcm.edu.cn (H.S.); gaofei@muc.edu.cn (F.G.)

† These authors contributed equally to this work.

Abstract: *Apocynum venetum* L. is a natural fiber and medicinal plant species with significant economic value. Jasmonic acid is an endogenous growth regulatory substance present in higher plants that participate in plant growth, development, and defense. As important endogenous single-stranded RNA molecules, microRNA (miRNA) plays an important role in the post-transcriptional regulation of plant genes. A combination of miRNA and mRNA sequencing techniques was used to systematically identify the methyl jasmonate miRNAs and mRNAs in *A. venetum*. Up to 135 conserved and 26 species-specific miRNAs have been identified in *A. venetum*. These miRNAs mainly target genes that encode transcription factors and enzymes. The expression levels of 23 miRNAs, including miR398 and miR482, significantly changed after MeJA treatment. A total of 1778 genes were differentially expressed under MeJA treatment, of which 825 were upregulated and 953 were downregulated. The main biological processes enriched in these differentially expressed genes were redox balance, secondary metabolism, photosynthesis, and plant hormone signal transduction. Joint analysis of the miRNAs and mRNA revealed that MeJA-responsive miRNAs function by forming regulatory modules, including miR398-CSD, miR482-NBS-LRR, miR156-SPL10, and miR164-NAC056, which further regulate multiple biological processes, including redox balance, disease resistance, and morphogenesis in *A. venetum*. This study provides important information to understand the biological roles of miRNAs in *A. venetum*.

Keywords: *Apocynum venetum*; leaf; methyl jasmonate; microRNA; miR398; miR482

Citation: Tan, J.; Han, X.; Liu, Q.; Dorjee, T.; Zhou, Y.; Sun, H.; Gao, F. Joint Analysis of Small RNA and mRNA Sequencing Unveils miRNA-Mediated Regulatory Network in Response to Methyl Jasmonate in *Apocynum venetum* L. *Horticulturae* **2024**, *10*, 173. https://doi.org/10.3390/horticulturae10020173

Academic Editor: Alessandra Carrubba

Received: 9 January 2024
Revised: 6 February 2024
Accepted: 9 February 2024
Published: 15 February 2024

1. Introduction

Apocynum venetum L., also known as LUOBUMA in China, is a perennial subshrub of the Apocynaceae family [1]. This plant is widely distributed in Xinjiang, Inner Mongolia, Gansu, Qinghai, and other regions of China, and there is large-scale artificial cultivation of *A. venetum* in Qinghai, Heilongjiang, and Xinjiang. *A. venetum* is a medicinal plant with multiple therapeutic and healthcare functions [2]. Research has shown that *A. venetum* has good pharmacological effects by lowering blood pressure, blood lipids, and acting as an antidepressant. The main active pharmaceutical ingredients of *A. venetum* include flavonoids, such as quercetin, hyperoside, and isoquercetin. *A. venetum* is a hemp crop. The fibers prepared from the stem skin of *A. venetum* are slender and soft and are often used to make underwear and other textiles. *A. venetum* can tolerate saline–alkali stress and often grows on desert edges and coastal areas; thus, it has important ecological value [3]. In addition, *A. venetum* blooms beautifully and can be used as an ornamental plant. At present, the research of *A. venetum* mainly focuses on the analysis of chemical composition and their pharmacological action, and there are few studies on hormone response in *A. venetum*.

Jasmonates (JAs) include jasmonic acid (JA), methyl jasmonate (MeJA), derivatives of volatile MeJA, and derivatives of JA amino acids, which are plant signaling molecules in the oxylipin family [4]. MeJA is a methylated form of JA. After entering plant cells through the stomata, MeJA undergoes a series of chemical reactions to form JA, which together participate in the induction of plant defense reactions. JA mediates tolerance to drought, salt, heat, and cold stress, as well as resistance to pathogens and insects, and induces embryogenesis, flowering, and senescence [5]. It also involves tripartite interactions between insect vectors, viruses, and plants [6]. Research has found that exogenous application of MeJA increased the antioxidant enzyme activity of banana, thereby increasing their resistance to *Fusarium oxysporum* [7]. Exogenous JA application can increase the ability of *Alyssum inflatum* to resist heavy metal stress by reducing H_2O_2 concentration and increasing photosynthetic pigments [8]. MeJA can enhance the ability of *Malus crabapple* to resist O_3 stress by increasing photosynthetic rate and antioxidant activity, and reducing lipid peroxidation [9]. However, little is known about the effects of MeJA on various biological processes and metabolic pathways in *A. venetum*.

Analyzing the JA response in *A. venetum* at the gene expression level helps understand the effects of JA on various biological processes and metabolic pathways in plants. In addition to classical transcriptome sequencing, analyzing the role of JA in miRNA regulation can provide a deeper understanding of JA-related transcriptional regulatory networks in plants. With an approximate length of 18 to 24 nt, microRNAs (miRNAs) are a class of endogenous single-stranded non-coding small RNAs that negatively regulate protein-coding genes at the posttranscriptional level. MiRNAs interact with their target mRNAs in two ways: cleavage induced degradation and translation inhibition. In plants, miRNA-targeted mRNAs are primarily cleaved and degraded by precise cleavage at the 10th or 11th base [10,11]. MiRNAs play key roles in regulating plant growth, development, and responses to environmental cues [12,13]. Emerging evidence has highlighted the responsiveness of miRNAs to various environmental stressors such as high salinity, drought, and low temperatures. Several studies have demonstrated that some miRNAs, such as miR156, miR166, miR398, miR482, and miR858, are involved in abiotic stress responses by targeting transcription factors and other functional genes, including SQUAMOSA Promoter-Binding Protein-Like (SPL) [14], HD-zip [15], Cu/Zn-superoxide dismutase (SOD; CSD) [16], NBS-LRR [17], and R2R3-MYB transcription factor [18].These miRNAs are believed to play crucial roles in the regulation of plant hormones, signal transduction, plant–pathogen interactions, and metabolic pathways. Plants contain both conserved and species-specific miRNAs. The systematic identification of all miRNAs in a specific species is of great value for understanding their biological characteristics and regulation. Many plant species have undergone systematic identification of miRNAs, including cabbage [19], jojoba [20], Chinese olive [21], tea plant [22], and ginkgo [23], but there have been no reports on miRNAs in *A. venetum*.

The present study employed high-throughput nucleic acid sequencing technology and bioinformatics methods to systematically identify conserved and species-specific miRNAs in *A. venetum*, as well as their corresponding target genes. Transcriptome sequencing was conducted to determine the gene profiling in leaves of *A. venetum* under MeJA treatment. Based on the joint analysis of the alterations in the expression levels of both miRNAs and mRNA, a regulatory network of the MeJA response mediated by miRNAs was constructed. This study advances our understanding of the molecular mechanisms associated with the MeJA response in *A. venetum* and the biological role of miRNAs in plants.

2. Materials and Methods

2.1. Plant Materials and Stress Treatments

Seeds of *A. venetum* were collected from Korla (E 85°88′, N 41°4′) in the Xinjiang Uygur Autonomous Region of China. *A. venetum* seeds were first washed with ultrapure water, sterilized with 75% ethanol for 30 s, washed several times with ultrapure water to remove residual ethanol, sterilized with 5% sodium hypochlorite (NaClO) for 5 min, washed three times with sterile water, and dried with filter paper to remove surface moisture. The sterilized seeds were immersed in ultrapure water at a temperature of 4 °C for 24 h. The seeds were germinated in 10 L pots, which consisted of a combination of nutrient soil and vermiculite in a 1:1 ratio.

The seedlings were grown at 25/20 °C (d/n) under a photosynthetic photon flux density of 150 μmol m^{-2} s^{-1} with long-day conditions (16/8 h light/dark cycle). Five months after gemination, seedlings that exhibited comparable heights were randomly allocated in four groups. The first group comprised seedlings that were unstressed and was denoted as CK. In the second group, 1 mM of MeJA (Sigma-Aldrich, Madrid, Spain) was topically applied to the foliage for 72 h (MeJA group). In the third group, the leaves were sprayed with 50 mM methyl viologen (MV) for 24 h (MV group). In the fourth group, leaves were sprayed with 1 mM MeJA for 72 h. Subsequently, a spray treatment with 50 mM MV was administered for 24 h (MeJA+MV group). The MeJA solution used in our experiment was prepared by dissolving it with 95% ethanol, and a total of 5 mL of MeJA solution was used for each seedling. Leaf samples of the three biological replicates were collected and rapidly cryopreserved in liquid nitrogen until RNA extraction.

2.2. Physiological and Biochemical Analysis

Relative electrolyte leakage (REL) was measured using the method described by Yasar and Yang, respectively [24,25]. Malondialdehyde (MDA) content and the activity of SOD in *A. venetum* leaves were detected using the MDA detection kit and SOD kit, respectively. These kits were manufactured by Jiancheng Bioengineering Ltd. (Nanjing, China). The accumulation of H_2O_2 and O_2^- in the leaves of *A. venetum* was analyzed using 3,3′-diaminobenzidine (DAB)-staining and nitro blue tetrazolium (NBT)-staining according to previously described methods [26]. For each sample, the leaves from the three *A. venetum* seedlings were taken for histochemical staining, and leaves from the MV and MeJA+MV groups were chosen at random for imaging.

2.3. Chromosomal Positioning of miRNA Precursor

Chromosomal mapping was conducted to identify and locate conserved and species-specific miRNA precursor sequences in the chromosomes of *A. venetum*. To accomplish this, the online platform MapGene2Chrom (http://mg2c.iask.in/mg2c_v2.1/, accessed on 2 August 2023) [27] was utilized.

2.4. Transcriptome Sequencing and Data Analysis

The total RNA of *A. venetum* leaves was extracted by the TRIzol method. RNA integrity was assessed using electrophoresis (Bio-Rad Laboratories, Hercules, CA, USA). The RNA samples were sent to Baimaike Gene Technology Co., Ltd. (http://www.biomarker.com.cn/archives/24629, accessed on 2 July 2023). (Hangzhou) for library (350 bp) construction, and the libraries were sequenced using the Illumina HiseqTM4000 platform.

Initially, clean reads were obtained using Cutadapt 4.7 software, which effectively eliminated adaptor and ow-quality reads from the dataset [28]. Furthermore, sequencing reads were aligned to the *A. venetum* genome (unpublished data) using Hisat2 2.2.1 software [29]. StringTie 1.3.6 software was used to assemble the reads [30]. Third, genes were annotated using Trinotate v4.0.0 software, which was compared with Non-Redundant Protein Sequences in NCBI (NR) [31], Nucleotide Collection (NT), Universal Protein (Uniprot) [32], Kyoto encyclopedia of genes and genomics database (KEGG) [33], Clusters of orthogonal groups of proteins (COG) [34], and Gene ontology (GO) [35] databases to obtain the functional annotation information of genes in *A. venetum*. The differentially expressed genes (DEGs) were identified using DESeq2 3.11 software [36], with a screening threshold of $|\log2\text{Ratio}| \geq 1$ and q value ≤ 0.05. GO and KEGG pathway enrichment analyses of DEGs in *A. venetum* was carried out by R 3.6.3 software.

2.5. Utilization of Small RNA Sequencing Techniques Coupled with Comprehensive Data Analysis

Small RNA libraries were prepared using a TruSeq Small RNA sample Preparation Kit (Illumina, San Diego, CA, USA) in accordance with established procedures. Subsequently, the six libraries were sequenced by single-end sequencing on the Illumina HiSeq 2500 platform (Baimaike Gene Technology [Hangzhou] Co., Ltd., Hangzhou, China).

Adaptor and low-quality sequences were removed using Trimmomatic v0.35, as described by Bolger [37]. The miRNA sequences and their corresponding precursors were identified usingMiRDeep2 and miRPlant software, as described by Friedlander [38,39]. Sequence fragments with mismatched bases < 4 and no base insertions or deletions were considered known miRNAs and were called conserved miRNAs. Unmatched sequences that were in line with the identification standard of "high reliability miRNA", were considered species-specific miRNA [40].

The differentially expressed miRNA (DEMs) in *A. venetum* were determined by employing DESeq2 software, and the criteria were set as $|\log2Ratio| \geq 1$ and q-value ≤ 0.05. The target genes of all miRNAs in *A. venetum* were predicted using psRNAtarget software (http://plantgrn.noble.org/psRNATarget/, accessed on 2 December 2023) with default parameters.

2.6. qRT-PCR Analysis of miRNAs and Their Targeting mRNAs

cDNA was obtained by reverse transcription of the extracted *A venetum* RNA, using the miRcute miRNA plus cDNA first-strand synthesis kit and the FastKing gDNA Displacing RT Supermix kit from TIANGEN. The cDNA reaction solution that was acquired underwent 100-fold dilution and was subsequently used as a template for qRT-PCR.

Forward, reverse, and universal reverse primers for miRNAs (Supplementary Table S1) were designed using Primer 3.0 (https://primer3.ut.ee/, accessed on 2 August 2023). The expression levels of both the target mRNAs and miRNAs were analyzed by qRT-PCR, and the internal references used were *eIF* and *18s* rRNA, respectively.

qRT-PCR amplification was performed on an Applied Biosystems QuantStudio 5 (Thermo Fisher Scientific, Waltham, MA, USA) using the miRcute plus miRNA fluorescence quantitative detection kit and the qPCR SuperMix kit (TIANGEN, China). The reaction system and qPCR procedures were set up according to the manufacturer's instructions. Each reaction was repeated three times independently, and the relative expression of each miRNA or target gene was analyzed using the $2^{-\Delta\Delta Ct}$ method [41].

2.7. Promoter Analysis of miRNAs in A. venetum

PLACE version 30.0 [42] software (https://www.dna.affrc.go.jp/PLACE/?action=newPlaceSite&site=S000453, accessed on 2 August 2023) was utilized to predict cis-acting elements situated approximately 1500 bp upstream of the precursor sequences of miRNAs.

2.8. Double-Luciferase Reporter Gene Test

The double-luciferase reporter assay test [43] was used to validate the targeting relationship between miRNAs and their predicted targets. The precursor sequence of miR398 was ligated to the pGreen II 62 SK vector, and the coding sequence of its target (CSD) was fused to the pGreen II 0800-LUC vector. The primers listed in Supplementary Table S1 were designed using Snapgene 3.2.1. The design process considered the precursor sequences of miR398, as well as the sequences of its target. Additionally, sequences of the pGreen II 62 SK and pGreen II 0800-LUC vectors were considered during the primer design process. After the protoplast of *Arabidopsis thaliana* was extracted [44], the luciferase signal was identified using the Bio-Lite TM luciferase system kit (Vazyme, Nanjing, China) 12 h later. Each luciferase signal was tested using five independent replicates.

3. Results

3.1. *Identification and Characterization of Conserved and Species-Specific miRNAs in A. venetum*

To identify the miRNAs within *A. venetum* and examine their responses to MeJA, we constructed six small RNA libraries, including three MeJA-treated groups and three control groups, followed by high-throughput sequencing. Each library yielded approximately 20.39 million raw reads. After eliminating adaptors, sequences of low quality, and sequences shorter than 18 nucleotides, every library yielded a range of 9.73 to 16.10 million clean reads. These clean reads had lengths varying from 19 to 24 nucleotides and high sequence quality (Supplementary Table S2).

A total of 135 conserved miRNAs were identified in *A. venetum*, belonging to 26 conserved miRNA families (Supplementary Table S3). The family with the highest number of miRNAs was miR166, which consisted of 25 members. The miR396 family had 13 members, the miR319 family had 12 members, and the miR171, miR167, and miR399 families had 10, 8, and 8 members, respectively. Five of the miRNA families contained only one miRNA member (Figure 1A).

All the miRNAs predicted by miRBase with more than four mismatches were classified as species-specific. The present investigation identified 26 miRNAs specific to *A. venetum*, utilizing a set of established criteria for discerning miRNA annotations with high and low confidence (Supplementary Table S4). Eight miRNAs were classified as high-confidence miRNAs, namely ave-miRN1, ave-miRN2, ave-miRN6, ave-miRN10, ave-miRN17, ave-miRN18, ave-miRN22, and ave-miRN23, which exhibited stable secondary structures (Figure 1B,C) and high expression levels.

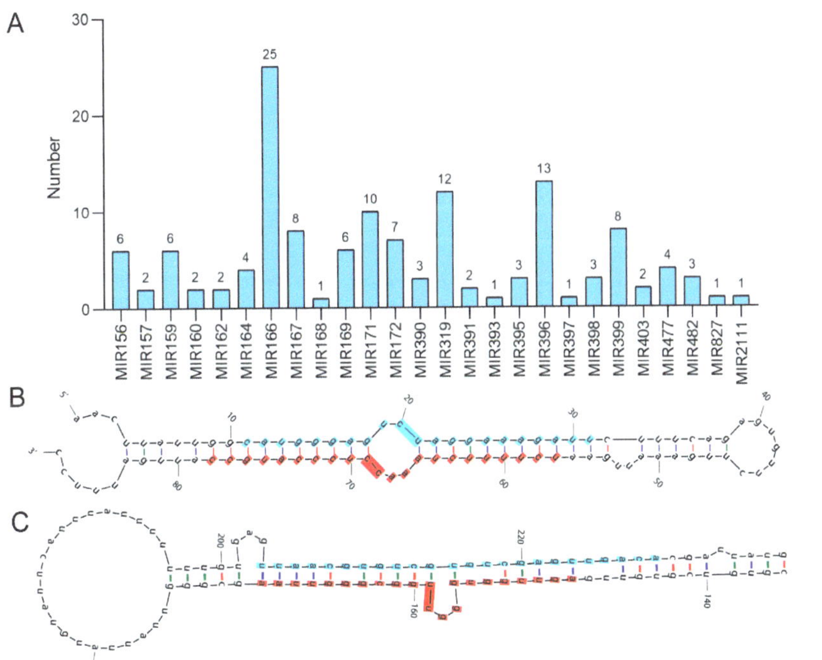

Figure 1. Identification of miRNAs in *A. venetum*. Secondary structures of precursors of two species-specific miRNA for ave-miRN1 (**A**) and ave-miRN21 (**B**). The structure diagrams were generated using the UNAFold web server. 5p strands were labeled in blue, and 3p strands were labeled in red. (**C**) Member quantity of conserved miRNA families in *A. venetum*. miRNA, microRNA.

activity', and 'molecular adaptor activity.' The cellular component terms that exhibited the highest level of enrichment were predominantly associated with 'cellular anatomical entity' and 'protein-containing complex' (Figure 3A). GO enrichment analysis in species-specific miRNAs in *A. venetum* showed that the enriched biological process terms include 'biological regulation', 'developmental process', 'pigmentation', 'carbon utilization', and 'sulfur utilization', and enriched GO molecular function terms include 'catalytic activity', 'antioxidant activity', 'transporter activity', and 'nutrient reservoir.' The most enriched cellular component terms were the same as those used in the GO enrichment analysis of the conserved miRNAs (Figure 3C).

The metabolic pathways enriched in the miRNA target genes were identified using KEGG analysis. The enriched metabolic pathways in the conserved miRNAs were related to the MAPK signaling pathway, ABC transporters, biosynthesis of secondary metabolites, and plant–pathogen interactions (Figure 3B). The enriched metabolic pathways in species-specific miRNAs in *A. venetum* were associated with ABC transporters, flavonoid biosynthesis, phenylalanine, tyrosine, tryptophan biosynthesis, anthocyanin biosynthesis, and betalain biosynthesis (Figure 3D).

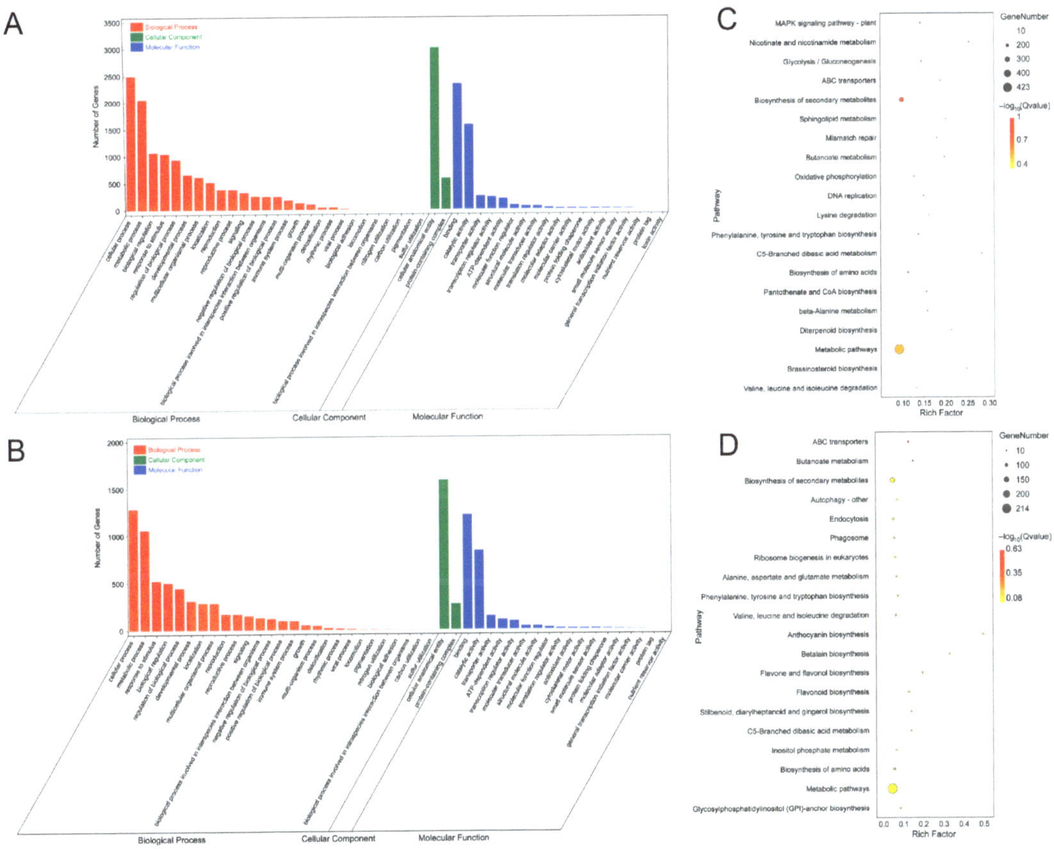

Figure 3. GO and KEGG enrichment analysis of the targets of conserved and species-specific miRNAs in *A. venetum*. (**A**,**C**) Display GO and KEGG enrichment analysis for target genes of conserved miRNA in *A. venetum*. (**B**,**D**) Depict GO and KEGG enrichment analysis for target genes of species-specific miRNA in *A. venetum*. GO, Gene Ontology; KEGG, Kyoto Encyclopedia of Genes and Genomics; miRNA, microRNA.

3.4. Identification of miRNAs Responding to MeJA Application in A. venetum

Twenty-three DEMs were identified in *A. venetum* by comparing the expression profiles of the miRNAs in the CK and MeJA groups (Figure 4A). Among them, 16 miRNAs were upregulated by MeJA treatment, including six conserved miRNAs and 10 species-specific miRNAs. Among the conserved miRNAs induced by MeJA, the upregulation of ave-miR171e-3p was the highest, since it was 3.18 times higher than that in the CK group. Among the species-specific miRNAs induced by MeJA, the upregulation of ave-miRN17-5p was the highest, since it was 3.55 times higher than that of the CK group (Figure 4B). In response to MeJA treatment, seven miRNAs in *A. venetum* were found to be downregulated. Among these, five miRNAs were conserved, while the remaining two were specific to *A. venetum* (Figure 4B).

Ten DEMs, including five conserved and five species-specific miRNAs, were randomly selected for qRT-PCR verification. The changes in miRNA expression under MeJA treatment, estimated based on small RNA sequencing data, were generally consistent with the results of qRT-PCR (Figure 4C), indicating that the results of the miRNA differential expression analysis in this study are reliable.

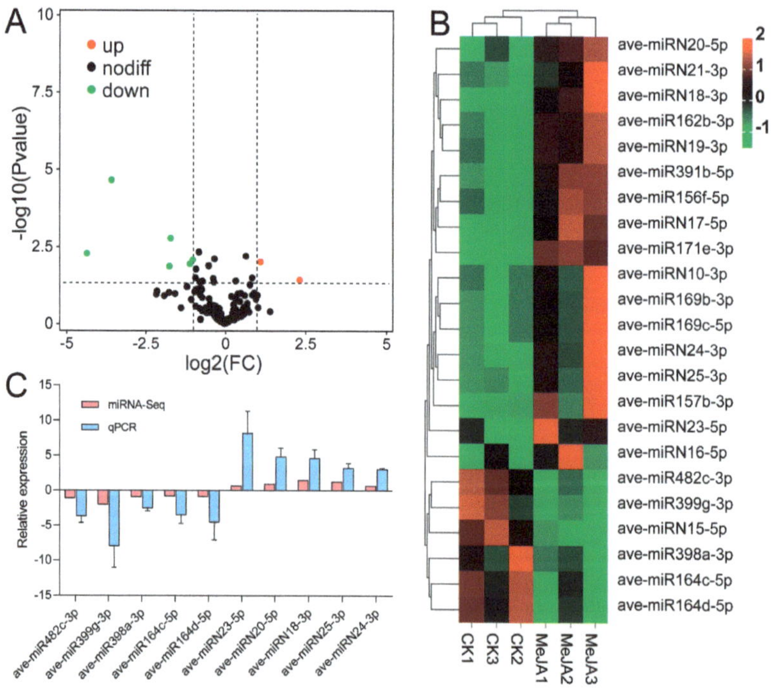

Figure 4. Differential expression analysis of identified miRNAs in *A. venetum* under MeJA treatment. (**A**) Volcano map of DEMs; red indicates upregulated genes, green indicates downregulated genes, and black indicates unchanged genes. (**B**) The expression of conserved and species-specific DEMs of *A. venetum* in response to MeJA. (**C**) qRT-PCR verification of MeJA-responsive miRNAs of *A. venetum*. The Y-axis shows the relative expression levels of conserved miRNAs and species-specific miRNAs. The numerical value is expressed by the average ± SD (n = 3). miRNA, microRNA; MeJA, methyl jasmonate; DEMs, differentially expressed miRNAs.

3.5. Promoter Analysis of MeJA-Responsive miRNAs

To understand the potential expression characteristics of the MeJA-responsive miRNAs in *A. venetum*, promoter analysis software was used to predict the presence and distribution of cis-acting elements in the promoter regions of these miRNAs. A large number of cis-acting elements are involved in the response to hormones and abiotic stress in the promoter regions of these miRNAs (Figure 5, Supplementary Table S7). Among them, ABRE is closely linked to the response of ABA, CGTCA and TGACG motifs play roles in the response to MeJA, GARE motif participates in GA response, MBS is related to drought response, LTR is associated with low-temperature response, and TC- rich repeats are involved in plant defense.

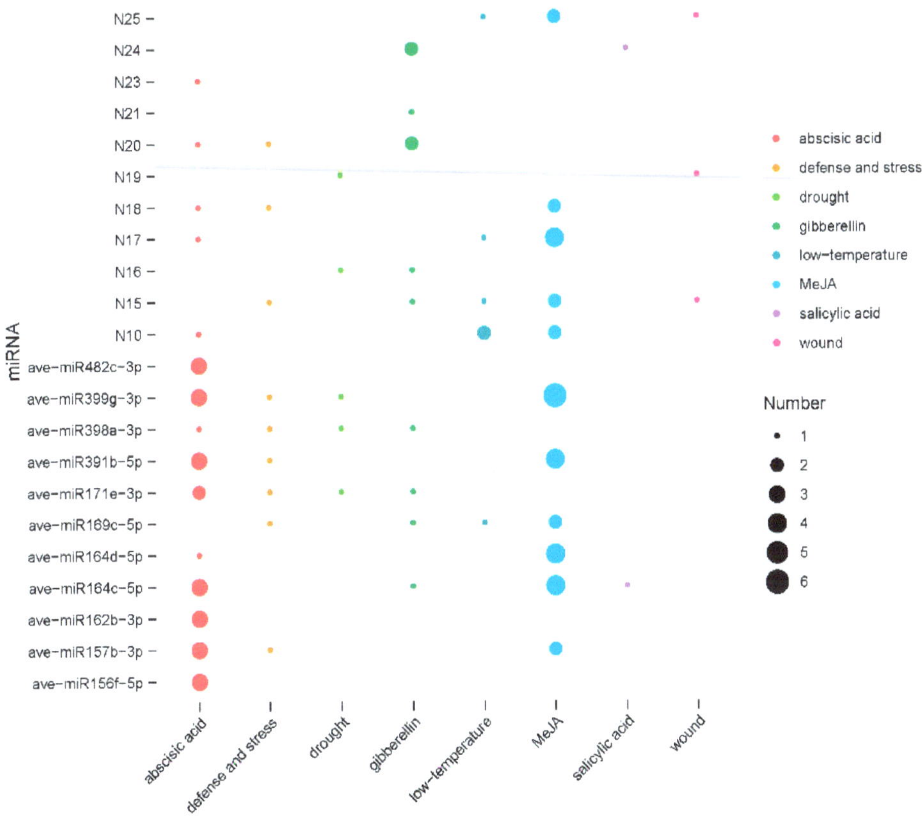

Figure 5. Promoter analysis of MeJA-responsive miRNAs in *A. venetum*. The X-axis represents the stress conditions and hormones that miRNAs may be related to, the Y-axis represents the DEMs, the circle color represents the stress conditions and hormones involved, and the size of the dots represents the degree of involvement. miRNA, microRNA; MeJA, methyl jasmonate; DEMs, differentially expressed miRNAs.

Not surprisingly, there are multiple copies of cis-acting elements involved in the MeJA response in the promoter regions of nearly half of the MeJA-responsive miRNAs. These miRNAs are both conserved and species-specific. In the promoter region of over 65% of miRNAs, cis-acting elements are involved in the ABA response, and the copy number of cis-acting elements carried by the promoter of the conserved miRNA is significantly higher than that of species-specific MeJA-responsive miRNAs.

3.6. Functional Analysis of Targets of MeJA-Responsive miRNAs in A. venetum

A total of 5664 gens (involving 5010 unique genes) were targeted by 23 MeJA-responsive miRNAs, including 2731 targets of conserved DEMs and 2405 targets of species-specific DEMs

GO enrichment analysis can help researchers understand the functions and biological processes in which a group of genes is enriched. The possible functions of these target genes in response to MeJA in *A. venetum* were clarified using GO enrichment analysis. GO enrichment analysis, in the conserved DEMs, showed that the response to MeJA in *A. venetum* was associated with biological processes, including 'metabolic process', 'response to stimulus', 'growth', and 'nitrogen utilization', and enriched GO molecular function terms were 'transporter activity', 'transcription regulator activity', 'antioxidant activity', and 'molecular adaptor activity'. Additionally, 'cellular anatomical entity', and 'protein-containing complex' were the two most enriched cellular component GO terms (Figure 6A). GO enrichment analysis in species-specific DEMs showed that the response to MeJA in *A. venetum* was associated with biological processes, including 'biological regulation', 'developmental process', 'pigmentation', 'carbon utilization', and 'sulfur utilization', and enriched GO molecular function terms were 'catalytic activity', 'antioxidant activity', 'transporter activity', and 'nutrient reservoir'. The most enriched cellular component terms were the same as those used in the GO enrichment analysis of the conserved miRNAs (Figure 6C).

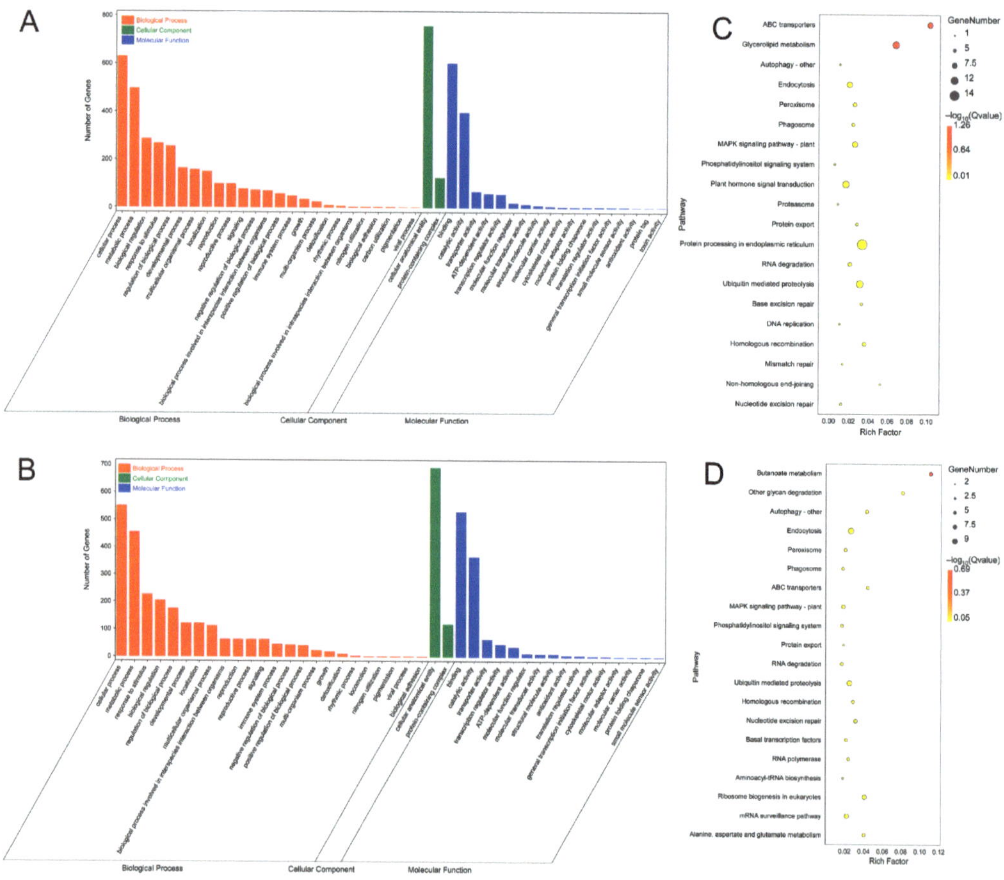

Figure 6. Functional enrichment analysis of the targets of MeJA-responsive miRNAs in *A. venetum*. (**A**,**C**) GO and KEGG enrichment analyses of the target genes of conserved DEMs in *A. venetum*. (**B**,**D**) GO and KEGG enrichment analyses of the target genes of species-specific DEMs in *A. venetum*. miRNA, microRNA; MeJA, methyl jasmonate; DEMs, differentially expressed miRNAs; GO, Gene Ontology; KEGG, Kyoto Encyclopedia of Genes and Genomics.

KEGG is helpful for studying genes and their expression information as a whole. It not only provides the metabolic pathways in which genes may participate, but also provides a comprehensive explanation of the catalytic reactions of enzymes produced by these genes. Metabolic pathways enriched in the target genes of the DEMs were investigated using KEGG annotation. KEGG enrichment analysis of conserved DEMs showed that the response to MeJA in *A. venetum* was related to ABC transporters, biosynthesis of secondary metabolites, oxidative phosphorylation, flavone and flavonol biosynthesis, and plant–pathogen interactions (Figure 6B). KEGG enrichment analysis of species-specific DEMs showed that the response to MeJA in *A. venetum* was related to anthocyanin biosynthesis, ribosome biogenesis in eukaryotes, phenylalanine, tyrosine and tryptophan biosynthesis, photosynthesis, and alanine, aspartate, and glutamate metabolism (Figure 6D).

3.7. Transcriptome Analysis of the Response to MeJA in A. venetum

To investigate changes in gene expression levels in response to MeJA in *A. venetum* leaves, six cDNA libraries were created and sequenced from *A. venetum* leaves cultivated under normal conditions and after MeJA application. In total, 37.08 Gb of sequencing data with a GC content of approximately 44% and a Q30 value of 93.04% were retrieved after low-quality sequences were removed (Supplementary Table S2). The expression levels of each mRNA under the CK and MeJA treatments were compared to identify the DEGs in *A. venetum*. In *A. venetum*, 1778 DEGs were identified, including 825 upregulated and 953 downregulated genes (Figure 7A,B).

To validate the gene expression calculated using the transcriptome data, ten DEGs were randomly chosen, and their expression levels were confirmed by qRT-PCR (Figure 7C). The expression patterns revealed by qRT-PCR analysis agreed with those observed by transcriptome sequencing, indicating the reliability of the transcriptome sequencing results.

For GO functional analysis, DEGs were segregated into 2862 GO categories. These categories encompassed 1881 classifications related to biological processes, 244 classifications associated with cellular components, and 737 classifications related to molecular functions. GO enrichment analysis in DEGs showed that the response to MeJA in *A. venetum* was associated with biological processes, including 'metabolic process', 'response to stimulus', 'developmental process', and 'signaling', and the enriched GO molecular function terms were 'catalytic activity', 'transporter activity', 'antioxidant activity', 'signal transducer activity', and 'transcription regulator activity'. The most enriched cellular component terms were 'cell', 'membrane', 'organelle', and 'extracellular region' (Figure 7E). KEGG enrichment analysis of the DEGs revealed 31 significantly enriched pathways, including plant hormones, signal transduction, porphyrin, secondary metabolism, and photosynthesis (Figure 7D).

3.8. Joint Analysis of miRNA and mRNA Sequencing in A. venetum under MeJA Application

To identify the miRNA–mRNA interaction pairs in *A. venetum* leaves under MeJA treatment, a comprehensive analysis was conducted by integrating datasets from small RNA and transcriptome sequencing. Consequently, 2819 miRNA-target pairs were identified, including 12 MeJA-responsive conserved miRNAs and 1614 targets, as well as 11 MeJA-responsive species-specific miRNAs and 1205 targets in *A. venetum* (Supplementary Tables S8 and S9). These miRNA–mRNA interaction pairs regulate many biological processes involved in the response to MeJA in *A. venetum*, including reactive oxygen species (ROS), homeostasis (miR398-CSD), plant defense (miR482-TIR-NBS-LRR), plant growth and development (miR156-SPL10, miR399-PHO2), and signal transduction (miR164a-5p-NAC056) (Figure 8).

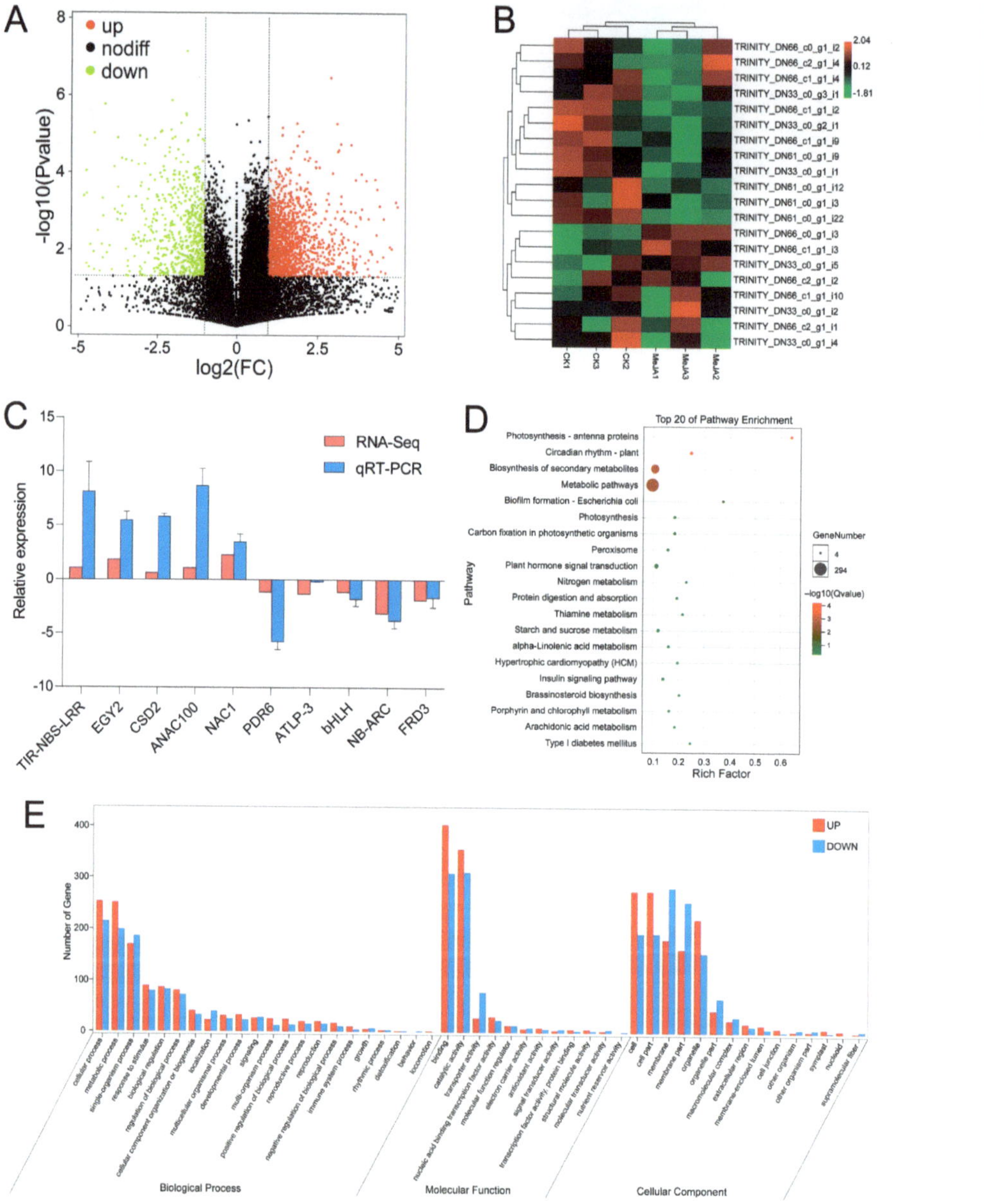

Figure 7. Transcriptomic analysis of the response to MeJA in *A. venetum*. (**A**) Volcano map of DEGs. Red represents upregulated genes, green represents downregulated genes, and black represents unchanged genes. (**B**) Heatmap of the top 20 DEGs. (**C**) qRT-PCR verification. The Y-axis shows the relative expression level of target genes, and *eIF* was used as internal reference. The relative expression level is averaged in three technical replicates. The numerical value is expressed by the average ± SD (n = 3). (**D**) KEGG pathway enrichment analysis of the DEGs. The X-axis represents the enrichment factor, and the Y-axis represents the enriched KEGG pathway. (**E**) GO enrichment analysis of the DEGs. GO, Gene Ontology; KEGG, Kyoto Encyclopedia of Genes and Genomics; MeJA, methyl jasmonate; DEGs, differentially expressed genes.

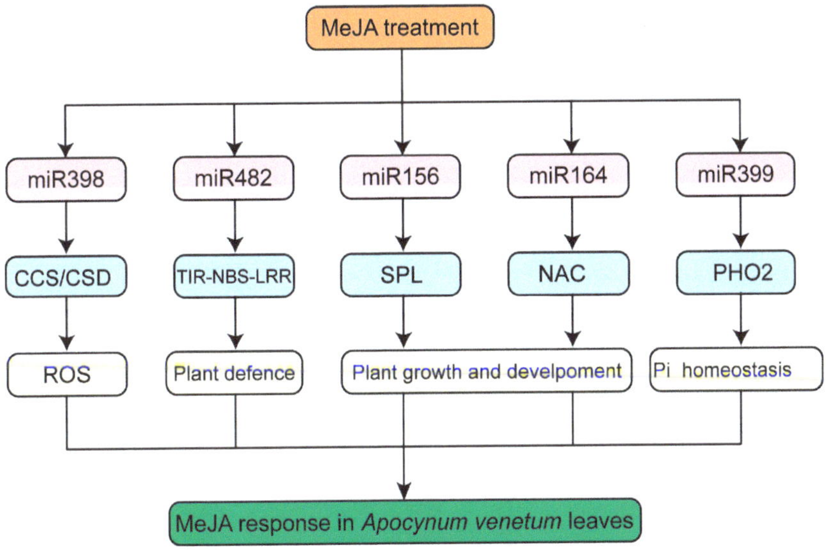

Figure 8. The miRNA-mediated gene regulation network in response to MeJA in *A. venetum* leaves. Red square represents miRNAs, blue square represents targets, yellow square represents biological process. miRNA, microRNA; MeJA, methyl jasmonate; ROS, reactive oxygen species.

3.9. Dual-Luciferase Reporter Analysis Proving the Targeting Relation between miR398 and CSD2

miRNA target prediction indicated that CSD2 (TRINITY_DN9077_c0_g1_i2) was the target of ave-miR398a-3p. The double-luciferase reporter method is a dependable experimental approach for identifying and confirming the specific interactions between miRNAs and their targets. Compared with 62 SK+CSD2 cells, the ratio of luciferase/Renilla in Pre-miR398+CSD2 cells decreased significantly. These results indicated that miR398 negatively regulates CSD expression in *Arabidopsis* protoplasts. qRT-PCR analysis showed that the expression profiles of ave-miR398a-3p and CSD2 (TRINITY_DN9077_c0_g1_i2) exhibited contrasting alterations in *A. venetum* (Figure 7C), which also supported the targeting relation between miR398 and CSD (Figure 9C).

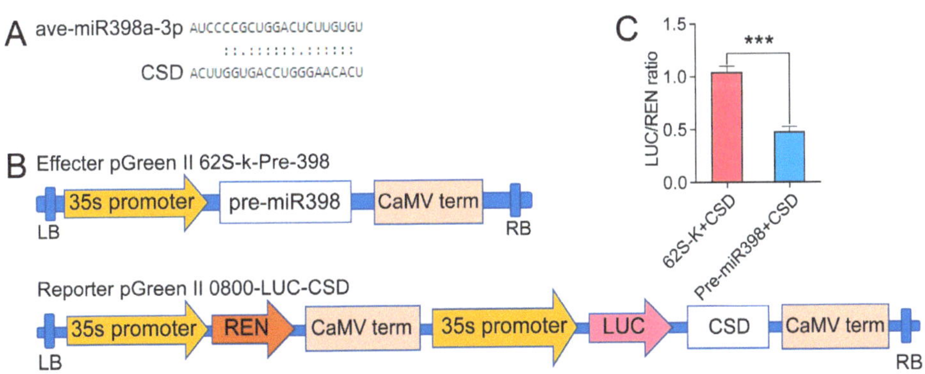

Figure 9. Experimental validation of the targeting relationship between miR398 and CSD. (**A**) The binding site of ave-miR398a-3p on CSD predicted by psRNAtarget. (**B**) Vector construction for double luciferase reporter gene vector assay. (**C**) The ratio of LUC/REN in 62-SK+CSD and ave-miR398a-3p+CSD cells. The numerical value is expressed by mean ± SD ($n = 5$). t test, *** $p < 0.01$. CSD, Cu/Zn-superoxide dismutase; miRNA, microRNA; LUC/REN, luciferase/Renilla.

3.10. MeJA Treatment Improved Tolerance of A. venetum to Oxidative Stress

MV treatment can lead to oxidative stress in plant tissues. After 24 h of MV treatment, all *A. venetum* leaves in the MV group showed significant curling, whereas after treating the leaves of *A. venetum* with MeJA and then MV for 24 h, only a few leaves in the MeJA+MV group showed slight curling (Figure 10A).

Figure 10. MeJA treatment improved the tolerance of *A. venetum* to oxidative stress. (**A**) Physiological changes in the leaves of *A. venetum* after MeJA treatment and MV treatment. (**B**) MDA content in the leaves of *A. venetum* under MeJA treatment and MV treatment. (**C**) REL values in the leaves of *A. venetum* with MeJA treatment and MV treatment. Different letters indicate significant differences between different groups, and the significance of differences was analyzed using one-way analysis of variance and Tukey's multiple range test ($p < 0.05$), $n = 5$. MeJA, methyl jasmonate; MDA, malondialdehyde; REL, relative electrolyte leakage; MV, methyl viologen.

MDA is a lipid peroxide decomposition substance produced by free radicals, which can hinder the normal metabolism of cells and even lead to death. REL indicates the degree of cell membrane damage, and the lower the value, the stronger the membrane stability.

Therefore, REL and MDA content, which are frequently employed as established markers for assessing cellular membrane impairment, can reflect the degree of lipid peroxidation in plants. REL and MDA levels were measured 24 h after MV treatment. The results showed that both the MeJA group and CK group exhibited no significant change in REL and MDA levels. After MV treatment, REL and MDA values in both the MeJA+MV and MV groups increased significantly. However, compared with that in the MV group, the magnitude of the increase in REL and MDA values was significantly smaller in the MeJA+MV group (Figure 10B,C). Collectively, these results indicate that MeJA improved the tolerance of *A. venetum* to MV-induced oxidative stress.

3.11. Tolerance to Oxidative Stress Induced by MeJA Is Related to the Elevation of SOD Enzyme Activity

DAB and NBT staining are effective methods for detecting the accumulation of H_2O_2 and O_2^- in plants. To investigate the effect of MeJA on ROS levels in *A. venetum* leaves, DAB and NBT staining were performed on leaves from the MV and MeJA+MV groups. The staining of *A. venetum* leaves in the MeJA+MV group was lighter than that in the MV group, indicating that MeJA treatment inhibited the MV-induced increase in ROS levels caused by MV (Figure 11A).

MeJA may affect the activity of the SOD enzyme through miR398. Therefore, the SOD enzyme activities in the CK, MeJA, MV, and MeJA+MV groups were detected. SOD enzyme activity in *A. venetum* leaves significantly increased after MeJA treatment (MeJA vs. CK and MeJA+MV vs. MV) (Figure 11B).

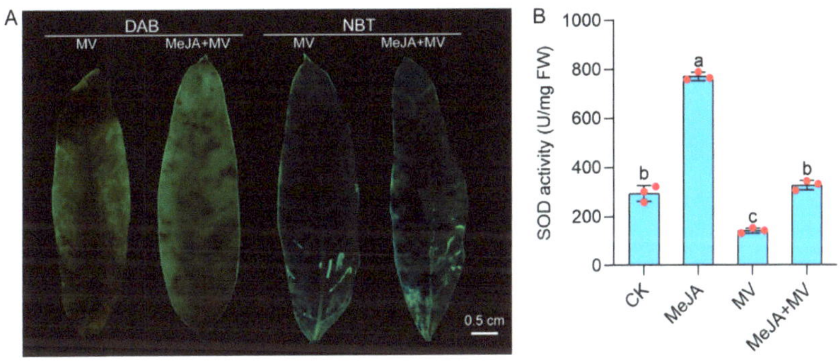

Figure 11. Alteration in ROS accumulation and antioxidant enzyme activities of *A. venetum* seedling leaves under MeJA application. (**A**) DAB staining showed the site of H_2O_2 accumulation in *A. venetum* leaves under MeJA. The darker the brown is, the more H_2O_2 content accumulates. NBT staining showed the site of O_2^- accumulation in *A. venetum* leaves under MeJA. The darker the blue is, the more O_2^- content accumulates. (**B**) The enzyme activities of SOD. Each experiment was performed using three independent biological replicates. Different letters indicate significant differences between different groups, and the significance of differences was analyzed using one-way analysis of variance and Tukey's multiple range test ($p < 0.05$). MeJA, methyl jasmonate; SOD, superoxide dismutase; DAB, 3,3′-diaminobenzidine; ROS, reactive oxygen species.

4. Discussion

A. venetum is an economically, industrially, and ecologically useful crop. MeJA improves environmental stress tolerance in many plants. MiRNAs are a class of endogenous, single-stranded regulatory RNA molecules that may play important regulatory roles in the MeJA response in *A. venetum*. In this study, we conducted a large-scale MeJA-responsive miRNA analysis in *A. venetum*, which contributes to the adaptation of organisms to biological and abiotic stresses. By predicting and annotating miRNAs, we identified conserved and species-specific miRNAs in *A. venetum*. Through the association analysis of small RNA sequencing data and transcriptome sequencing data, high-confidence miRNA target genes were predicted, miRNA–mRNA interaction pairs involved in the MeJA response were obtained, and interaction networks between miRNAs and different target genes were constructed.

In this study, 135 conserved miRNAs belonging to 26 gene families were identified in *A. venetum*. Based on the degree of conservation, these miRNA families can be classified into five groups. Group 1 comprises ubiquitous miRNAs, including miR156, miR166, miR167, miR168, and miR172. Group 2 miRNAs are present in most taxonomic groups, including miR159, miR160, miR162, miR164, miR169, miR171, miR319, and miR390. Group 3 miRNAs were poorly enriched in monocots, including miR391, miR482, and miR2111. Group 4 miRNAs were enriched in dicots, including miR403. Group 5 miRNAs [45], including miR395, miR398, miR399, miR408, and miR827, were enriched in angiosperms.

A total of 26 species-specific miRNAs were identified in *A. venetum*, eight of which were identified as high-confidence species-specific miRNAs because they met the criteria for novel miRNA identification [46]. The abundances of the eight high-confidence species-specific miRNAs ranged from hundreds to tens of thousands. Among them, ave-miRN1 was the most abundant (71903 reads), and miRN9 was the least abundant (18 reads). The remaining 18 possible new miRNAs had stable hairpin structures, but their expression levels were insufficient for their identification as highly reliable species-specific miRNAs.

MeJA is a plant hormone that has the potential to combat infections, salt stress, drought stress, low temperatures, and heavy metal stress. The application of MeJA can promote plant growth, induce the accumulation of active substances, and affect physiological and biochemical characteristics, such as endogenous hormone levels. In this study, transcriptome sequencing revealed that 1778 DEGs were involved in the response of *A. venetum* leaves to MeJA. Enrichment analysis of these DEGs revealed that they were mainly involved in biological processes, including antioxidants, plant–pathogen interactions, and flavonoid biosynthesis.

There were 24 DEGs encoding different antioxidant families in *A. venetum* leaves (Supplementary Table S10). Many genes encode antioxidant enzymes, such as copper/zinc SOD, ascorbic acid oxidase, glutathione peroxidase (GSH-Px), and catalase (CAT). Seven of the eleven differentially expressed SOD genes, four of the five differentially expressed GSH-Px genes, and five of the eight differentially expressed CAT genes were upregulated by MeJA treatment. These results indicate that antioxidant enzyme genes play important roles in the response of *A. venetum* to MeJA, possibly protecting cells from oxidative damage caused by ROS by clearing ROS.

Plant–pathogen interactions also play an important role in the MeJA response of *A. venetum*. In this study, 131 DEGs were involved in plant–pathogen interactions, and 63% of these genes were upregulated (Supplementary Table S10). Several pathogenic genes involved in the MeJA response, such as PR1, RPS2, and CC-NBS-LRR, were upregulated.

It is worth noting that this study identified 16 DEGs encoding key enzymes in flavonoid biosynthesis, including ANS, DFR, CHS, C4H, HCT, FLS, CCoAOMT1, F3H, and F3′H. The differential expression of these genes may lead to changes in flavonoid biosynthesis, and changes in the concentrations of these flavonoid substances may further affect the tolerance of *A. venetum* to abiotic stress. A related study has indicated that MeJA stimulates changes in the concentration of flavonoids in pear callus tissue [47].

In the present study, a batch of DEMs belonging to the miR156, miR164, miR399, and miR482 families were found to respond to MeJA in *A. venetum*. These DEMs may regulate the biosynthesis and physiological metabolism of *A. venetum* in response to MeJA by forming a regulatory network with DEGs identified by transcriptome sequencing.

The functional diversity of the SPL family of transcription factors allows them to regulate several essential elements of plant growth and development, such as branching [48], leaf initiation rate, blooming time [49], and vegetative phase shift [50]. It has been proven that miR156 regulates these functions by inhibiting the expression of SPL in *A. thaliana* [51]. Six miRNA members of the ave-MIR156 family have been identified in *A. venetum*, the most abundant of which is ave-miR156f. In total, 179 genes were targeted by ave-miR156f, including the SPL transcription factor genes. Furthermore, ave-miR156f was significantly upregulated following MeJA treatment. A combined analysis of small RNA sequencing and RNA sequencing data revealed a negative correlation between the expression levels of ave-miR156f and some of its predicted targets, including TRINITY_DN807_c0_g2_i15-SPL10, TRINITY_DN807_c0_g2_i26-SPL2, TRINITY_DN6812_c0_g1_i2-SPL4, and TRINITY_DN6812_c0_g1_i2-SPL9. Studies have shown that MeJA application induces the expression of SPL4, SPL5, SPL7, and SPL15 in alfalfa [52]. Our results suggest that miR156 mediates the regulation of SPL transcription factors by MeJA in *A. venetum*.

PHO2, which encodes the ubiquitin-binding E2 enzyme 24, regulates the degradation of PHO1 in plant membranes to maintain Pi homeostasis [53]. miR399 affects organic phosphate (Pi) homeostasis by regulating PHO2 expression. The miR399-PHO2 module regulates Pi homeostasis in *Brassica napus* [54] and flowering time in response to varying ambient temperatures in *Arabidopsis thaliana* [55]. In this study, eight miRNA members of the ave-MIR399 family were identified. Ave-miR399g was the most abundant member of the miR399 family. Up to 154 genes were predicted as targets of ave-miR399g. One of the ave-miR399g targets, TRINITY_DN4408_c0_g1_i9, encodes PHO2. The expression level of ave-miR399g was downregulated after MeJA application, whereas the PHO2 expression level displayed opposite patterns, as revealed by RNA-seq. The opposite relation was validated using qRT-PCR. Studies proposed that JA affects the Pi activation response in *B. napus* [56]. Our study provides new evidence for understanding the relation between the MeJA signaling pathway and Pi homeostasis in *A. venetum*.

Studies have shown that miR164 targets the NAC family of transcription factors, which is highly conserved in a wide range of plant species and is involved in the growth of lateral roots in *A. thaliana* [57], negative regulation of drought tolerance in rice [58], and the ripening of kiwifruit [59]. In this study, four miRNA members of the ave-MIR164 family were identified using short RNA sequencing. Ave-miR164c was the most abundant member of the ave-MIR164 family. Up to 157 mRNAs were predicted to be ave-miR164c targets using the psRNA target. TRINITY_DN8710_c1_g1_i3, an anave-miR164c target, encodes a transcription factor belonging to the NAC family. The expression level of NAC showed opposite expression patterns in small RNA sequencing compared with the expression level of ave-miR164c. qRT-PCR was performed to validate the inverse expression relationship between TRINITY_DN8710_c1_g1_i3 and miR164c. A related study showed that MeJA treatment induces the expression of NAC2/3 in kiwifruit [60].

The high-throughput sequencing data from this study showed that, after MeJA treatment, the expression of miRNA482 was downregulated in *A. venetum* leaves, whereas the expression of some miRNA482-targeted TIR-NBS-LRR genes increased. Furthermore, qRT-PCR analysis confirmed the relationship between miRNA482 and its target genes in *A. venetum*. Diseases and pests are the main biological stresses affecting the growth of *A. venetum* [3], including anthrax caused by *Gloeosporium apocyni* [61], various pests and diseases, including leaf spot disease caused by *Phyllostachys oleifera* leaf spot mold [62,63], which seriously affect the cultivation of *A. venetum*. Plant TIR-NBS-LRR genes play important roles in plant defense [64]. Consistent with our findings, it has been reported that the TIR-NBS-LRR defense gene is involved in regulating biological stress by targeting miR482 in alfalfa [65], tomatoes [66], and cotton [67]. These data indicate that in *A. venetum*, MeJA treatment can regulate the disease resistance gene TIR-NBS-LRR through miR482, thereby affecting the defense of *A. venetum* against pests and diseases.

Under abiotic stress conditions, ROS accumulation in plant cells leads to severe oxidative damage, ultimately resulting in cell death. Increasing evidence has shown that plant defense and stress response pathways are significantly influenced by ROS [68]. Plants have developed effective antioxidant systems, including SOD, ascorbic acid peroxidase, CAT, GSH-Px, monodehydroascorbate reductase, dehydroascorbate reductase, glutathione reductase, glutathione S-transferase, and peroxiredoxin, which are distributed in different parts of plant cells to maintain redox homeostasis [69].

MV treatment causes significant oxidative stress in plants. In this study, MV treatment resulted in the curling of *A. venetum* leaves, with significantly increased levels of REL and MDA. However, the tolerance of *A. venetum* treated with MeJA to MV-induced oxidative stress caused by MV treatment was significantly enhanced (Figure 10). Compared with untreated *A. venetum* leaves, the increase in ROS levels in *A. venetum* leaves pretreated with MeJA was smaller (Figure 11A), indicating that MeJA treatment can inhibit the excessive accumulation of ROS caused by oxidative stress. The small RNA sequencing and RNA-seq data in this study indicated that MeJA treatment led to changes in the expression levels of some miRNAs and many genes, with the expression level of miR398 being downregulated and the target gene CSD2 being upregulated. These alterations were further validated using qRT-PCR and SOD enzyme activity analyses. Similar studies showed that miR398 participated in controlling abiotic stress by targeting CSD1/CSD2 in rice [70], grapevine [71], and tomatoes [72]. Therefore, we speculate that MeJA treatment could increase the expression of CSD2 and further enhance the activity of SOD through miR398 in *A. venetum*, thereby enhancing the tolerance of *A. venetum* plants to various environmental stresses.

5. Conclusions

A. venetum is an economically important plant, and jasmonic acid is a key plant growth regulator. Understanding the MeJA response of *A. venetum* is important. A total of 135 conserved miRNAs and 26 species-specific miRNAs were identified using small RNA sequencing. Transcriptome analysis showed that 1778 genes were differentially expressed under MeJA treatment, of which 825 were upregulated and 953 were downregulated. Transcriptome and qRT-PCR analyses revealed a broad negative regulatory relationship between MeJA-responsive miRNAs and most of their target genes. A total of 2819 miRNA-target pairs were identified through combined analyses of RNA-seq and small RNA, and it was found that they regulate a variety of biological processes, including ROS homeostasis (miR398-CSD2), plant defense (miR482-TIR-NBS-LRR), and plant growth and development (miR156-SPL10, miR164a-5p-NAC056/NAC2). These data contribute to our understanding of the regulatory network of gene expression in the MeJA response of plants and advance our understanding of the biological functions of plant miRNAs. In the next step of our research, we will further analyze the biological functions of these MeJA-responsive miRNAs using transgenic systems in *A. thaliana* or *A. venetum*.

Supplementary Materials: The following supporting information can be downloaded at: https://www.mdpi.com/article/10.3390/horticulturae10020173/s1, Table S1: Primers used in this study; Table S2: Summary of transcriptome and small RNA sequencing data; Table S3: The conserved miRNAs identified in *A. venetum*; Table S4: The species-specific miRNAs identified in *A. venetum*; Table S5: The predicted targets of the conserved miRNA in *A. venetum*; Table S6: The predicted targets of the species-specific miRNA in *A. venetum*; Table S7: Predicted cis-acting elements in the promoters of MeJA-responsive miRNAs in *A. venetum*; Table S8: Targets of the MeJA-responsive conserved miRNA in *A. venetum*; Table S9: Targets of the MeJA-responsive species-specific miRNA in *A. venetum*; Table S10: DEGs involved in flavonoid synthesis, ROS scavenging, and plant–pathogen interactions in *A. venetum*.

Author Contributions: Conceptualization, Y.Z. and F.G.; formal analysis, J.T., X.H., Q.L., T.D., H.S. and F.G.; funding acquisition, X.H., Y.Z. and F.G.; investigation, J.T., Q.L. and H.S.; supervision, Y.Z.; writing—original draft, J.T., H.S. and F.G.; writing—review and editing, F.G. All authors have read and agreed to the published version of the manuscript.

Funding: This work was supported by the Scientific and Technological Research Projects of Colleges and Universities of Hebei Province (grant number ZD2020111), S&T Program of Hebei (grant number 22326418D), Hebei Provincial Administration of Traditional Chinese Medicine (grant number 2023355 and 2023351), the National Natural Science Foundation of China (grant numbers 31670335 and 31770363), and the Industrial Innovation Team Program of Traditional Chinese Medicine of Modern Agricultural Industry Technology System of Hebei Province (grant number HBCT2018060205).

Data Availability Statement: The raw data of transcriptome and small RNA sequencing were deposited in the GenBank Short Read Archive (SRA) database under accession number PRJNA961898.

Conflicts of Interest: The authors declare no conflicts of interest.

References

1. Xie, W.; Zhang, X.; Wang, T.; Hu, J. Botany, traditional uses, phytochemistry and pharmacology of *Apocynum venetum* L. (Luobuma): A review. *J. Ethnopharmacol.* **2012**, *141*, 1–8. [CrossRef]
2. Lu, C.; Zhang, W.; Peng, X.; Gu, G.; Chen, M.; Tang, Z. Development of randomly amplified polymorphic DNA-Sequence characterized amplified region marker for identification of *Apocynum venetum* LINN. from *A. pictum* SCHRENK. *Biol. Pharm. Bull.* **2010**, *33*, 522–526. [CrossRef]
3. Thevs, N.; Zerbe, S.; Kyosev, Y.; Rozi, A.; Tang, B.; Abdusalih, N.; Novitskij, Z. *Apocynum venetum* L. and *Apocynum pictum* Schrenk (Apocynaceae) as multi-functional and multi-service plant species in central asia: A review on biology, ecology, andutilization. *J. Appl. Bot. Food Qual.* **2012**, *85*, 159–167. Available online: https://hdl.handle.net/10863/178 (accessed on 2 December 2023).
4. Yu, X.; Zhang, W.; Zhang, Y.; Zhang, X.; Lang, D.; Zhang, X. The roles of methyl jasmonate to stress in plants. *Funct. Plant Biol.* **2018**, *46*, 197–212. [CrossRef] [PubMed]
5. Wasternack, C.; Feussner, I. The oxylipin pathways: Biochemistry and function. *Annu. Rev. Plant Biol.* **2018**, *69*, 363–386. [CrossRef] [PubMed]
6. Zheng, X.; Chen, Y.; Zhao, L.; Chen, Y.; Zheng, L.; Zheng, K.; Zhang, J. Tripartite interactions between jasmonic/salicylic acid pathways, western flower thrips, and thrips-transmitted tomato zonate spot virus infection in *Capsicuum annuum*. *Arthropod-Plant Interact.* **2019**, *13*, 289–297. [CrossRef]
7. Sun, D.; Lu, X.; Hu, Y.; Li, W.; Hong, K.; Mo, Y.; Xie, J. Methyl jasmonate induced defense responses increase resistance to *Fusarium oxysporum* f. sp. cubense race 4 in banana. *Sci. Hortic.* **2013**, *164*, 484–491. [CrossRef]
8. Karimi, N.; Ghasempour, H.R. Salicylic acid and jasmonic acid restrains nickel toxicity by ameliorating antioxidant defense system in shoots of metallicolous and non-metallicolous *Alyssum inflatum* Náyr. *Populations. Plant Physiol. Biochem.* **2019**, *135*, 450–459. [CrossRef]
9. Qiu, Y.; An, K.; Sun, J.; Chen, X.; Gong, X.; Ma, L.; Wang, Y. Investigating the effect of methyl jasmonate and melatonin on resistance of *Malus crabapple* 'Hong Jiu'to ozone stress. *Environ. Sci. Pollut. Res.* **2019**, *26*, 27761–27768. [CrossRef]
10. Kasschau, K.D.; Xie, Z.; Allen, E.; Llave, C.; Chapman, E.J.; Krizan, K.A.; Carrington, J.C. P1/HC-Pro. a viral suppressor of RNA silencing, interferes with *Arabidopsis* development and miRNA function. *Dev. Cell* **2002**, *4*, 205–217. [CrossRef]
11. Llave, C.; Xie, Z.; Kasschau, K.D.; Carrington, J.C. Cleavage of scarecrow-like mRNA targets directed by a class of *Arabidopsis* miRNA. *Science* **2002**, *297*, 2053–2056. [CrossRef]
12. Song, X.; Li, Y.; Cao, X.; Qi, Y. MicroRNAs and their regulatory roles in plant-environment interactions. *Annu. Rev. Plant Biol.* **2019**, *70*, 489–525. [CrossRef]
13. Zhu, M.; Wang, X.; Zhou, Y.; Tan, J.; Zhou, Y.; Gao, F. Small RNA sequencing revealed that miR4415, a legume-specific miRNA, was involved in the cold acclimation of *Ammopiptanthus nanus* by targeting an L-ascorbate oxidase gene and regulating the redox state of apoplast. *Front. Genet.* **2022**, *13*, 870446. [CrossRef]
14. Stief, A.; Altmann, S.; Hoffmann, K.; Pant, B.D.; Scheible, W.R.; Baurle, I. *Arabidopsis* mir156 regulates tolerance to recurring environmental stress through spl transcription factors. *Plant Cell* **2014**, *26*, 1792. [CrossRef]
15. Williams, L. Regulation of *Arabidopsis* shoot apical meristem and lateral organ formation by microRNA miR166g and its AtHD-ZIP target genes. *Development* **2005**, *132*, 3657–3668. [CrossRef]
16. Val-Torregrosa, B.; Bundó, M.; San Segundo, B. Crosstalk between nutrient signaling pathways and immune responses in rice. *Agriculture* **2021**, *11*, 747. [CrossRef]
17. Liao, L.; Xie, B.; Guan, P.; Jiang, N.; Cui, J. New insight into the molecular mechanism of miR482/2118 during plant resistance to pathogens. *Front. Plant. Sci.* **2022**, *13*, 1026762. [CrossRef]
18. Sumbur, B.; Gao, F.; Liu, Q.; Feng, D.; Bing, J.; Dorjee, T.; Zhou, Y. The Characterization of R2R3-MYB Genes in *Ammopiptanthus nanus* uncovers that the miR858-AnaMYB87 Module mediates the accumulation of anthocyanin under osmotic stress. *Biomolecules* **2023**, *13*, 1721. [CrossRef]
19. Sajad, S.; Dai, Q.; Yang, J.; Song, J. Identification of miRNAs involved in male fertility and pollen development in *Brassica oleracea* var. *capitata* L. by High-Throughput Sequencing. *Horticulturae* **2023**, *9*, 515. [CrossRef]
20. Zheng, L.; Wu, W.; Gao, Y.; Wu, Y.; Xu, Y.; Zhang, G.; Zhou, Y. Integrated transcriptome, small RNA and degradome analysis provide insights into the transcriptional regulatory networks underlying cold acclimation in jojoba. *Sci. Hortic.* **2022**, *299*, 111050. [CrossRef]
21. Lai, R.; Guan, Q.; Shen, C.; Feng, X.; Zhang, Y.; Chen, Y.; Wu, R. Integrated sRNA-seq and RNA-seq analysis reveals the regulatory roles of miRNAs in the low-temperature responses of *Canarium album*. *Horticulturae* **2022**, *8*, 667. [CrossRef]
22. Cao, D.; Li, J.; Ma, L.; Liu, Y.; Huang, J.; Jin, X. Genome-Wide identification of selenium-responsive microRNAs in tea Plant (*Camellia sinensis LO Kuntze*). *Horticulturae* **2023**, *9*, 1278. [CrossRef]
23. Lu, J.; Mao, X.; Xu, Y.; Liu, S.; Wang, L. MicroRNA identification and integrated network analyses for age-dependent flavonoid biosynthesis in *Ginkgo biloba*. *Forests* **2023**, *14*, 1706. [CrossRef]

24. Yang, G.; Rhodes, D.; Joly, R.J. Effects of high temperature on membrane stability and chlorophyll fluorescence in glycinebetaine-deficient and glycinebetaine-containing maize lines. *Funct. Plant Biol.* **1996**, *23*, 437–443. [CrossRef]
25. Yasar, F.; Ellialtioglu, S.; Yildiz, K. Effect of salt stress on antioxidant defense systems, lipid peroxidation, and chlorophyll content in green bean. *Russ. J. Plant Physiol.* **2008**, *55*, 782–786. [CrossRef]
26. Zhang, X.; Wang, L.; Meng, H.; Wen, H.; Fan, Y.; Zhao, J. Maize ABP9 enhances tolerance to multiple stresses in transgenic *Arabidopsis* by modulating ABA signaling and cellular levels of reactive oxygen species. *Plant Mol. Biol.* **2022**, *75*, 365–378. [CrossRef]
27. Chao, J.; Kong, Y.; Wang, Q.; Sun, Y.; Gong, D.; Lv, J.; Liu, G. MapGene2Chrom, a tool to draw gene physical map based on Perl and SVG languages. *Yi Chuan Hered.* **2015**, *37*, 91–97. [CrossRef]
28. Martin, M. Cutadapt removes adapter sequences from high-throughput sequencing reads. *EMBnet. J.* **2011**, *17*, 10–12. [CrossRef]
29. Kim, D.; Langmead, B.; Salzberg, S.L. HISAT: A fast spliced aligner with low memory requirements. *Nat. Methods* **2015**, *12*, 357–360. [CrossRef]
30. Pertea, M.; Pertea, G.M.; Antonescu, C.M.; Chang, T.C.; Mendell, J.T.; Salzberg, S.L. StringTie enables improved reconstruction of a transcriptome from RNA-seq reads. *Nat. Biotechnol.* **2015**, *33*, 290–295. [CrossRef]
31. Hyatt, D.; Chen, G.L.; LoCascio, P.F.; Land, M.L.; Larimer, F.W.; Hauser, L.J. Integrated nr database in protein annotation system and its localization. *Nat. Commun.* **2010**, *6*, 1–8.
32. Apweiler, R.; Bairoch, A.; Wu, C.H.; Barker, W.C.; Boeckmann, B.; Ferro, S.; Yeh, L.S.L. UniProt: The universal protein knowledgebase. *Nucleic Acids Res.* **2004**, *32*, D115–D119. [CrossRef]
33. Kanehisa, M.; Araki, M.; Goto, S.; Hattori, M.; Hirakawa, M.; Itoh, M.; Yamanishi, Y. KEGG for linking genomes to life and the environment. *Nucleic Acids Res.* **2007**, *36*, D480–D484. [CrossRef]
34. Tatusov, R.L.; Galperin, M.Y.; Natale, D.A.; Koonin, E.V. The COG database: A tool for genome-scale analysis of protein functions and evolution. *Nucleic Acids Res.* **2000**, *28*, 33–36. [CrossRef] [PubMed]
35. Ashburner, M.; Ball, C.A.; Blake, J.A.; Botstein, D.; Butler, H.; Cherry, J.M.; Sherlock, G. Gene ontology: Tool for the unification of biology. *Nat. Genet.* **2000**, *25*, 25–29. [CrossRef] [PubMed]
36. Love, M.I.; Huber, W.; Anders, S. Moderated estimation of fold change and dispersion for RNA-seq data with DESeq2. *Genome Biol.* **2014**, *15*, 550. [CrossRef] [PubMed]
37. Bolger, A.M.; Lohse, M.; Usadel, B. Trimmomatic: A flexible trimmer for Illumina sequence data. *Bioinformatics* **2014**, *30*, 2114–2120. [CrossRef] [PubMed]
38. Friedländer, M.R.; Mackowiak, S.D.; Li, N.; Chen, W.; Rajewsky, N. miRDeep2 accurately identifies known and hundreds of novel microRNA genes in seven animal clades. *Nucleic Acids Res.* **2012**, *40*, 37–52. [CrossRef]
39. An, J.; Lai, J.; Sajjanhar, A.; Lehman, M.L.; Nelson, C.C. miRPlant: An integrated tool for identification of plant miRNA from RNA sequencing data. *BMC Bioinform.* **2014**, *15*, 275. [CrossRef]
40. Kozomara, A.; Griffiths-Jones, S. miRBase: Annotating high confidence microRNAs using deep sequencing data. *Nucleic Acids Res.* **2014**, *42*, D68–D73. [CrossRef]
41. Li, W.; Zhao, F.A.; Fang, W.; Xie, D.; Hou, J.; Yang, X.; Lv, S. Identification of early salt stress responsive proteins in seedling roots of upland cotton (*Gossypium hirsutum* L.) employing iTRAQ-based proteomic technique. *Front. Plant Sci.* **2015**, *6*, 732. [CrossRef]
42. Higo, K.; Ugawa, Y.; Iwamoto, M.; Higo, H. PLACE: A database of plant cis-acting regulatory DNA elements. *Nucleic Acids Res.* **1998**, *26*, 358–359. [CrossRef]
43. Clément, T.; Salone, V.; Rederstorff, M. Dual luciferase gene reporter assays to study miRNA function. *Methods Mol Biol.* **2015**, *1296*, 187–198.
44. Yoo, S.D.; Cho, Y.H.; Sheen, J. *Arabidopsis* mesophyll protoplasts: A versatile cell system for transient gene expression analysis. *Nat. Protoc.* **2007**, *2*, 1565–1572. [CrossRef]
45. Chavez, M.R.A.; Rosas-Cárdenas, D.F.F.; De Paoli, E.; Accerbi, M.; Rymarquis, L.A.; Mahalingam, G.; De Folter, S. Sample sequencing of vascular plants demonstrates widespread conservation and divergence of microRNAs. *Nat. Commun.* **2014**, *5*, 3722. [CrossRef]
46. Wang, X.; Zhang, J.; Li, F.; Gu, J.; He, T.; Zhang, X.; Li, Y. MicroRNA identification based on sequence and structure alignment. *Bioinformatics* **2005**, *21*, 3610–3614. [CrossRef] [PubMed]
47. Premathilake, A.T.; Ni, J.; Shen, J.; Bai, S.; Teng, Y. Transcriptome analysis provides new insights into the transcriptional regulation of methyl jasmonate-induced flavonoid biosynthesis in pear calli. *BMC Plant Biol.* **2020**, *20*, 388. [CrossRef]
48. Yant, L.; Mathieu, J.; Dinh, T.T.; Ott, F.; Lanz, C.; Wollmann, H.; Schmid, M. Orchestration of the floral transition and floral development in *Arabidopsis* by the bifunctional transcription factor APETALA2. *Plant Cell* **2010**, *22*, 2156–2170. [CrossRef] [PubMed]
49. Yu, N.; Cai, W.J.; Wang, S.; Shan, C.M.; Wang, L.J.; Chen, X.Y. Temporal control of trichome distribution by microRNA156-targeted SPL genes in *Arabidopsis thaliana*. *Plant Cell* **2010**, *22*, 2322–2335. [CrossRef] [PubMed]
50. Xing, S.; Salinas, M.; Höhmann, S.; Berndtgen, R.; Huijser, P. miR156-targeted and nontargeted SBP-box transcription factors act in concert to secure male fertility in *Arabidopsis*. *Plant Cell* **2010**, *22*, 3935–3950. [CrossRef]
51. Wang, J.W.; Park, M.Y.; Wang, L.J.; Koo, Y.; Chen, X.Y.; Weigel, D.; Poethig, R.S. miRNA control of vegetative phase change in trees. *PLoS Genet.* **2011**, *7*, e1002012. [CrossRef] [PubMed]

52. Wang, Y.; Ruan, Q.; Zhu, X.; Wang, B.; Wei, B.; Wei, X. Identification of *Alfalfa* SPL gene family and expression analysis under biotic and abiotic stresses. *Sci. Rep.* **2023**, *13*, 84. [CrossRef] [PubMed]
53. Lin, S.I.; Chiang, S.F.; Lin, W.Y.; Chen, J.W.; Tseng, C.Y.; Chiou, W.T.J. Regulatory network of microrna399 and pho2 by systemic signaling. *Plant Physiol.* **2008**, *147*, 732–746. [CrossRef] [PubMed]
54. Du, K.; Yang, Y.; Li, J.; Wang, M.; Jiang, J.; Wu, J.; Wang, Y. Functional Analysis of Bna-miR399c-PHO2 Regulatory Module Involved in Phosphorus Stress in *Brassica napus*. *Life* **2023**, *13*, 310. [CrossRef] [PubMed]
55. Kim, W.; Ahn, H.J.; Chiou, T.J.; Ahn, J.H. The role of the miR399-PHO2 module in the regulation of flowering time in response to different ambient temperatures in *Arabidopsis thaliana*. *Mol. Cells* **2011**, *32*, 83–88. [CrossRef]
56. Li, Y.; Yang, X.; Li, X.; Wang, C.; Ding, G.; Xu, F.; Shi, L. Jasmonic acid participating in the systemic regulation of phosphate starvation response in *Brassica napus*. *Plant Soil* **2023**. [CrossRef]
57. Guo, H.S.; Xie, Q.; Fei, J.F.; Chua, N.H. MicroRNA directs mRNA cleavage of the transcription factor NAC1 to downregulate auxin signals for *Arabidopsis* lateral root development. *Plant Cell* **2005**, *17*, 1376–1386. [CrossRef]
58. Fang, Y.; Xie, K.; Xiong, L. Conserved miR164-targeted NAC genes negatively regulate drought resistance in rice. *J. Exp. Bot.* **2014**, *65*, 2119–2135. [CrossRef]
59. Wang, W.Q.; Wang, J.; Wu, Y.Y.; Li, D.W.; Allan, A.C.; Yin, X.R. Genome-wide analysis of coding and non-coding RNA reveals a conserved miR164-NAC regulatory pathway for fruit ripening. *N. Phytol.* **2020**, *225*, 1618–1634. [CrossRef]
60. Wu, Y.Y.; Liu, X.F.; Fu, B.L.; Zhang, Q.Y.; Tong, Y.; Wang, J.; Yin, X.R. Methyl jasmonate enhances ethylene synthesis in kiwifruit by inducing NAC genes that activate ACS1. *J. Agric. Food Chem.* **2020**, *68*, 3267–3276. [CrossRef]
61. Ellis, J.B.; Everhart, B.M. The North American species of Gloeosporium. *J. Mycol.* **1885**, *1*, 109–119. [CrossRef]
62. Martin, G. The Phyllostictas of North America. *J. Mycol.* **1886**, *2*, 13–20. [CrossRef]
63. Horst, R.K. *Field Manual of Diseases on Garden and Greenhouse Flowers*; Springer: Dordrecht, The Netherlands, 2013.
64. González, V.M.; Müller, S.; Baulcombe, D. Evolution of NBS-LRR gene copies among dicot plants and its regulation by members of the miR482/2118 superfamily of miRNAs. *Mol. Plant* **2015**, *8*, 329–331. [CrossRef]
65. Zhai, J.; Jeong, D.H.; De Paoli, E.; Park, S.; Rosen, B.D.; Li, Y.; Meyers, B.C. MicroRNAs as master regulators of the plant NB-LRR defense gene family via the production of phased, trans-acting siRNAs. *Gene Dev.* **2011**, *25*, 2540–2553. [CrossRef]
66. Shivaprasad, P.V.; Chen, H.M.; Patel, K.; Bond, D.M.; Santos, B.A.; Baulcombe, D.C. A microRNA superfamily regulates nucleotide binding site–leucine-rich repeats and other mRNAs. *Plant Cell* **2012**, *24*, 859–874. [CrossRef]
67. Zhu, Q.H.; Fan, L.; Liu, Y.; Xu, H.; Llewellyn, D.; Wilson, I. miR482 regulation of NBS-LRR defense genes during fungal pathogen infection in cotton. *PLoS ONE* **2013**, *8*, e84390. [CrossRef]
68. You, J.; Chan, Z. ROS regulation during abiotic stress responses in crop plants. *Front. Plant Sci.* **2015**, *6*, 1092. [CrossRef]
69. Dreyer, A.; Dietz, K.J. Reactive oxygen species and the redox-regulatory network in cold stress acclimation. *Antioxidants* **2018**, *7*, 169. [CrossRef]
70. Lu, Y.; Feng, Z.; Bian, L.; Xie, H.; Liang, J. miR398 regulation in rice of the responses to abiotic and biotic stresses depends on CSD1 and CSD2 expression. *Funct. Plant Biol.* **2010**, *38*, 44–53. [CrossRef]
71. Leng, X.; Wang, P.; Zhu, X.; Li, X.; Zheng, T.; Shangguan, L.; Fang, J. Ectopic expression of CSD1 and CSD2 targeting genes of miR398 in grapevine is associated with oxidative stress tolerance. *Funct. Integr. Genomics* **2017**, *17*, 697–710. [CrossRef]
72. Yan, G.; Hua, Y.; Jin, H.; Huang, Q.; Zhou, G.; Xu, Y.; Zhu, Z. Sly-miR398 participates in cadmium stress acclimation by regulating antioxidant system and cadmium transport in tomato (*Solanum lycopersicum*). *Int. J. Mol. Sci.* **2023**, *24*, 1953. [CrossRef] [PubMed]

horticulturae

MDPI

Article

Mediator Subunit RhMED15a Regulates Drought Tolerance in Rose

Xiaoman Shang [1,†], Nanxin Xie [1,†], Yalin Li [1], Zixin Zhao [1], Ping Luo [1], Yongyi Cui [1], Xianlong Rao [2,*] and Wen Chen [1,*]

1 Key Laboratory of Quality and Safety Control for Subtropical Fruit and Vegetable, Ministry of Agriculture and Rural Affairs, Collaborative Innovation Center for Efficient and Green Production of Agriculture in Mountainous Areas of Zhejiang Province, College of Horticulture Science, Zhejiang A&F University, Hangzhou 311300, China
2 School of Urban Construction, Zhejiang Shuren University, Hangzhou 310015, China
* Correspondence: 601226@zjsru.edu.cn (X.R.); askcw2005@zafu.edu.cn (W.C.)
† These authors contributed equally to this work.

Abstract: Mediator is a multiprotein complex integral to the transcription machinery, mediated by RNA polymerase II. Some Mediator subunits have been found to have critical functions in plants' responses to abiotic stresses. However, the role of plant Mediator subunits in drought responses remains largely enigmatic. Here, we identified a Mediator subunit, *RhMED15a*, in roses (*Rosa hybrida*). Its expression was greatly and swiftly induced by dehydration treatment in the root. The promoter sequence of *RhMED15a* contains *cis*-acting elements that respond to abscisic acid (ABA) and methyl jasmonate (MeJA). In addition, the expression of *RhMED15a* was significantly up-regulated with ABA treatment and inversely down-regulated with MeJA treatment. Silencing *RhMED15a* using virus-induced gene silencing (VIGS) in roses significantly reduced drought tolerance in rose plants. This resulted in a significant increase in the malondialdehyde (MDA) level and a decreased survival rate in comparison to TRV controls. Moreover, we found that the expression of five drought-related genes, including *dehydration responsive element binding factor 1B* (*DREB1B*), *responsive to desiccation stress 29A* (*RD29A*), *responsive to desiccation stress 29B* (*RD29B*), *early response to dehydration 14* (*ERD14*), and *9-cis-epoxycarotenoid dioxygenase 1* (*NCED1*), was considerably suppressed in *RhMED15a*-silenced plants during drought stress. Taken together, our results present that the Mediator tail module subunit *RhMED15a* serves as an enhancer of drought tolerance in rose, probably through the modulation of the expression of some drought-related genes.

Keywords: *Rosa hybrida*; drought; mediator; RhMED15a; VIGS

Citation: Shang, X.; Xie, N.; Li, Y.; Zhao, Z.; Luo, P.; Cui, Y.; Rao, X.; Chen, W. Mediator Subunit RhMED15a Regulates Drought Tolerance in Rose. *Horticulturae* **2024**, *10*, 84. https://doi.org/10.3390/horticulturae10010084

Academic Editor: Jinzhi Zhang

Received: 19 December 2023
Revised: 8 January 2024
Accepted: 8 January 2024
Published: 15 January 2024

1. Introduction

The rose (*Rosa hybrida*), one of the world's most popular flowers, holds significant global economic value. Nonetheless, rose plants frequently encounter abiotic stresses throughout their lifespan, including drought, salinity, extreme temperatures, and varying nutrient availabilities. Among these, drought is a particularly common stressor worldwide due to climate change and rapid industrialization [1–3]. This factor poses severe limitations to crop productivity and distribution [2]. In rose seedlings, harsh drought stress leads to leaf wilting and curling, diminished photosynthetic performance, nutrient assimilation reduction, and stunted plant growth, culminating in substantial economic loss [2,4].

To counteract drought stress, plants have developed various intricate response strategies. Numerous studies have shown that drought can trigger responses such as proline and trehalose production, antioxidant defense system enhancement for redox homeostasis maintenance, and peroxidase enzyme use to control cellular damage and sustain membrane integrity [2,5–7]. Phytohormones like abscisic acid (ABA) and jasmonic acid (JA) play a role in coordinating such adaptive responses. These hormones are paramount for managing environmental stimuli responses and regulating plant growth and development [2,8,9]. Their synthesis and signal pathways also participate in regulating osmotic adjustment and other processes, heightening drought tolerance in plants [2,9,10]. On the other hand, the intricate signaling transduction pathways that perceive and react to drought conditions often drive changes in gene expression, achieved through transcriptional regulation [11,12]. In eukaryotic cells, this regulation involves an interplay of active agents like RNA polymerase II (Pol II), general transcription factors (TFs), transcriptional repressors or activators, and co-regulators such as Mediator [12].

Mediator, a multi-protein complex, facilitates as a molecular bridge linking DNA-binding TFs to the Pol II enzyme, playing a vital role in numerous stages of transcription, such as initiation, elongation, and splicing [13,14]. The Mediator complex incorporates an estimated 34 subunits in plants, around 31 subunits in mammals, and roughly 25 subunits in yeast [13,15]. In eukaryotes, it comprises three core modules: the head, middle, and tail modules, along with a separable module, cyclin-dependent kinase 8 (CDK8) [13,14,16]. The head module instigates the principal interactions with Pol II, while the middle module performs a structural function, interacting with Pol II when Mediator's conformation changes after the initial interaction. The tail module is significant for interacting with gene-specific TFs [14].

Each Mediator subunit in plants is believed to perform varying functions due to the subunit composition of Mediator diverging throughout evolution and responding to environmental stimuli and tissue-specific inputs [14]. Several Mediator subunits contribute to basic cellular processes, such as cell proliferation and growth, organ growth, developmental timing, and phytohormone responses [14]. Recent genetic analyses have shown that Mediator subunits also participate in plants' responses to diverse abiotic stresses such as salinity, cold, and drought [17,18]. Notably, particular Mediator subunits like MED14, MED16, and MED25 are implicated in multiple abiotic stress responses. MED25 specifically, a vital part of stress-response signaling pathways, influences responses to salt and drought stresses. It was observed that *Arabidopsis med25* showcased sensitivity to salt stress but resistance to drought [19–21]. Genetic analyses suggest that MED25 positively regulates the JA signaling *via* interaction with MYC2, while it negatively regulates ABA signaling and response through interaction with ABI5 [22]. MED14 and MED16 are crucial for cold acclimation-induced freezing tolerance through modulating Pol II's recruitment to cold-regulated CBF target genes and their transcription [23–26]. Meanwhile, MED14 also influences plant responses to high temperatures by activating heat-sensitive genes [27]. MED16 has a role in regulating nutrient uptake and homeostasis, such as iron and phosphate [28]. Moreover, it can interact with MED25, mutations to which alter responses to iron deficiency [29]. Other Mediator subunits show individual functionalities in abiotic stress responses too. For instance, *Arabidopsis* seedlings of *med8* mutant display increased tolerance to oxidative stress and enhanced transcriptional activation of defense-related genes, predominantly those in the salicylic acid (SA) and JA-related pathways [30]. Recently, CDK8 was identified as a positive regulator of ABA signaling and drought response through association with RAP2.6 and SnRK2.6 [31]. Although the roles of the Mediator complex in *Arabidopsis* have been increasingly identified, the understanding of its functions in crop plants, particularly ornamental plants like roses, remains limited.

The molecular regulatory mechanism for water stress response in roses is still largely unexplored. Past studies identified crucial genes for dehydration/drought tolerance in roses (*Rosa hybrida*). Transcription factors RhNAC2 and RhNAC3, for example, were found to enhance dehydration tolerance through the regulation of genes associated with cell-wall structure and osmotic adjustment, respectively, in rose flowers [32,33]. The regulatory module RhABF2/RhFer1 can influence Fe levels in rose petals, thus increasing dehydration tolerance [34]. RhHB1, a transcription factor induced by dehydration, directly suppresses the expression of the JA biosynthesis gene *RhLOX4*, leading to a decreased JA-Ile level and enhanced rose flower dehydration tolerance [9]. Furthermore, transcriptomic analyses of rose plants have identified drought-responsive genes [4]. Considering the significant and varied roles of Mediator in plant stress response, this study focused on the function and mechanism of Mediator subunits in roses. In this study, the expression pattern of nine *RhMED* genes was explored in roses during dehydration treatment. Among them, *RhMED15a*, encoding a subunit in the tail module of the Mediator complex, was found to be rapidly and significantly triggered by dehydration stress. It was demonstrated that *RhMED15a* plays a crucial role in the rose plant's drought tolerance, partially by regulating the expression of certain drought-responsive genes. This provides new details regarding the function of Mediator subunits in plant drought responses.

2. Materials and Methods

2.1. Plant Materials and Treatments

Rosa hybrida cv. Samantha plantlets were propagated *via* tissue culture as per previously established techniques [35]. Following growth in a combination of peat moss and vermiculite (1:1), the rooted rose and tobacco (*Nicotiana benthamiana*) plants were subjected to conditions of approximately 60% relative humidity at 22 ± 1 °C and a 16/8 h light/dark photoperiod.

For dehydration treatments, the roots of rose plant were washed and submerged in deionized water for 24 h. Subsequently, the rose plants were dehydrated by horizontally positioning them on the laboratory bench at 22 °C and approximately 60% relative humidity, under a light intensity of 100 μmol m^{-2} s^{-1} for periods of 0, 1, 6, 12, and 24 h. The control plants remained submerged in water under identical conditions. At each time point, leaves and roots of both groups were collected and preserved at −80 °C for subsequent analysis.

For the ABA treatment, the leaves of 1-month-old rose plantlets were sprayed with a solution of 100 μM ABA and 0.05% Tween. The mock group was treated with 0.1% ethanol, excluding any phytohormones. Each seedling was sprayed with 10 mL of the solution. Post-treatment of 12 h, leaf and root samples from these roses were collected respectively.

Leaf discs were submerged in 20 mL of solutions of either 100 μM methyl jasmonate (MeJA) or 100 μM salicylic acid (SA) and 0.05% Tween for 24 h, for the JA and SA treatments, respectively. The mock group, in the case of MeJA treatment, received a treatment of 0.1% ethanol along with 0.05% Tween, while mock samples of the SA treatment were treated with 0.1% DMSO and 0.05% Tween.

2.2. Sequence Analysis of RhMED15a

The conserved domains of the RhMED15a amino acid sequence were scrutinized utilizing the Conserved Domain Database (CDD) of the National Center for Biotechnology Information (NCBI). ProtParam (https://www.expasy.org/, accessed on 16 May 2022) was employed to probe the molecular weight (MW) and the isoelectric point (pI). The task of multiple sequence alignment was fulfilled through Bioedit using the ClustalW algorithm as its foundation. The phylogenetic associations between RhMED15a and its homologs from other plant species were conducted through MEGA 7.0 by implementing the neighbor-joining algorithm, and bootstrap values were obtained through 1000 iterations.

2.3. RNA Extraction and Quantitative RT-PCR

RNA extractions from the roots and leaves of rose plants were performed using an RNA rapid extraction kit (BOLAZ, Nanjing, China). The conversion of 1 μg of total RNA into complementary DNAs (cDNAs) was accomplished using the PrimeScript™ RT reagent Kit along with the gDNA Eraser (Takara, Shiga, Japan). This was followed by quantitative RT-PCR, carried out using 2 μL of cDNA as a template, on the StepOne Real-Time PCR System (Applied Biosystems, Foster City, CA, USA) in conjunction with the Kapa SYBR Fast Universal qPCR Kit (Kapa Biosystems, Boston, MA, USA). The primers used for this process were available in Table S1, while *RhUBI2* served as an internal control. Three biological replicates formed the basis of all reactions, and representative results are presented. The statistical testing was performed using the SPSS version 17.0.

2.4. Protein Subcellular Localization Assays

The open-reading frame (ORF) of *RhMED15a* was amplified by PCR and then inserted into a modified binary vector pCAMBIA2300 harboring 35S and a GFP label using Clon-Express II One Step Cloning Kit (Vazyme, Nanjing, China). The resultant pCAMBIA2300-*GFP-RhMED15a* construct was then transferred into GV3101 *Agrobacterium* competent cells. The tomato bushy stunt virus p19 protein served to inhibit gene silencing during the transient expression procedure in tobacco leaves, as previously outlined [35]. Negative control was provided by p35S::*GFP*, with p35S::*NF-YA4-mCherry* serving as a nuclear marker. *Agrobacterium* cultures containing the pCAMBIA2300-*GFP-RhMED15a* construct (or negative control construct), nuclear marker, and the p19 plasmid were mixed at OD_{600} of 0.5:0.5:0.3. Cells were pelleted and resuspended in infiltration medium (10 mM $MgCl_2$, 10 mM MES, and 100 mM acetsyringone). After incubation for 2 h, the suspension was filtrated into leaves of 1-month old tobacco (*Nicotiana benthamiana*) plants. Three days post-infiltration, laser confocal fluorescence microscopy was employed to observe the fluorescence signal (FV1000; Olympus, Tokyo, Japan).

2.5. Virus-Induced Gene Silencing

Virus-induced gene silencing (VIGS) was used to silence *RhMED15a* in rose plants, following previous methodology with minor amendments [32]. *Tobacco rattle virus* (TRV) vectors were employed in VIGS. TRV is a bipartite RNA virus with the TRV1 and TRV2 genomes. pTRV1 encodes the replication and movement functions while pTRV2 harbors the coat protein and the sequence used for VIGS. A specific 350 bp fragment in *RhMED15a* ORF was used to generate pTRV2-*RhMED15a*. *Agrobacterium* strain EHA105, containing pTRV1, pTRV2, and pTRV2-*RhMED15a*, was cultured in Luria-Bertani medium with 10 mM MES, 20 μM acetosyringone, and appropriate antibiotic selections. The Agrobacterium cells were collected and suspended in infiltration buffer (10 mM $MgCl_2$, 200 μM acetosyringone, and 10 mM MES, pH 5.6) to achieve an approximate OD_{600} of 1.5. Agrobacterium cultures carrying pTRV1 and pTRV2 or its derivatives were mixed in equal ratios (v/v) and left at room temperature for 4 h. Rose plantlets were then immersed in the bacterial suspension solution, exposed to −25 KPa vacuum twice, washed with deionized water, and potted for analysis.

For the drought treatment, infected rose plants were planted in a combination of peat moss and vermiculite (1:1) in a growth room maintained at a steady 22 ± 1 °C, approximately 60% relative humidity, and subjected to a 16 h/8 h light/dark cycle. After 40 days of growth, these conditions were altered: watering was ceased for 20 days, then resumed for an additional 20 days. Plant phenotypes were periodically checked. When a compound leaf's leaflets display over 50% wrinkling, the leaf is categorized as wilted. The rate of leaf wilting is computed by dividing the quantity of wilted compound leaves by the compound leaves' total number.

2.6. Malondialdehyde (MDA) Content Determination

The concentration of MDA was ascertained using an MDA content assay kit (Comin Biotechnology Co., Ltd., Suzhou, China). MDA is capable of participating readily in a nucleophilic addition reaction with 2-thiobarbituric acid (TBA), resulting in the creation of a red-colored MDA:TBA adduct. Subsequently, the colorimetric methodology is utilized to quantify the adduct content, from which the MDA content can then be further determined. Fresh leaves from each plant, either prior to the drought treatment or on the 12th day of drought treatment, were ground and homogenized along with a 10% (w/v) extracting solution. The absorbance was then calculated at 532 nm and 600 nm wavelengths using a spectrophotometer (T6 New Century, Purkinje General, Beijing, China). Based on its extinction coefficient, 155 $mM^{-1} \cdot cm^{-1}$, the MDA concentration was then derived.

3.2. Chromosomal Localization of Identified miRNAs in *A. venetum*

Chromosomal localization analysis of the miRNAs showed that the miRNA precursors were distributed on all 11 chromosomes (Figure 2). Many miRNAs are distributed on chromosomes 2, 8, and 3, with 23, 18, and 17 miRNA precursors, respectively. However, only one species-specific miRNA was found on chromosome 9, and no conserved miRNAs were found on this chromosome. Notably, several conserved miRNAs were clustered on the chromosomes, including miR156, miR157, miR159, miR160, miR162, miR164, miR166, miR167, miR169, miR171, miR172, miR390, miR391, miR395, miR398, miR399, miR403, miR477, and miR482. The three largest clusters comprised eight miR399, five miR395, and five miR167.

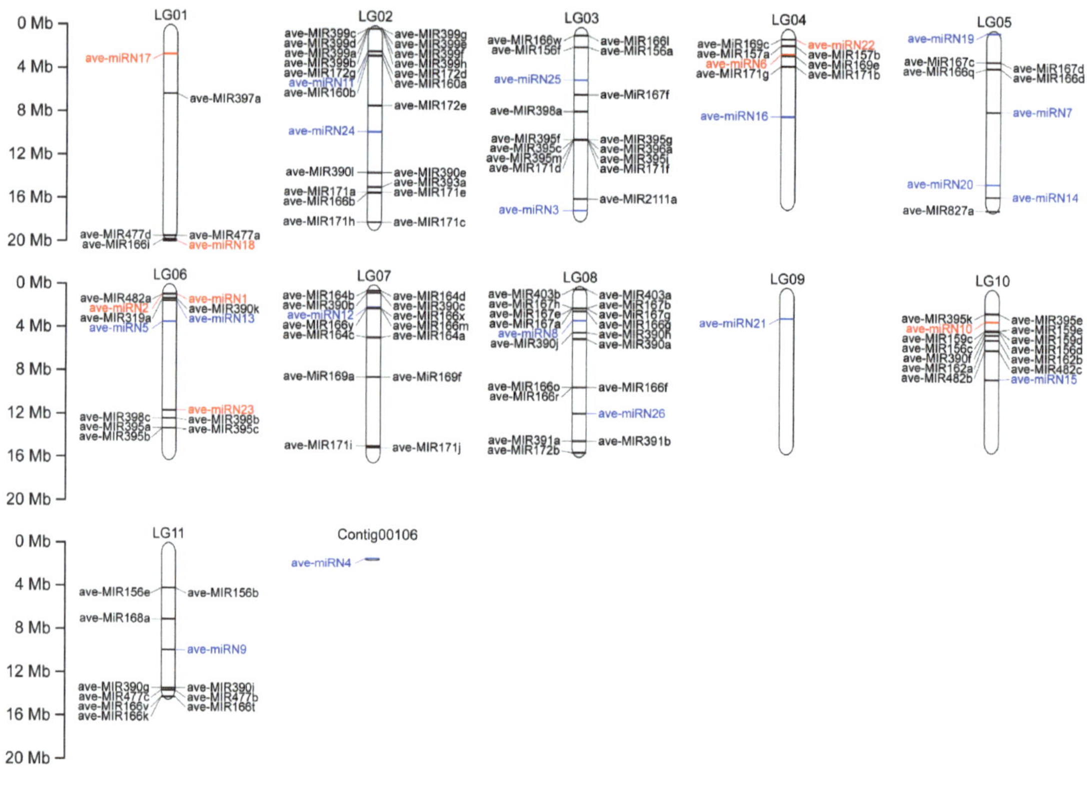

Figure 2. Chromosomal localization of conserved and species-specific miRNAs in *A. venetum*. Black letters indicate conserved miRNAs, red letters indicate species-specific miRNAs with high confidence, and blue letters indicate potential species-specific miRNAs in *A. venetum*. miRNA, microRNA.

3.3. Prediction and Functional Analysis of Target Genes of miRNA in *A. venetum*

The miRNA target gene prediction software psRNAtarget was used to predict the target genes of both conserved and species-specific miRNAs in *A. venetum*. A total of 43,484 target genes, encompassing 15,938 unique genes, were identified. Among these, 11,450 genes were targeted by conserved miRNAs, while 5834 genes were targeted by species-specific miRNAs in *A. venetum* (Supplementary Tables S5 and S6).

To gain insights into the potential functions of the predicted target genes in *A. venetum*, GO enrichment analysis was conducted. GO enrichment analysis in conserved miRNA in *A. venetum* showed that the enriched biological process terms include 'metabolic process', 'response to stimulus', 'growth', and 'nitrogen utilization', and enriched GO molecular function terms include 'transporter activity', 'transcription regulator activity', 'antioxidant

2.7. *Statistical Analysis*

Each experiment was performed independently at least three times. At least three plants were used in each independent experiment. In each biological replicate of *quantitative RT-PCR*, three technical replicates of each PCR reaction were run. SPSS 16 was used for statistical analysis. Significant differences between the experimental data were analyzed using Student's t test or one-way ANOVA with Duncan's multiple range tests (* $p < 0.05$; ** $p < 0.01$; *** $p < 0.001$).

3. Results

3.1. *RhMED15a Was Induced by Dehydration Treatment in Roses*

To investigate the role of Mediator subunits in the drought response of rose plants, nine differentially expressed *RhMED*s from the transcriptome data of dehydration-treated flowers (unpublished data) were selected. Their expression profiles in rose seedlings during dehydration were confirmed using quantitative real-time PCR. Furthermore, the expression of the well-characterized water deficit-inducible gene, *dehydration responsive element binding factor 1B* (*RhDREB1B*), was detected as a positive control [36]. As depicted in Figure 1, both the leaves and roots expressed a significant increase in *RhDREB1B* during dehydration, suggesting the rose plants were subjected to moderate water stress. When compared to the control, the expression of six *RhMED* genes (*RhMED12*, *RhMED13*, *RhMED19a*, *RhMED23*, *RhMED25*, and *RhMED32*) significantly declined at least at three time points within the 24 h of dehydration in both the leaf and root. This indicates that dehydration treatment inhibits the expression of these *RhMED*s in rose plants. Among the remaining *RhMED*s, *RhMED8*'s expression profile differed between the leaf and root; its expression significantly increased in the root after six hours of dehydration treatment compared to the control, while concurrently, it substantially declined in the leaf. *RhMED26b*'s expression was not up-regulated until after 12 h of dehydration treatment in both the leaf and root. Notably, among all the tested *RhMED*s, *RhMED15a* experienced the most significant dehydration-induced increase in root expression. Its abundance markedly increased at 1 h of dehydration treatment and reached an approximately two-fold value at 12 h compared to the control, despite no significant alteration in leaf expression during dehydration.

3.2. *Isolation and Sequence Analysis of RhMED15a*

The full-length cDNA of *RhMED15a* was isolated from rose plants and its structural features were analyzed. Its open reading frame is 4086 bp in length and codes for an inferred amino acid sequence of 1361 residues. The predicted relative molecular mass of these residues is 150.4 kDa, with a theoretical pI of 9.4. When the inferred amino acid sequences of RhMED15a were aligned with homologs from other plant species, high homology was observed Figures 2B and S1). Moreover, the RhMED15a protein was found to contain a KIX domain at the amino terminus. This domain has previously been seen as pivotal for the interaction between Med15 (a Mediator tail module subunit) and various transcription activators in yeast and other plants [37]. Phylogenetic analysis revealed that MED15a homologs from the Rosaceae form a clearly defined and substantially supported branch, distinct from homologs in other families (Figure 2A). In Arabidopsis, the closest ortholog to RhMED15a was found to be AtMED15 (NP_001321150.1), which shares 53.3% identity with the protein sequence of RhMED15a. Such results suggest that RhMED15a is a part of the tail module of the Mediator complex.

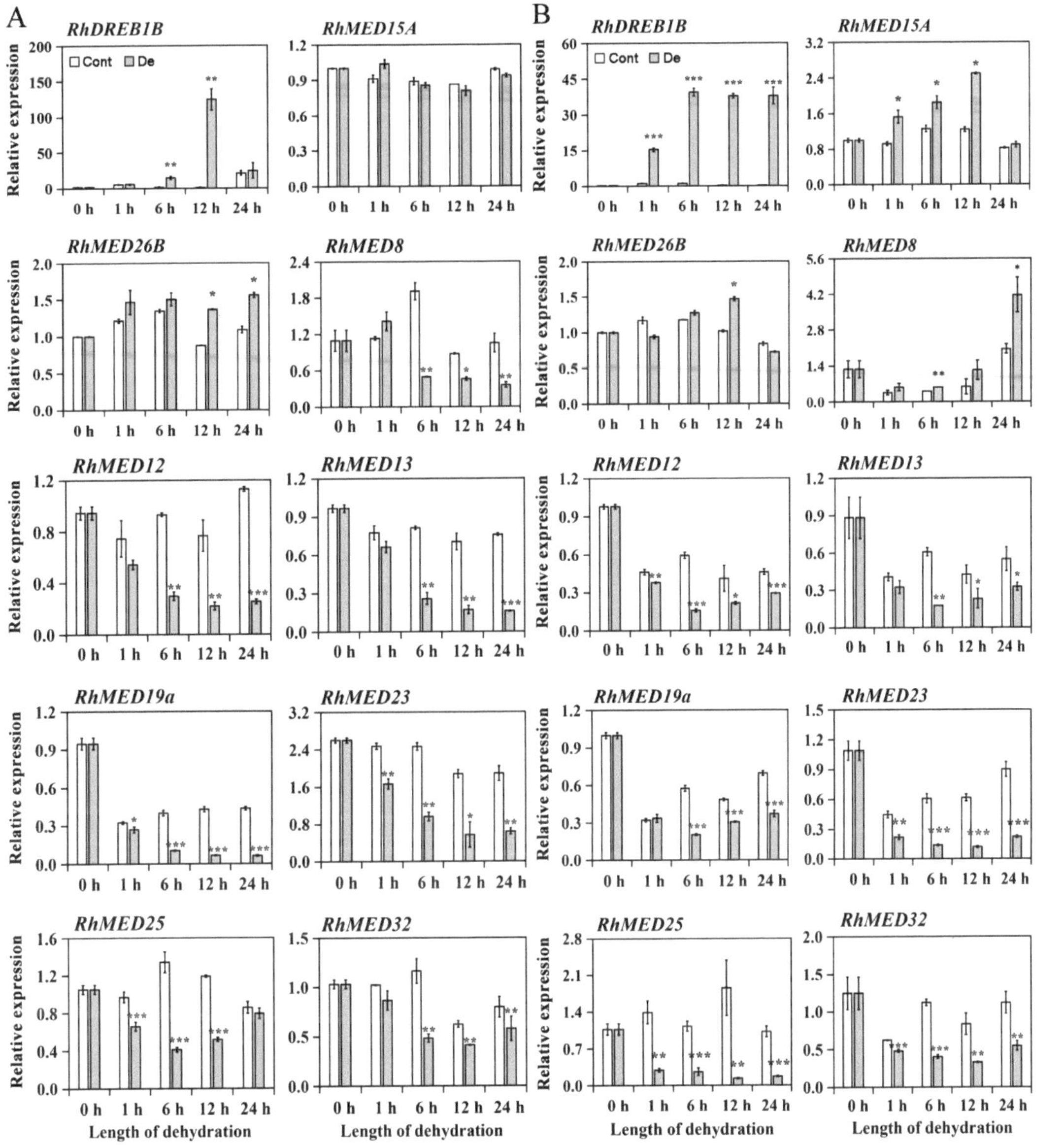

Figure 1. The expression profiles of *RhMEDs* in roses under dehydration stress. (**A**) The expression profiles of *RhMEDs* in rose leaf during dehydration stress. (**B**) The expression profiles of *RhMEDs* in rose root during dehydration stress. The samples for quantitative RT-PCR analysis were treated by dehydration for 0, 1, 6, 12, and 24 h, respectively. *RhUBI2* was used as an internal control. The expression level of *RhMEDs* at 0 h was defined as 1. Values are means ± SD (n = 3). Cont, control; De, dehydration treatment. Asterisks indicate significant differences calculated with a t-test (* $p < 0.05$; ** $p < 0.01$; *** $p < 0.001$).

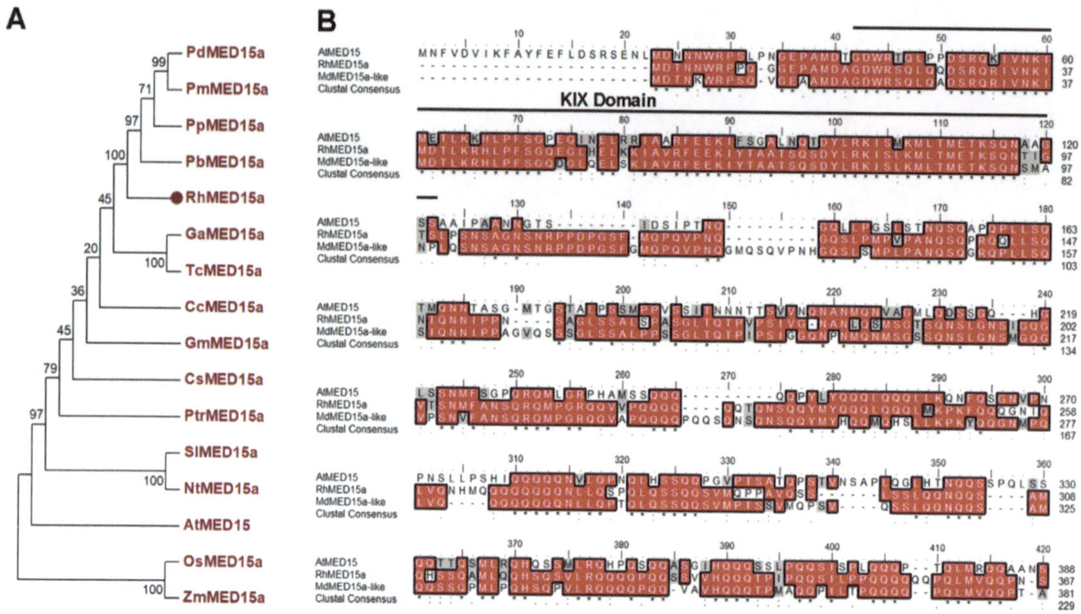

Figure 2. Evolutionary analysis of RhMED15a. (**A**) Phylogenetic relationships of deduced RhMED15a amino acid sequences with homologs from other plants. The phylogenetic tree file was produced by MEGA7. RhMED15a is indicated by the red solid circle. Bootstrap values indicate the divergence of each branch, and the scale indicates branch length. PdMED15a (*Prunus dulcis*, XP_034219548.1), PbMED15a (*Pyrus × bretschneideri*, XP_048426092.1), PmMED15a (*Prunus mume*, XP_008218603.1), RhMED15a (*Rosa hybrida*, XM_024301679.2), PpMED15a (*Prunus persica*, XP_007205180.2), GmMED15a (*Glycine max*, XP_014623364.1), CcMED15a (*Citrus clementina*, XP_024036137.1), PtrMED15a (*Populus trichocarpa*, XP_024449454.1), GaMED15a (*Gossypium arboreum*, XP_017648934.1), SlMED15a (*Solanum lycopersicum*, XP_004236821.1), CsMED15a (*Camellia sinensis*, XP_028120173.1), TcMED15a (*Theobroma cacao*, XP_007023331.2), NtMED15a (*Nicotiana tomentosiformis*, XP_009616506.1), AtMED15 (*Arabidopsis thaliana*, NP_001321150.1), OsMED15a (*Oryza sativa*, XP_015649820.1), ZmMED15a (*Zea mays*, XP_035820115.1). (**B**) Multiple sequence alignment between N-terminal of RhMED15a, AtMED15, and MdMED15a-like (*Malus domestica*, XP_028955279.1). The red mark is the identical amino acid, and the grey mark is the similar amino acid. KIX domain was indicated by the line above the amino acids.

3.3. Expression of RhMED15a Was Significantly Up-Regulated by ABA but Down-Regulated by MeJA

To anticipate the expression profiles of *RhMED15a*, we analyzed nearly a 3kb promoter sequence of *RhMED15a*. PlantCare database was used to conduct bioinformatics analysis of this promoter sequence. As revealed in Figure 3A, the sequence exhibited typical regulatory elements found in plant gene promoters. These elements include those that are responsive to plant hormones and abiotic stress factors. There were four *cis*-acting motifs in the *RhMED15a* promoter that interact with two phytohormones: ABA and MeJA. This suggests the possibility of *RhMED15a* having a role in the stress response mediated by MeJA and ABA. In terms of abiotic stress, the sense strand of the promoter lacks any stress-response elements, although two LTRs (CCGAAA), which are regulatory elements for low-temperature stress, were found on the other strand. This indicates that the expression of *RhMED15a* in dehydration stress situations may be regulated by the stress hormones ABA or MeJA, both of which have been identified as potent mediators of plant response to water stress [9,38].

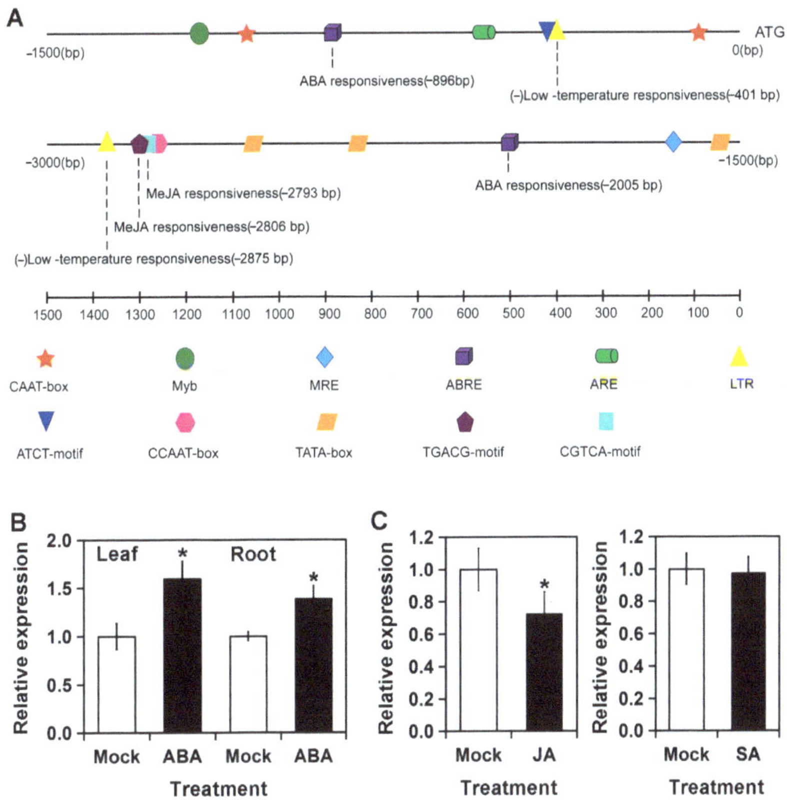

Figure 3. Effect of plant hormones on the expression of *RhMED15a*. (**A**) *Cis*-acting elements analysis of *RhMED15a* promoter. The negative numbers indicate the upstream ends of the sequence before the ATG. The *cis*-acting elements on the antisense strand are indicated by a negative sign in parentheses. Myb, MYB transcription factor recognition sequence; MRE, MYB binding site involved in light responsiveness; ABRE, *cis*-acting element involved in abscisic acid responsiveness; ARE, *cis*-acting element essential for anaerobic induction; LTR, *cis*-acting element involved in low-temperature responsiveness. (**B**) *RhMED15a* expression in rose leaf (left) and root (right) in response to exogenous ABA treatment. Rose plantlets were sprayed with 100 μM ABA for 12 h. (**C**) *RhMED15a* expression in rose leaf in response to exogenous MeJA and SA treatments. Leaf discs from rose plant were soaked in 100 μM MeJA or 100 μM SA solution for 24 h. The expression level of *RhMED15a* in Mock was defined as 1. *RhUBI2* was used as an internal control. Values are means ± SD (n = 3). Asterisks indicate significant differences calculated with a t-test (* $p < 0.05$).

We further analyzed *RhMED15a's* response to plant hormones using quantitative RT-PCR analysis. The focus was on ABA and MeJA, as these phytohormones were identified from the analysis of cis-elements in the RhMED15a promoter. We also studied the impact of SA on the expression of *RhMED15a*, since *AtMED15* in *Arabidopsis* is known to play a part in the SA-mediated defense response [39]. As shown in Figure 3B, we found that ABA treatment led to a significant increase in the expression of *RhMED15a* by approximately 1.6-fold in leaves and 1.4-fold in roots after 12 h of exposure. Conversely, MeJA was found to suppress *RhMED15a* expression by about 27.4% in leaf tissue at 24 h stimulation when compared with the controls (Figure 3C). This result hints at *RhMED15a's* potential involvement in ABA and JA-regulated development and abiotic stress response in roses. A 24 h treatment with SA did not significantly change *RhMED15a* expression in leaf tissue, aligning with our initial bioinformatics analysis of the *RhMED15a* promoter that found no SA-responsive *cis*-elements (Figure 3A).

3.4. *RhMED15a Was Localized in the Nucleus*

We explored the subcellular location of RhMED15a in the epidermal cells of Nicotiana benthamiana leaves by executing a fusion of the ORF sequence of *RhMED15a* to the C-terminal side of the GFP reporter gene. The resultant construct, along with an mCherry-marked nuclear marker, was co-introduced into tobacco leaves. As delineated in Figure 4, green fluorescence signals from RhMED15a were located in the nucleus, overlapping significantly with the red fluorescence signal emanating from mCherry. This observation aligns with prior studies on *Arabidopsis* MED15 that pinpoint RhMED15a as a nuclear-localized protein [40].

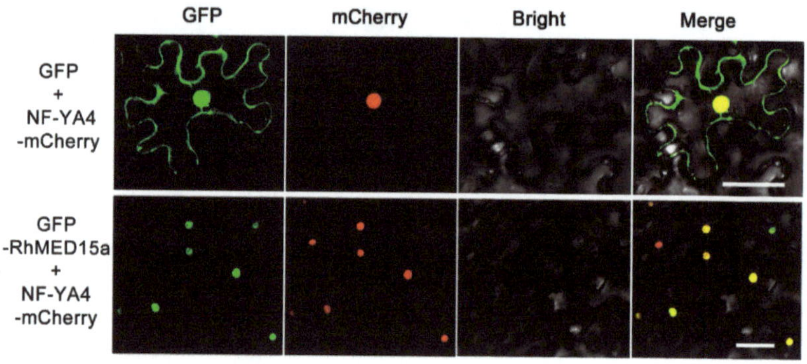

Figure 4. The subcellular localization of RhMED15a in the epidermis cells of *Nicotiana benthamiana* leaves. GFP-RhMED15a was colocalized with a mCherry-labelled nuclear marker (NF-YA4-mCherry). The experiment was performed independently three times, and representative results are shown. Bars = 50 μm.

3.5. *Silencing of RhMED15a Decreased Drought Tolerance in Rose Plants*

To investigate the role of *RhMED15a* in response to water deficiency, we employed the VIGS approach using Tobacco Rattle Virus (TRV) to silence this gene in rose plants, following which, drought tolerance was gauged as outlined previously [35]. We leveraged a 350 bp gene-specific fragment of *RhMED15a* from the ORF region for pTRV2-*RhMED15a* assembly. Rose plantlets infiltrated with *Agrobacterium* incorporating either TRV control or TRV-*RhMED15a* were subsequently exposed to drought for 20 days and later reverted to standard watering regimes for an additional 20 days. Demonstrated in Figure 5B, we clinched a total of 7 successfully silenced plants exhibiting considerably lower expression of *RhMED15a* relative to the TRV control. Under standard growth conditions, there were no significant variances in plant height, leaf count, or leaf surface area between the TRV controls and *RhMED15a*-silenced plants, suggesting a likely non-involvement of *RhMED15a* in the regulation of plant growth in roses (Figure S2).

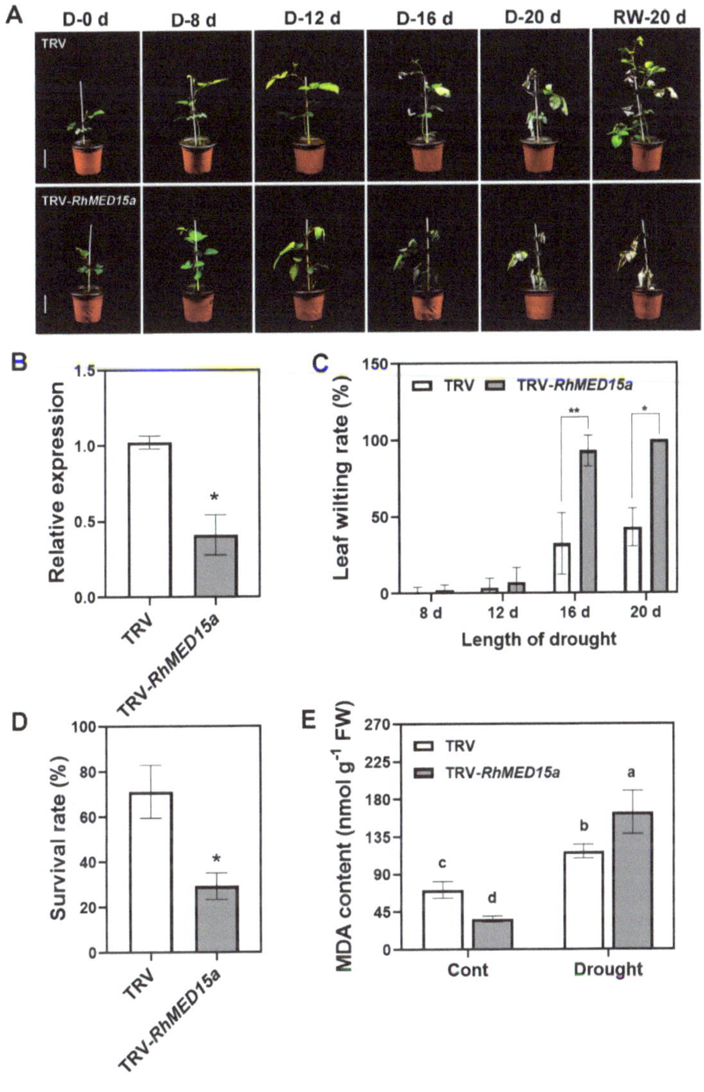

Figure 5. Silencing of *RhMED15a* reduced drought tolerance in rose plants. (**A**) Morphology of *RhMED15a*-silenced plants and TRV controls during drought stress. *RhMED15a*-silenced plants and TRV controls were grown for 40 days under normal conditions. Then, water was withheld for 20 days, followed by 20 days of re-watering. Scale bar, 4 cm. D-0 d, D-8 d, D-12 d, D-16 d, and D-20 d indicate drought treatment for 0, 8, 12, 16, and 20 days, respectively; RW-20 d indicates re-watering for 20 days. (**B**) Quantitative RT–PCR of *RhMED15a* in leaves of TRV and TRV- *RhMED15a*-infected plants at the 40th day after infiltration. (**C**) Leaf wilting rates of *RhMED15a*-silenced plants and TRV plants during drought stress. The leaf wilting rate was monitored every fourth day. (**D**) Survival rates of *RhMED15a*-silenced plants and TRV plants on 20th day of the re-watering treatment. The survival rate was calculated based on data from three independent experiments. (**E**) MDA contents of *RhMED15a*-silenced plants and TRV plants under normal conditions (control) and on the 12th day of the drought stress treatment. Asterisks indicate significant differences calculated with a *t*-test (* $p < 0.05$; ** $p < 0.01$). Lower-case letters indicate significant differences according to Duncan's multiple range test ($p < 0.05$).

When subjected to drought stress, rose plants manifested drought-related phenotypes, including yellowing and curvature of leaf blades, wilted petioles, defoliation, and inhibited growth (Figure 5A). TRV plants began to wilt on the 16th day of drought treatment, becoming more severe by the 20th day. Contrastingly, *RhMED15a*-silenced plants exhibited noticeable wilting by the 12th day of drought treatment, with most leaves dried and defoliated by the 20th day (Figure 5A). *RhMED15a*-silenced plants demonstrated a significantly higher leaf wilting rate than the TRV controls at the 16th and 20th days of drought treatment (Figure 5C). Following re-watering, about 71.2% of the TRV control plants recovered, whereas only 29.2% of *RhMED15a*-silenced plants survived the drought stress (Figure 5D). MDA content was tested, as it is a product of peroxidation of unsaturated fatty acids in phospholipids and is commonly used as an indicator of cellular membrane lipid peroxidation [41]. The test results revealed that MDA levels in both groups displayed a rising trend post-drought stress (Figure 5E). On the 12th day of drought treatment, MDA levels of *RhMED15a*-silenced plants increased by approximately 4.5-fold, while TRV controls only showed about 1.6-fold induction, suggesting *RhMED15a*-silenced plants endured more severe damage during drought treatment.

Considering Mediator's role in transcriptional regulation, we hypothesized that *RhMED15a* could impact plant tolerance to drought stress by controlling the expression of genes involved in drought response. Accordingly, we analyzed six drought-responsive genes, including *responsive to desiccation stress 29A (RD29A)*, *responsive to desiccation stress 29A (RD29B)*, *9-cis-epoxycarotenoid dioxygenase 1 (NCED1)*, *DREB1B*, *early response to dehydration 14 (ERD14)*, and *Δ1-pyrroline-5-carboxylate synthetase (P5CS)* [42–45], *via* quantitative RT-PCR analysis on the 12th day of drought treatment in the leaves of TRV-*RhMED15a* plants and TRV controls. The findings revealed significant reductions in the expression levels of five genes (*RD29A*, *RD29B*, *NCED1*, *DREB1B*, and *ERD14*) in *RhMED15a*-silenced rose plants compared to TRV controls under drought stress. However, *P5CS* expression showed no significant difference (Figure 6). Particularly, in *RhMED15a*-silenced plants, expression levels of *RD29A*, *NCED1*, and *DREB1B* were markedly inhibited by roughly 97%, 96.5%, and 83.7%, respectively, compared to TRV controls. These results suggest that *RhMED15a* positively regulates the expression of these drought-responsive genes under drought conditions.

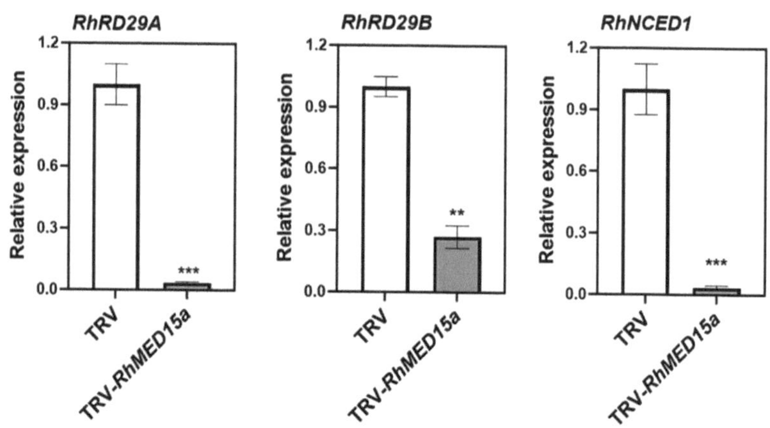

Figure 6. *Cont.*

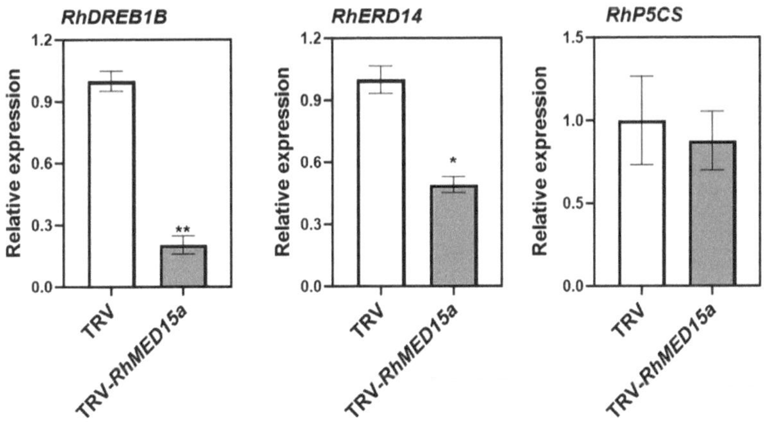

Figure 6. Expression levels of drought-related genes in *RhMED15a*-silenced plants and TRV controls under drought stress by quantitative RT–PCR. Relative expression was normalized against the reference gene of *RhUBI2*. The expression level of the indicated genes in the TRV controls under drought treatment was set as 1. The mean values ± SD from three biological replicates ($n = 3$) are shown. Asterisks indicate significant differences calculated with a *t*-test (* $p < 0.05$; ** $p < 0.01$; *** $p < 0.001$).

4. Discussion

The Mediator complex in plants functions as a transcription cofactor, and is involved in various growth, developmental, and stress-response processes [14]. The necessity for individual Mediator subunits typically varies within the complex, indicating distinct functions for different subunits in these processes [14,46,47]. Currently, research on plant Mediator subunits is primarily confined to *Arabidopsis* and rice [46,48]. Thus, it is crucial to identify potential Mediator subunits in crops that can influence yield, product quality, and stress tolerance [46]. In this study, we examined the expression profiles of nine Mediator subunits in roses during a water deficit scenario, with particular emphasis on the Mediator tail module subunit *RhMED15a*, which exhibited notable up-regulation following dehydration treatment. We explored *RhMED15a*'s role in drought tolerance in roses using VIGS, and evaluated the expression of drought-responsive genes within *RhMED15a*-silenced plants.

4.1. Dehydration Stress Responsive Mediator Subunits in Roses

The established role of the plant Mediator complex in immunity is well-recognized, with a few Mediator subunits implicated in responses to abiotic stresses [12,20,31,49]. It is evident that key *Arabidopsis* MEDs partake in the regulation of abiotic stress signaling pathways. For instance, *AtMED16*, *AtMED14*, and *AtMED2* oversee the regulation of *COLD ON-REGULATED* (*COR*) genes, demonstrating sensitivity to cold stress [26]. *AtMED25* and *AtMED19a* contribute to plant resistance against drought stress and control ABA-responsive transcription [19,22,49]. Evaluating *RhMEDs*' expression profiles during water deficit situations sheds light on their functional roles within the drought response mechanism in rose plants. Nine *RhMED* expressions were confirmed during dehydration treatment in rose seedlings in this study. Six of the observed *RhMEDs* showed significant down-regulation in both leaf and root under dehydration, while three exhibited significant up-regulation in either leaf or root at varying degrees (Figure 1). The varied impacts of dehydration on *RhMED* expressions point to the diverse responsiveness of *RhMEDs* to water stress across Mediator subunits and plant organs. Notably, *RhMED26b* was induced at 12 h of dehydration treatment in both leaves and roots, in alignment with previous reports on tomato *SlMED26b*, suggesting a potential role in the regulation of plant dehydration tolerance [50]. Interestingly, *RhMED25* demonstrated rapid and significant down-regulation in both leaf and root during dehydration treatment. Since *AtMED25* can negatively regulate drought tolerance in *Arabidopsis* [19,22], such results hint that *RhMED25* may perform a role similar to *AtMED25* in plant drought tolerance. Another finding was the significant down-regulation of *RhMED19a* in both organs under dehydration, a surprising observation as *AtMED19a* reportedly bolsters plant resistance to drought stress [49]. This could potentially hint at the diverse roles MEDs play across different species and the decoupling of gene expression from its function due to multi-level molecular mechanisms interfacing with the regulatory process in stress responses [51,52]. Among the observations, the rapid rise of *RhMED15a* expression, significantly quicker and greater than other *MED*s in the root, is notable, suggesting a potential role in the rose's water stress response.

4.2. Insights into the Role of RhMED15a in Drought Stress Responses in Roses

The *MED15* subunit of the mediator tail module in plants is reported to be an essential component involved in the transcriptional regulation of various processes, including floral transition [39], lipid biosynthesis in seeds [40], and plant immunity [39,53]. Existing studies reveal how *AtMED15* in *Arabidopsis* functions downstream of *NPR1* (*Non-expresser-of Pathogenesis-Related-1*) to regulate the SA-mediated defense response [39]. AtMED15 also activates glycolysis-related and fatty acid biosynthesis genes by interacting with the plant-specific transcription factor, Wrinkled 1 (WRI1) [40]. In rice plants, OsMED15a connects OsNAC024 and OsNAC02 to their target genes, modulating grain size and weight [54]. However, the role of MED15 in the drought stress response of plants remains undefined. Our study demonstrated that compared to controls, the silencing of RhMED15a through the VIGS approach induced quicker and more severe leaf wilting, increased MDA content, and decreased the survival rate of rose plants during drought stress (Figure 5). These findings genetically validate the conducive role of *RhMED15a* in the regulation of drought tolerance in roses, potentially revealing a new aspect of MEDs' involvement in the plant response to drought stress. This implies that the MED15 mediator subunit might be instrumental in addressing both biotic and abiotic stresses in plants.

Additionally, we observed that the expression of five drought-responsive genes, including one ABA-independent gene (*DREB1B*) [55], three stress/ABA responsive genes (*RD29A*, *RD29B*, and *ERD14*) [56,57], and one ABA synthesis gene (*NCED1*) [58], were considerably down-regulated in *RhMED15a*-silenced plants under drought stress compared to TRV controls (Figure 6). This suggests that *RhMED15a*, as a transcriptional co-regulator, may enhance roses' resistance to drought stress by regulating the expression of drought-stress-related genes. Thus, further studies, such as RNA-seq of TRV controls and *RhMED15a*-silenced plants, would help explicate the underlying molecular mechanisms of the role of *RhMED15a* in drought resistance in roses.

In addition, under normal conditions, we noted no statistically significant difference in plant height, the number of compound leaves, and leaf area between *RhMED15a*-silenced and TRV control plants on the 42nd day after infection (Figure S2). This suggests that *RhMED15a* may not regulate the growth rate of rose plants. However, compared to the wild type in *Arabidopsis*, the *med15/nrb4-4* variant was smaller and had fewer flowers but more additional stems after bolting [39]. The phenotypic discrepancy between *RhMED15a*-silenced roses and *Arabidopsis med15/nrb4-4* could be traced back to *MED15*'s diverse roles in different plant species, the various developmental periods of plants, and the effect of gene silencing. Consequently, the impact of *RhMED15a* on rose growth and development necessitates further investigation.

Alterations in hormonal content and corresponding signaling networks in plants are triggered by drought stress. These modifications aim to enhance tolerance or stimulate programed cell death within distinct cells, tissues, or organs, promoting survival under unfavorable conditions [10,59]. Certain phytohormones, such as ABA and JA, are crucial for regulating plant responses to water stress [38]. ABA, a primary stress-signaling phytohormone, tends to accumulate during drought periods and predominantly acts as a hormonal signaling molecule addressing drought stress [60,61]. In contrast, JA, a lipid-derived plant hormone, fine-tunes plant responses to various biotic and abiotic stresses through global transcriptional activation of JA-responsive genes [62]. Both the internally produced and externally introduced JAs contribute to drought stress tolerance in plants [63]. Specific Mediator subunits, including MED25, AtMED19a, and CDK8, critically participate in JA and/or ABA signaling. MED25 positively regulates JA signaling *via* interaction with MYC2, while negatively regulating ABA signaling and response through interacting with ABI5 [22]. AtMED19a acts as a positive regulator in ABA responses mediated by ABI5 [49]. CDK8 has been implicated as a critical regulator in the ABA signaling and drought response pathways. The *cdk8* mutant demonstrated diminished sensitivity to ABA, compromised stomatal aperture, and hypersensitivity to drought stress compared to control [31]. In tomatoes, *SlMED17*, *SlMED21*, and *SlMED23* were observed to have been up-regulated more than twice following 8 h of MeJA treatment, while the expression of *SlMED18* significantly decreased in response to ABA treatment [50]. Our study found that *RhMED15a* displayed increased mRNA accumulation under ABA treatment and reduced accumulation under MeJA treatment (Figure 3B,C). Given its expression levels following ABA and MeJA treatments and the presence of ABA- and MeJA-related *cis*-acting elements in the promoter region (Figure 3A), we propose that *RhMED15a* could serve as a regulator of ABA and MeJA signaling in roses. Recently, it was reported that JA-Ile levels elevate in rose petals during dehydration, limiting the osmotic adjustment capability in petal cells and causing a decline in dehydration tolerance [9,35]. Nullifying *RhMED15a* function may deteriorate drought tolerance in rose plants due to its involvement in JA signaling, although this hypothesis demands more robust genetic evidence.

Considering the nature of the Mediator's tail module, RhMED15a could potentially play a vital role in drought stress response *via* directed interactions with hormone signaling components and numerous TFs. For instance, the Mediator tail module subunit MED25 has been found to regulate drought tolerance negatively upon interacting with drought stress-induced proteins, including DREB2A, ZINC FINGER HOMEODOMAIN 1 (ZFHD1), and PHOSPHATE STARVATION RESPONSE LIKE1 (PHL1) [19,64]. Additionally, MED15a homologs were discovered interacting with multiple TFs [37]. In *Arabidopsis*, the AtMED15a interactome entailed 45 proteins interacting with the KIX domain of AtMED15a, including 11 TFs, three distinct nucleic acid-binding proteins, and a splicing factor [37]. A similar finding was recorded for RhMED15a, where the KIX domain was observed in its respective deduced protein (Figure 2B and Figure S1). Thus, there is a need to identify the interacting TFs of RhMED15a in roses that may contribute to water stress tolerance. In addition, Mediator subunits can interact to form flexible and diverse Mediator complexes that regulate the transcription of select genes. The specificity of these combinations is determined by both the nature of the environmental stimuli and the identity of the induced gene [14]. For instance, in *Arabidopsis*, the recruitment of RNA polymerase II to cold-regulated genes responsive to CBF requires the MED16, MED2, and MED14 subunits during cold acclimation-induced freezing tolerance [26]. These three subunits also manage the expression of certain cold-responsive genes, which are not recognized CBF targets, during low-temperature induction. Genes induced by darkness also necessitate MED16 but require a different Mediator subunit combination for their expression compared to the genes induced by cold. Consequently, it is worthwhile to further comprehend the unique Mediator complex composed of RhMED15a and other subunits affected by drought stress, as well as the specific downstream genes they regulate.

5. Conclusions

Drought stress profoundly impacts the growth and development of the rose plant, making it crucial to uncover its regulatory mechanisms. This research characterizes a rose-derived Mediator subunit gene, *RhMED15a*, providing insights into the role played by Mediator subunits in drought response in roses. *RhMED15a* operates as a positive regulator of drought tolerance as well as being a crucial transcriptional regulator mediating gene expression alterations associated with drought stress. This study contributes to understanding the role of Mediator subunits in plant responses to abiotic stresses and offers significant clues for investigating drought tolerance in roses.

Supplementary Materials: The following supporting information can be downloaded at: https://www.mdpi.com/article/10.3390/horticulturae10010084/s1, Figure S1: The alignment of the predicted amino acid sequences of RhMED15a and other homologs with the highest similarity from different plant species; Figure S2: The silencing of RhMED15a in roses does not affect growth; Table S1: Primers used in RT-PCR analysis and vector construction.

Author Contributions: Conceptualization, W.C., X.R. and Y.C.; performed the experiments, X.S., N.X., Y.L., Z.Z. and P.L.; writing and original draft preparation, W.C. and X.S.; approved the final manuscript for publication and agreed to be accountable for all aspects of the manuscript, W.C.; funding acquisition, W.C. All authors have read and agreed to the published version of the manuscript.

Funding: This work was funded by the National Natural Science Foundation of China, grant number 32072612.

Data Availability Statement: All data generated or analyzed during this study are included in this article and its Supplementary Information files.

Conflicts of Interest: The authors declare no conflicts of interest.

References

1. Gleick, P.H. The world's water, 2000–2001: The biennial report on freshwater resources. *Electron. Green J.* **2002**, *1*, 210–212. [CrossRef]
2. Gupta, A.; Rico-Medina, A.; Caño-Delgado, A.I. The physiology of plant responses to drought. *Science* **2020**, *368*, 266–269. [CrossRef] [PubMed]
3. Zhu, X.; Cao, B.; Zhao, S.; Wang, X.; Zhang, H.; Gao, D.; Duan, Y. Selection and evaluation of suitable tree species in dry and dusty mining areas of Northwest China. *J. For. Res.* **2022**, *33*, 1817–1828. [CrossRef]
4. Li, W.; Fu, L.; Geng, Z.; Zhao, X.; Liu, Q.; Jiang, X. Physiological characteristic changes and full-length transcriptome of rose (*Rosa chinensis*) roots and leaves in response to drought stress. *Plant Cell Physiol.* **2021**, *61*, 2153–2166. [CrossRef] [PubMed]
5. Dinneny, J.R. Developmental responses to water and salinity in root systems. *Annu. Rev. Cell Dev. Biol.* **2019**, *35*, 239–257. [CrossRef]
6. Rellán-Álvarez, R.; Lobet, G.; Dinneny, J.R. Environmental control of root system biology. *Annu. Rev. Plant Biol.* **2016**, *67*, 619–642. [CrossRef] [PubMed]
7. Bailey-Serres, J.; Parker, J.E.; Ainsworth, E.A.; Oldroyd, G.E.D.; Schroeder, J.I. Genetic strategies for improving crop yields. *Nature* **2019**, *575*, 109–118. [CrossRef]
8. Sah, S.K.; Reddy, K.R.; Li, J. Abscisic acid and abiotic stress tolerance in crop plants. *Front Plant Sci.* **2016**, *7*, 571. [CrossRef]
9. Fan, Y.; Liu, J.; Zou, J.; Zhang, X.; Jiang, L.; Liu, K.; Lü, P.; Gao, J.; Zhang, C. The *RhHB1/RhLOX4* module affects the dehydration tolerance of rose flowers (*Rosa hybrida*) by fine-tuning jasmonic acid levels. *Hortic. Res.* **2020**, *7*, 74. [CrossRef]
10. Llanes, A.S.; Andrade, A.M.; Alemano, S.G.; Luna, M.V. Alterations of endogenous hormonal levels in plants under drought and salinity. *Am. J. Plant Sci.* **2016**, *7*, 1357–1371. [CrossRef]
11. Riechmann, J.L.; Heard, J.; Martin, G.; Reuber, L.; Jiang, C.; Keddie, J.; Adam, L.; Pineda, O.; Ratcliffe, O.J.; Samaha, R.R.; et al. *Arabidopsis* transcription factors: Genome-wide comparative analysis among eukaryotes. *Science* **2000**, *290*, 2105–2110. [CrossRef] [PubMed]
12. Crawford, T.; Karamat, F.; Lehotai, N.; Rentoft, M.; Blomberg, J.; Strand, A.; Björklund, S. Specific functions for Mediator complex subunits from different modules in the transcriptional response of *Arabidopsis thaliana* to abiotic stress. *Sci. Rep.* **2020**, *10*, 5073. [CrossRef] [PubMed]
13. Allen, B.L.; Taatjes, D.J. The Mediator complex: A central integrator of transcription. *Nat. Rev. Mol. Cell Biol.* **2015**, *16*, 155–166. [CrossRef] [PubMed]
14. Buendía-Monreal, M.; Gillmor, C.S. Mediator: A key regulator of plant development. *Dev. Biol.* **2016**, *419*, 7–18. [CrossRef] [PubMed]
15. Samanta, S.; Thakur, J.K. Importance of Mediator complex in the regulation and integration of diverse signaling pathways in plants. *Front Plant Sci.* **2015**, *6*, 757. [CrossRef] [PubMed]
16. Robinson, P.J.; Trnka, M.J.; Bushnell, D.A.; Davis, R.E.; Mattei, P.J.; Burlingame, A.L.; Kornberg, R.D. Structure of a complete mediator-RNA polymerase II pre-Initiation complex. *Cell* **2016**, *166*, 1411–1422. [CrossRef] [PubMed]
17. Yang, Y.; Li, L.; Qu, L.J. Plant Mediator complex and its critical functions in transcription regulation. *J. Integr. Plant Biol.* **2016**, *58*, 106–118. [CrossRef]

18. Chen, J.; Yang, S.; Fan, B.; Zhu, C.; Chen, Z. The Mediator Complex: A central coordinator of plant adaptive responses to environmental stresses. *Int. J. Mol. Sci.* **2022**, *23*, 6170. [CrossRef]
19. Elfving, N.; Davoine, C.; Benlloch, R.; Blomberg, J.; Brännström, K.; Müller, D.; Nilsson, A.; Ulfstedt, M.; Ronne, H.; Wingsle, G.; et al. The *Arabidopsis thaliana* Med25 mediator subunit integrates environmental cues to control plant development. *Proc. Natl. Acad. Sci. USA* **2011**, *108*, 8245–8250. [CrossRef]
20. An, C.; Li, L.; Zhai, Q.; You, Y.; Deng, L.; Wu, F.; Chen, R.; Jiang, H.; Wang, H.; Chen, Q.; et al. Mediator subunit MED25 links the jasmonate receptor to transcriptionally active chromatin. *Proc. Natl. Acad. Sci. USA* **2017**, *114*, E8930–E8939. [CrossRef]
21. Liu, Y.; Du, M.; Deng, L.; Shen, J.; Fang, M.; Chen, Q.; Lu, Y.; Wang, Q.; Li, C.; Zhai, Q. MYC2 regulates the termination of jasmonate signaling *via* an autoregulatory negative feedback loop. *Plant Cell* **2019**, *31*, 106–127. [CrossRef] [PubMed]
22. Chen, R.; Jiang, H.; Li, L.; Zhai, Q.; Qi, L.; Zhou, W.; Liu, X.; Li, H.; Zheng, W.; Sun, J.; et al. The *Arabidopsis* mediator subunit MED25 differentially regulates jasmonate and abscisic acid signaling through interacting with the MYC2 and ABI5 transcription factors. *Plant Cell* **2012**, *24*, 2898–2916. [CrossRef] [PubMed]
23. Knight, H.; Veale, E.L.; Warren, G.J.; Knight, M.R. The *sfr6* mutation in *Arabidopsis* suppresses low-temperature induction of genes dependent on the CRT/DRE sequence motif. *Plant Cell* **1999**, *11*, 875–886. [CrossRef] [PubMed]
24. Knight, H.; Mugford, S.G.; Ulker, B.; Gao, D.; Thorlby, G.; Knight, M.R. Identification of SFR6, a key component in cold acclimation acting post-translationally on CBF function. *Plant J.* **2009**, *58*, 97–108. [CrossRef] [PubMed]
25. Boyce, J.M.; Knight, H.; Deyholos, M.; Openshaw, M.R.; Galbraith, D.W.; Warren, G.; Knight, M.R. The *sfr6* mutant of *Arabidopsis* is defective in transcriptional activation via CBF/DREB1 and DREB2 and shows sensitivity to osmotic stress. *Plant J.* **2003**, *34*, 395–406. [CrossRef] [PubMed]
26. Hemsley, P.A.; Hurst, C.H.; Kaliyadasa, E.; Lamb, R.; Knight, M.R.; De Cothi, E.A.; Steele, J.F.; Knight, H. The *Arabidopsis* mediator complex subunits MED16, MED14, and MED2 regulate mediator and RNA polymerase II recruitment to CBF-responsive cold-regulated genes. *Plant Cell* **2014**, *26*, 465–484. [CrossRef]
27. Ohama, N.; Moo, T.L.; Chua, N.H. Differential requirement of MED14/17 recruitment for activation of heat inducible genes. *New Phytol.* **2020**, *229*, 3360–3376. [CrossRef]
28. Zhang, Y.; Wu, H.; Wang, N.; Fan, H.; Chen, C.; Cui, Y.; Liu, H.; Ling, H.Q. Mediator subunit 16 functions in the regulation of iron uptake gene expression in *Arabidopsis*. *New Phytol.* **2014**, *203*, 770–783. [CrossRef]
29. Yang, Y.; Ou, B.; Zhang, J.; Si, W.; Gu, H.; Qin, G.; Qu, L.J. The *Arabidopsis* Mediator subunit MED16 regulates iron homeostasis by associating with EIN3/EIL1 through subunit MED25. *Plant J.* **2014**, *77*, 838–851. [CrossRef]
30. He, H.; Denecker, J.; Van Der Kelen, K.; Willems, P.; Pottie, R.; Phua, S.Y.; Hannah, M.A.; Vertommen, D.; Van Breusegem, F.; Mhamdi, A. The *Arabidopsis* mediator complex subunit 8 regulates oxidative stress responses. *Plant Cell* **2021**, *33*, 2032–2057. [CrossRef]
31. Zhu, Y.; Huang, P.; Guo, P.; Chong, L.; Yu, G.; Sun, X.; Hu, T.; Li, Y.; Hsu, C.C.; Tang, K.; et al. CDK8 is associated with RAP2.6 and SnRK2.6 and positively modulates abscisic acid signaling and drought response in *Arabidopsis*. *New Phytol.* **2020**, *228*, 1573–1590. [CrossRef] [PubMed]
32. Dai, F.; Zhang, C.; Jiang, X.; Kang, M.; Yin, X.; Lü, P.; Zhang, X.; Zheng, Y.; Gao, J. *RhNAC2* and *RhEXPA4* are involved in the regulation of dehydration tolerance during the expansion of rose petals. *Plant Physiol.* **2012**, *160*, 2064–2082. [CrossRef] [PubMed]
33. Jiang, X.; Zhang, C.; Lü, P.; Jiang, G.; Liu, X.; Dai, F.; Gao, J. RhNAC3, a stress-associated NAC transcription factor, has a role in dehydration tolerance through regulating osmotic stress-related genes in rose petals. *Plant Biotechnol. J.* **2014**, *12*, 38–48. [CrossRef] [PubMed]
34. Liu, J.; Fan, Y.; Zou, J.; Fang, Y.; Wang, L.; Wang, M.; Jiang, X.; Liu, Y.; Gao, J.; Zhang, C. A RhABF2/Ferritin module affects rose (*Rosa hybrida*) petal dehydration tolerance and senescence by modulating iron levels. *Plant J.* **2017**, *92*, 1157–1169. [CrossRef]
35. Zhang, S.; Feng, M.; Chen, W.; Zhou, X.; Lu, J.; Wang, Y.; Li, Y.; Jiang, C.Z.; Gan, S.S.; Ma, N.; et al. In rose, transcription factor PTM balances growth and drought survival via PIP2;1 aquaporin. *Nat. Plants* **2019**, *5*, 290–299. [CrossRef]
36. Wang, T.; Tong, Z.; Ma, N.; Gao, J. Isolation and expression analysis of *Rh-DREB1s* gene in cut roses (*Rosa hybrida*) under ethylene treatment and water deficit stress. *Acta Hortic. Sin.* **2009**, *36*, 65–72. [CrossRef]
37. Kumar, V.; Waseem, M.; Dwivedi, N.; Maji, S.; Kumar, A.; Thakur, J.K. KIX domain of AtMed15a, a Mediator subunit of *Arabidopsis*, is required for its interaction with different proteins. *Plant Signal. Behav.* **2018**, *13*, e1428514. [CrossRef]
38. Verma, V.; Ravindran, P.; Kumar, P.P. Plant hormone-mediated regulation of stress responses. *BMC Plant Biol.* **2016**, *16*, 86. [CrossRef]
39. Canet, J.V.; Dobón, A.; Tornero, P. Non-recognition-of-*BTH4*, an *Arabidopsis* mediator subunit homolog, is necessary for development and response to salicylic acid. *Plant Cell* **2012**, *24*, 4220–4235. [CrossRef]
40. Kim, M.J.; Jang, I.C.; Chua, N.H. The Mediator complex MED15 subunit mediates activation of downstream lipid-related genes by the WRINKLED1 transcription factor. *Plant Physiol.* **2016**, *171*, 1951–1964. [CrossRef]
41. Sharma, P.; Jha, A.B.; Dubey, R.S.; Pessarakli, M. Reactive oxygen species, oxidative damage, and antioxidative defense mechanism in plants under stressful conditions. *J. Bot.* **2012**, *2012*, 1–26. [CrossRef]
42. Yamaguchi-Shinozaki, K.; Shinozaki, K. A novel *cis*-acting element in an *Arabidopsis* gene is involved in responsiveness to drought, low-temperature, or high-salt stress. *Plant Cell* **1994**, *6*, 251–264. [CrossRef] [PubMed]

43. Liu, Q.; Kasuga, M.; Sakuma, Y.; Abe, H.; Miura, S.; Yamaguchi-Shinozaki, K.; Shinozaki, K. Two transcription factors, DREB1 and DREB2, with an EREBP/AP2 DNA binding domain separate two cellular signal transduction pathways in drought- and low-temperature-responsive gene expression, respectively, in *Arabidopsis*. *Plant Cell* **1998**, *10*, 1391–1406. [CrossRef]
44. Narusaka, Y.; Nakashima, K.; Shinwari, Z.K.; Sakuma, Y.; Furihata, T.; Abe, H.; Narusaka, M.; Shinozaki, K.; Yamaguchi-Shinozaki, K. Interaction between two *cis*-acting elements, ABRE and DRE, in ABA-dependent expression of *Arabidopsis rd29A* gene in response to dehydration and high-salinity stresses. *Plant J.* **2003**, *34*, 137–148. [CrossRef] [PubMed]
45. Lehr, P.P.; Hernández-Montes, E.; Ludwig-Müller, J.; Keller, M.; Zrb, C. Abscisic acid and proline are not equivalent markers for heat, drought and combined stress in grapevines. *Aust. J. Grape Wine Res.* **2022**, *28*, 119–130. [CrossRef]
46. Mathur, S.; Vyas, S.; Kapoor, S.; Tyagi, A.K. The Mediator complex in plants: Structure, phylogeny, and expression profiling of representative genes in a dicot (*Arabidopsis*) and a monocot (rice) during reproduction and abiotic stress. *Plant Physiol.* **2011**, *157*, 1609–1627. [CrossRef]
47. Larivière, L.; Seizl, M.; Cramer, P. A structural perspective on Mediator function. *Curr. Opin. Cell Biol.* **2012**, *24*, 305–313. [CrossRef]
48. Bäckström, S.; Elfving, N.; Nilsson, R.; Wingsle, G.; Björklund, S. Purification of a plant mediator from *Arabidopsis thaliana* identifies PFT1 as the Med25 subunit. *Mol. Cell* **2007**, *26*, 717–729. [CrossRef]
49. Li, X.; Yang, R.; Gong, Y.; Chen, H. The *Arabidopsis* mediator complex subunit MED19a is involved in ABI5-mediated ABA responses. *J. Plant Biol.* **2018**, *61*, 97–110. [CrossRef]
50. Wang, Y.; Liang, H.; Chen, G.; Liao, C.; Wang, Y.; Hu, Z.; Xie, Q. Molecular and phylogenetic analyses of the mediator subunit genes in *Solanum lycopersicum*. *Front. Genet.* **2019**, *10*, 1222. [CrossRef]
51. Floris, M.; Mahgoub, H.; Lanet, E.; Robaglia, C.; Menand, B. Post-transcriptional regulation of gene expression in plants during abiotic stress. *Int. J. Mol. Sci.* **2009**, *10*, 3168–3185. [CrossRef] [PubMed]
52. Guo, J.; Wang, S.; Valerius, O.; Hall, H.; Zeng, Q.; Li, J.F.; Weston, D.J.; Ellis, B.E.; Chen, J.G. Involvement of *Arabidopsis* RACK1 in protein translation and its regulation by abscisic acid. *Plant Physiol.* **2011**, *155*, 370–383. [CrossRef] [PubMed]
53. Wang, C.; Du, X.; Mou, Z. The mediator complex subunits MED14, MED15, and MED16 are involved in defense signaling crosstalk in *Arabidopsis*. *Front Plant Sci.* **2016**, *7*, 1947. [CrossRef]
54. Dwivedi, N.; Maji, S.; Waseem, M.; Thakur, P.; Kumar, V.; Parida, S.K.; Thakur, J.K. The mediator subunit OsMED15a is a transcriptional co-regulator of seed size/weight-modulating genes in rice. *BBA Gene Regul. Mech.* **2019**, *1862*, 194432. [CrossRef]
55. Wei, T.; Deng, K.; Gao, Y.; Liu, Y.; Yang, M.; Zhang, L.; Zheng, X.; Wang, C.; Song, W.; Chen, C.; et al. *Arabidopsis* DREB1B in transgenic *Salvia miltiorrhiza* increased tolerance to drought stress without stunting growth. *Plant Physiol. Biochem.* **2016**, *104*, 17–28. [CrossRef] [PubMed]
56. Kiyosue, T.; Yamaguchi-Shinozaki, K.; Shinozaki, K. Characterization of two cDNAs (*ERD10* and *ERD14*) corresponding to genes that respond rapidly to dehydration stress in *Arabidopsis thaliana*. *Plant Cell Physiol.* **1994**, *35*, 225–231. [CrossRef]
57. Msanne, J.; Lin, J.; Stone, J.M.; Awada, T. Characterization of abiotic stress-responsive *Arabidopsis thaliana RD29A* and *RD29B* genes and evaluation of transgenes. *Planta* **2011**, *234*, 97–107. [CrossRef]
58. Lee, S.U.; Mun, B.G.; Bae, E.K.; Kim, J.Y.; Kim, H.H.; Shahid, M.; Choi, Y.I.; Hussain, A.; Yun, B.W. Drought stress-mediated transcriptome profile reveals NCED as a key player modulating drought tolerance in *Populus davidiana*. *Front. Plant Sci.* **2021**, *12*, 755539. [CrossRef]
59. De Ollas, C.; Arbona, V.; Gómez-Cadenas, A.; Dodd, I.C. Attenuated accumulation of jasmonates modifies stomatal responses to water deficit. *J. Exp. Bot.* **2018**, *69*, 2103–2116. [CrossRef]
60. Seki, M.; Umezawa, T.; Urano, K.; Shinozaki, K. Regulatory metabolic networks in drought stress responses. *Curr. Opin. Plant Biol.* **2007**, *10*, 296–302. [CrossRef]
61. Chen, K.; Li, G.J.; Bressan, R.A.; Song, C.P.; Zhu, J.K.; Zhao, Y. Abscisic acid dynamics, signaling, and functions in plants. *J. Integr. Plant Biol.* **2020**, *62*, 25–54. [CrossRef] [PubMed]
62. Howe, G.A.; Major, I.T.; Koo, A.J. Modularity in jasmonate signaling for multistress resilience. *Annu. Rev. Plant Biol.* **2018**, *69*, 387–415. [CrossRef] [PubMed]
63. Wang, J.; Song, L.; Gong, X.; Xu, J.; Li, M. Functions of jasmonic acid in plant regulation and response to abiotic stress. *Int. J. Mol. Sci.* **2020**, *21*, 1446. [CrossRef] [PubMed]
64. Zhai, Q.; Deng, L.; Li, C. Mediator subunit MED25: At the nexus of jasmonate signaling. *Curr. Opin. Plant Biol.* **2020**, *57*, 78–86. [CrossRef]

horticulturae

Article

Proteomic Analysis of the Cold Stress Response of *Ammopiptanthus mongolicus* Reveals the Role of AmCHIA in Its Cold Tolerance

Xuting Li [1,2,3,†], Qi Liu [1,2,3,†], Rongqi Wu [1,2,3], Jie Bing [4], Lamei Zheng [1,2,3], Batu Sumbur [1,2,3], Yijun Zhou [1,2,3,*] and Fei Gao [1,2,3,*]

1 Key Laboratory of Mass Spectrometry Imaging and Metabolomics, Minzu University of China, National Ethnic Affairs Commission, Beijing 100081, China; 15210636339@163.com (X.L.); bio_67@163.com (Q.L.); 15272960705@163.com (R.W.); zhenglm100@163.com (L.Z.); baikaishuisbebt@163.com (B.S.)
2 Key Laboratory of Ecology and Environment in Minority Areas, Minzu University of China, National Ethnic Affairs Commission, Beijing 100081, China
3 College of Life and Environmental Sciences, Minzu University of China, Beijing 100081, China
4 College of Life Sciences, Beijing Normal University, Beijing 100080, China; bingjie@bnu.edu.cn
* Correspondence: zhouyijun@muc.edu.cn (Y.Z.); gaofei@muc.edu.cn (F.G.); Tel.: +86-68932633 (Y.Z. & F.G.)
† These authors contributed equally to this work.

Abstract: *Ammopiptanthus mongolicus*, a traditional ethnic medicinal herb, is a rare broad-leaved evergreen shrub in the arid region of central Asia that can survive under extremely low temperatures during winter. In this study, we investigated the cold stress response of *A. mongolicus* leaves using physiological and proteomic approaches. Cold stress treatment increased the relative electrolyte leakage, proline, and soluble sugar levels and decreased the chlorophyll content in *A. mongolicus* leaves. Moreover, 93 differentially abundant proteins (DAPs) were identified using two-dimensional electrophoresis, of which 79 were further confirmed via tandem mass spectrometric analysis. The predicted functions of DAPs were mainly associated with photosynthesis in chloroplasts, reactive oxygen species scavenging, defense, and protein synthesis, folding, and degradation. *A. mongolicus* chitinase A (AmCHIA) is a cold-induced apoplast protein whose transcription is upregulated under cold, osmotic, high-salinity, and mechanical stresses. Recombinant AmCHIA expressed in *Escherichia coli* exhibits chitinase activity. Here, *AmCHIA* expression enhanced the cold tolerance of *E. coli* cells, suggesting that it may contribute to the cold adaptation of *A. mongolicus* after cold treatment. The present study not only provides important data for understanding the cold stress responses in plants but also serves as a basis for further exploration of the biological functions of cold-induced proteins in *A. mongolicus* using genetic approaches.

Keywords: *Ammopiptanthus mongolicus*; leaf; cold stress; proteomics; chitinase

Citation: Li, X.; Liu, Q.; Wu, R.; Bing, J.; Zheng, L.; Sumbur, B.; Zhou, Y.; Gao, F. Proteomic Analysis of the Cold Stress Response of *Ammopiptanthus mongolicus* Reveals the Role of AmCHIA in Its Cold Tolerance. *Horticulturae* **2023**, *9*, 1114. https://doi.org/10.3390/horticulturae9101114

Academic Editor: Roberto Berni

Received: 25 August 2023
Revised: 20 September 2023
Accepted: 21 September 2023
Published: 9 October 2023

1. Introduction

Plants cannot escape the harmful environmental factors as they are anchored in the soil. Low temperature is a key limiting factor affecting plant growth, development, and production [1]. Plants are vulnerable to temperature fluctuations, and most plants cannot tolerate low temperatures during the reproductive period, which can lead to plant infertility or a significant reduction in crop yield [2,3]. To adapt to low-temperature stress, plants have evolved countermeasures via many complex physiological and biochemical changes, including the activation of primary metabolism to produce more energy, increase in the antioxidant enzyme activity, and maintenance of osmotic balance by changing the membrane structure [4,5].

Ammopiptanthus mongolicus is an evergreen broad-leaved shrub of the family, *Leguminosae*, that has a height of 1.5–2 m. *A. mongolicus* is a traditional ethnic medicinal herb mainly used by Mongolians, whose branches and leaves can be used for medicinal purposes, including the dispelling of wind and dampness, promotion of blood circulation, dispersion of blood stasis, and pain relief. It can also be used externally to treat chilblains, particularly in people with rheumatoid arthritis. *A. mongolicus* plays essential roles in maintaining the desert ecosystem and preventing further desertification [6]. *A. mongolicus* can survive under extreme drought, salinity, and low temperature (up to −30 °C) conditions [7]. The extreme tolerance of this plant species to harsh environments makes it a valuable model to identify stress tolerance genes, especially those involved in cold tolerance, and explore their underlying mechanisms. Extensive large-scale transcriptional analyses using amplified fragment length polymorphisms [8] and transcriptome sequencing (RNA-seq) [9] have been performed to identify cold stress-responsive genes. Several cold-induced genes, including *AmCIP* [10], *AmLEA*, *AmEBP1* [11], and *AmCaM* [12], have been isolated and functionally characterized via ectopic expression. In addition, several antifreeze proteins proposed to contribute to the freezing tolerance of *A. mongolicus* have been purified, and their antifreeze activities have been determined [13,14]. These studies demonstrate the efficacy of using *A. mongolicus* to identify cold stress tolerance-related genes.

Proteomic studies identify protein markers, whose abundance changes are associated with changes in biological processes, thus providing important data to understand the metabolic adjustments in response to environmental stresses. In recent years, proteomics has been widely used as an efficient approach to study the plant responses to environmental stresses [15,16]. Proteomic analysis generally involves the separation of the protein mixtures via two-dimensional electrophoresis (2-DE), which is followed by the identification of protein spots on the 2-DE gel using mass spectrometry. Proteomic analysis of the cold stress response using 2-DE or quantitative proteomics with stable isotopic labeling has been performed in various plant species, such as rice [17], chickpea [18], peanut [19], jojoba [20], and *Medicago sativa*. However, to date, no large-scale screening for cold-responsive proteins in *A. mongolicus* has been reported at the protein level despite its importance in abiotic stress tolerance.

In the present study, changes in the *A. mongolicus* leaf proteome at low temperatures were investigated using 2-DE and tandem mass spectrometric analysis. A batch of cold-responsive proteins was identified, and the corresponding biological processes regulated under cold stress were determined. Additionally, biological functions of the cold-induced protein, *A. mongolicus* chitinase A (AmCHIA), were evaluated via expression profiling and ectopic expression in *Escherichia coli*. Our results provide important data to understand the cold stress response of *A. mongolicus* at the protein level.

2. Materials and Methods

2.1. Plant Materials, Growth Conditions, and Stress Treatments

Seeds of *A. mongolicus*, collected from the dryland in Zhongwei City (Ningxia Autonomous Region, China) were surface-sterilized with ethanol, soaked in water for 48 h at 25 °C, and sown in 30 cm diameter pots containing vermiculite, perlite, and nutritional supplements (3:1:1, *w*/*w*/*w*) in a greenhouse at approximately 25 °C and 35% relative humidity under a photosynthetic photon flux density of 120 µmol m^{-2} s^{-1} and a photoperiod of 16 h/8 h light/dark. The seedlings were watered at three-day intervals with half-strength Hoagland's solution. Eight weeks after germination, when four pairs of leaves had developed, seedlings of similar heights were used for cold stress treatment. The seedlings were randomly divided into four groups. The first group was harvested just before treatment and served as the control (0 h), whilst the second (1 h), third (24 h), and fourth (72 h) groups were kept at 4 °C (16 h/8 h light/dark) in a plant chamber (Percival I-36VL; Percival, Inc., Perry, IA, USA) for low temperature treatment for 1, 24, and 72 h, respectively. Leaf samples from all groups were snap-frozen immediately in nitrogen and stored at −80 °C until further processing. Three independent replicates of 15 plants were harvested for each sample. Leaf samples collected at 0, 1, and 72 h were further used for proteomic analysis.

2.2. Measurement of the Relative Electrolyte Leakage (REL) and Proline Levels

REL is used to evaluate cell membrane damage [21]. Here, leaves were cut into 1 cm segments and washed thrice with ultrapure water. The segments were placed in tubes containing 5 mL of ultrapure water and incubated at 28 °C. After 2 h, the electrical conductivity of the bathing solution (Lt) was measured. Tubes were first incubated at 100 °C for 20 min and subsequently at 28 °C for 1 h, and the electrical conductivity (L0) was measured again. REL was calculated using the following formula: $Lt/L0 \times 100\%$. Five replicates were used for each sample.

Free proline levels were measured as previously described [21]. In total, 0.5 g (fresh weight) of spring wheat leaves was homogenized in 3% sulfosalicylic acid and boiled for 10 min. The supernatant was treated with acetic acid and acid ninhydrin and boiled for 30 min. After cooling to room temperature (25 °C), the mixture was eluted with benzene, and the absorbance was read at 520 nm using the BioTek Epoch spectrophotometer.

Chlorophyll content was measured as previously described [22]. Additionally, the soluble sugar content was determined as described by Tang [23].

2.3. Protein Extraction and Quantification

Total soluble proteins were extracted as previously described with some modifications. Leaf samples (300 mg) were transferred to a mortar and ground into a fine powder with 10% polyvinyl pyrrolidone in liquid nitrogen. The powder was suspended in 2 mL of 10% (*w*/*v*) trichloroacetic acid/acetone containing 0.07% dithiothreitol (DTT) and kept at −20 °C for at least 1 h. The homogenate was centrifuged at 12,000× *g* for 30 min at 4 °C. The pellet was washed with chilled (−20 °C) acetone containing 0.07% DTT and centrifuged at 12,000× *g* for 30 min at 4 °C. The above-mentioned steps were repeated until the supernatant became colorless, and the pellet was air-dried and stored at −80 °C until use. The protein powder was weighed and suspended in a lysis buffer containing 8 mol/L urea, 2% (*w*/*v*) 3-[(3-cholamidopropyl) dimethylammonio]-1-propanesulfonate hydrate (CHAPS), 0.5% (*v*/*v*) ampholyte, and 20 mmol/L DTT, and shaken vigorously at 4 °C for at least 1 h. Insoluble material was removed via centrifugation at 12,000× *g* for 30 min at 4 °C [24]. After extraction, protein concentration was quantified using the Bradford method.

2.4. 2-DE Analysis and Protein Identification

Immobilized pH gradient (IPG) strips (24 cm, pH 4–7, linear) were rehydrated overnight with 450 mL of rehydration buffer (8 mol/L urea, 2% (*w*/*v*) CHAPS, 20 mmol/L DTT, and 0.5% (*v*/*v*) ampholyte) in a reswelling tray (Amersham Biosciences, Uppsala, Sweden) at room temperature [24]. Isoelectric focusing (IEF) was conducted at room temperature using the Ettan IPGphore-3 (GE Healthcare, Chicago, IL, USA) Dry Strip kit (Amersham Biosciences, Uppsala, Sweden). The operating conditions were as follows: 100 V for 1 h, 500 V for 1.5 h, 1500 V for 2 h, which was followed by a stepwise increase to 8000 V for approximately 70 kVh. The focused strips were equilibrated twice for 15 min in 15 mL of the equilibration solution. The first equilibration was performed in a solution containing 6 mol/L urea, 30% (*w*/*v*) glycerol, 2% (*w*/*v*) sodium dodecyl sulfate (SDS), 1% (*w*/*v*) DTT, and 0.05 mol/L Tris-HCl buffer, pH 8.8. The second equilibration was performed in a solution modified by the replacement of DTT by 4% (*w*/*v*) iodoacetamide. For SDS–polyacrylamide gel electrophoresis (PAGE), the equilibrated strips were positioned on the stacking gel and sealed with 0.5% agarose solution. The second dimension was run in the gel apparatus in 26 × 20 cm homogeneous 12.5% SDS-PAGE gels. Electrophoresis was performed in a standard Tris-glycine running buffer at a constant voltage of 500 V. Gels were stained with silver, and gel images were acquired using UMAX Powerlook 2011XL (GE Healthcare, Chicago, IL, USA). The data were analyzed using Image Master 2D Platinum 7.0 software (Amersham Bioscience). For quantitative analysis, the volume of each spot was normalized against the total number of valid spots. Spots from three biological replicates with more than a 1.5-fold change among the protein samples and p value < 0.05 were considered as differentially abundant proteins (DAPs).

Spots showing significant and reproducible changes were excised, digested with trypsin (Promega, http://www.promega.com/, accessed on 2 August 2022), and desalted, and the resulting protein fragments were analyzed using a 4800 Plus matrix-assisted laser desorption tandem time-of-flight instrument (AB Sciex, http://www.absciex.com/, accessed on 26 August 2022) (C11). Database searches, including peptide mass fingerprinting and tandem mass spectrometry (MS/MS) analyses, were performed using MASCOT (http://www.matrixscience.com, accessed on 26 August 2022). A customized protein database was set up based on the transcriptome dataset of *A. mongolicus*. Green plants were selected as the taxonomic categories. The other parameters for searching were an enzyme of trypsin, one missed cleavage, fixed modifications of carbamidomethyl group (C) and variable modifications of oxidation (M), peptide tolerance of 100 ppm, MS/MS tolerance of 0.4 Da, peptide charge of 1 +, and being monoisotopic. According to the MASCOT probability analysis ($p \leq 0.05$), only significant hits were accepted for the identification of protein samples [25].

2.5. Functional Classification and Annotation

Next, we used AgriGO (http://bioinfo.cau.edu.cn/agriGO/, accessed on 26 August 2022) for GO annotation enrichment analysis and KOBAS (http://kobas.cbi.pku.edu.cn/, accessed on 26 August 2022) for Kyoto Encyclopedia of Genes and Genomes (KEGG) pathway analysis of DAPs. When no GO annotation was available, proteins were manually annotated based on literature searches and closely related homologs. As a single classification of GO or KEGG does not really represent the function of the protein, we continued to classify these proteins with high confidence with reference to the Bevan standard [26].

2.6. Protein Subcellular Location and Protein–Protein Interaction (PPI) Analysis

Subcellular locations of the identified proteins were predicted using five internet tools: YLoc [27] (confidence score ≥ 0.4), LocTree3 [28] (expected accuracy $\geq 80\%$), Plant-mPLoc [29], ngLOC [30] (probability $\geq 80\%$), and TargetP [31] (reliability class ≤ 3). Only consistent predictions from at least two tools were considered reliable. If the method was still unable to determine subcellular localization, GO analysis of cell components and a literature search were used for predictions.

To determine the relationship among DAPs under cold stress, PPI networks were generated using the web tool, STRING v10 (http://string-db.org, accessed on 18 August 2022) [32], and edited using Cytoscape 2.8.3. The minimum required interaction score for PPI network construction is 0.7 (high confidence). DAP homologs in *Arabidopsis* were identified using sequence BLAST on the TAIR10 database.

2.7. Quantitative Reverse Transcription-Polymerase Chain Reaction (qRT-PCR) Analysis

Total RNA was extracted using the TRIzol reagent (Invitrogen, Burlington, ON, Canada), according to the manufacturer's protocol. The quality and purity of RNA were assessed by determining the absorbance at 280, 260, and 230 nm using the NanoDrop 2000 spectrophotometer (Thermo Scientific, Wilmington, NC, USA). Approximately 1 μg of DNase I-treated total RNA was converted into single-stranded cDNA using M-MLV Reverse Transcriptase (TIANGEN, Beijing, China). The cDNA products were diluted with deionized water before being used as templates for real-time PCR. The quantitative reaction was performed using TIANGEN SYBR Premix Ex Taq II on an MyiQ2 two-color real-time PCR detection system (Bio-Rad Laboratories, Hercules, CA, USA). The reaction mixture (10 μL) contained 2× qPCR Master Mix, 0.3 μL each of the forward and reverse primers, and 1 μL of template cDNA. PCR amplification was performed under the following conditions: 95 °C for 1 min, followed by 40 cycles of 95 °C for 5 s, 55 °C for 10 s, and 60 °C for 10 s. Three independent biological replicates for each sample and three technical replicates of each biological replicate were analyzed via qRT-PCR analysis. Expression levels of the selected genes were normalized against that of an internal reference gene, *eIF1*. Relative gene expression was calculated using the $2^{-\Delta\Delta Ct}$ method [33]. All primers used in this study are listed in Table S1.

2.8. Sequence Alignment and Phylogenetic Analysis

BLAST was used to run homologous sequence queries on the National Center of Biotechnology Information website (www.ncbi.nlm.nih.gov, accessed on 16 August 2022) to obtain the amino acid sequences of AmCHIA orthologs from 11 other plant species. Plant species and proteins include maize ZmCHIA (XP_008652265.1), *A. thaliana* AtCHIA (BAA21870.1), chickpea CaCHIA (XP_004496499.1), *Lotus japonicus* LjCHIA (AFK40655.1), *Acacia koa* AkCHIA (AFY08286.1), *Rhododendron irroratum* RiCHIA (ADF47473.1), rice OsCHIA (XP_015622676.1), *Lupinus angustifolius* LaCHIA (XP_019453315.1), *Trifolium pratense* TpCHIA (PNX81307.1), *Vigna angularis* VaCHIA (XP_017439504.1), and *Vigna unguiculata* VuCHIA (XP_027931365.1). MAFFT v7 [34] was used to perform multiple sequence alignments of AmCHIA and its orthologs in other plant species. MEGA-X [35] was used to build a phylogenetic tree using the neighbor-joining method.

2.9. Promoter Analysis of AmCHIA

The online promoter analysis software, PLACE [36] (http://www.dna.affrc.go.jp/htdocs/PLACE/, accessed on 26 August 2022), was used to predict and analyze the *cis*-acting element of the AmCHIA promoter region (2000 bp upstream of the start codon).

2.10. Prokaryotic Expression and Purification of AmCHIA and Chitinase Activity Measurement

AmCHIA was expressed in *E. coli* as previously described [37]. The pET-22b(+) vector with *Bam*HI and *Hind*III digestion sites and *E. coli* BL21(DE3) competent cell line were used for this experiment. Briefly, 0.8 mM IPTG was added to the culture medium, incubated at 25 °C for 6 h, and *E. coli* cells were harvested via centrifugation. Then, AmCHIA was purified. Purification and chitinase activity measurements were conducted as previously described [38].

2.11. Evaluation of the Cold Tolerance of E. coli Cells via Growth Curve Analysis

When *E. coli* grew to the logarithmic phase, the culture medium was transferred to a 4 or 12 °C incubator for cold treatment. *E. coli* cultured at 37 °C was used as the control. The OD600 value of the medium was measured every 2 h to plot the growth curve. Three independent biological replicates were used for each experiment.

3. Results

3.1. Physiological Changes in Cold-Stressed A. mongolicus Leaves

To estimate the cold-induced membrane damage in *A. mongolicus*, REL was measured in *A. mongolicus* leaves after exposure to cold stress for 1, 24, and 72 h (Figure 1A). REL increased gradually during cold stress treatment (Figure 1A). Proline protects the protein structures from abiotic stress damage. Here, the proline content of *A. mongolicus* leaves increased with exposure to low temperatures and reached a peak at 72 h (Figure 1B). The accumulation of soluble sugars reduces the osmotic potential of plant cells. The soluble sugar content in the leaves of *A. mongolicus* rapidly increased after 1 h of low-temperature treatment and further increased with the extension of the low-temperature treatment (Figure 1C). In addition, the chlorophyll content was significantly reduced in the leaves of *A. mongolicus* under low-temperature stress (Figure 1D). These results suggest that cold stress causes significant time-dependent damage to the leaf cells of *A. mongolicus*, and osmolytes accumulate in the leaves of *A. mongolicus* in response to this cold stress.

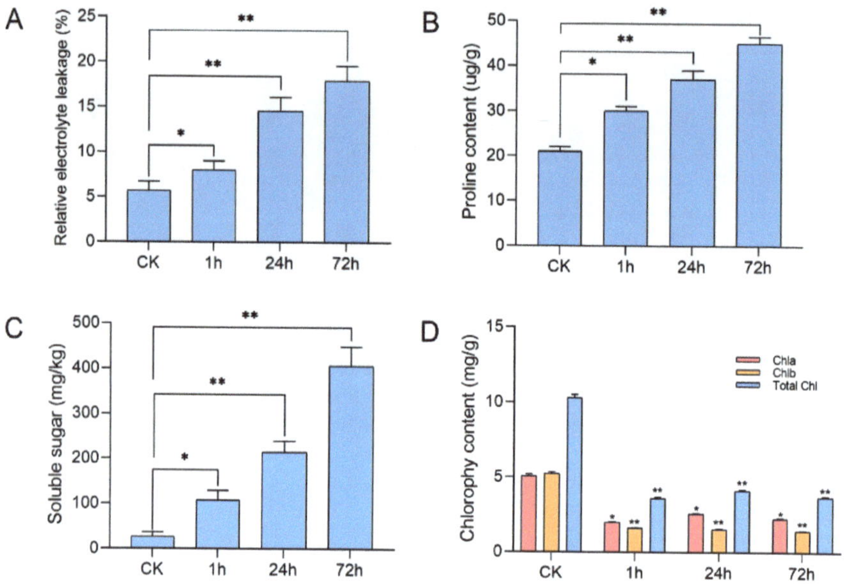

Figure 1. Physiological changes in the leaves of *A. mongolicus* under cold stress. (**A**) Relative electrolyte leakage in the leaves of *A. mongolicus* under cold stress. (**B**) Proline content in the leaves of *A. mongolicus* under cold stress. (**C**) Soluble sugar levels in the leaves of *A. mongolicus* under cold stress. (**D**) Chlorophyll content in the leaves of *A. mongolicus* under cold stress. * $p \leq 0.05$ and ** $p \leq 0.01$; $n = 5$.

3.2. 2-DE Analysis of Leaf Proteins in A. mongolicus under Cold Stress

To uncover the molecular mechanism underlying the physiological response of leaves to temporal and short-term cold stresses, alterations in the *A. mongolicus* leaf proteome after exposure to cold treatment for 1 and 72 h were examined using 2-DE. Triplicate gels were obtained from three independent biological experiments, and the gels from the different experiments showed high reproducibility. Representative gels from the control (0 h) and cold-treated (1 and 72 h) plants are shown in Figure 2. Approximately 1500 protein spots were reproducibly detected using ImageMaster 2-D Elite software version 4.01 (Amersham Biosciences, Freiburg, Germany) on silver-stained gels, and all gels showed similar protein distribution patterns in the 2D image. Quantitative image analysis revealed that the intensities of 93 protein spots changed significantly ($p \leq 0.05$) by more than 1.5-fold compared with the control and emerged at least at one time-point. The three typical regions are shown in Figure 3.

Among the 93 DAP spots, 60 were identified at both the 1 and 72 h time-points, whereas the remaining 33 were detected only at the 1 or 72 h time-point (Figure 4). We observed 26 upregulated spots, 33 downregulated spots, and five new spots after 1 h of cold treatment. After 72 h of cold treatment, 23 protein spots were upregulated, 65 were downregulated, and a new spot was observed.

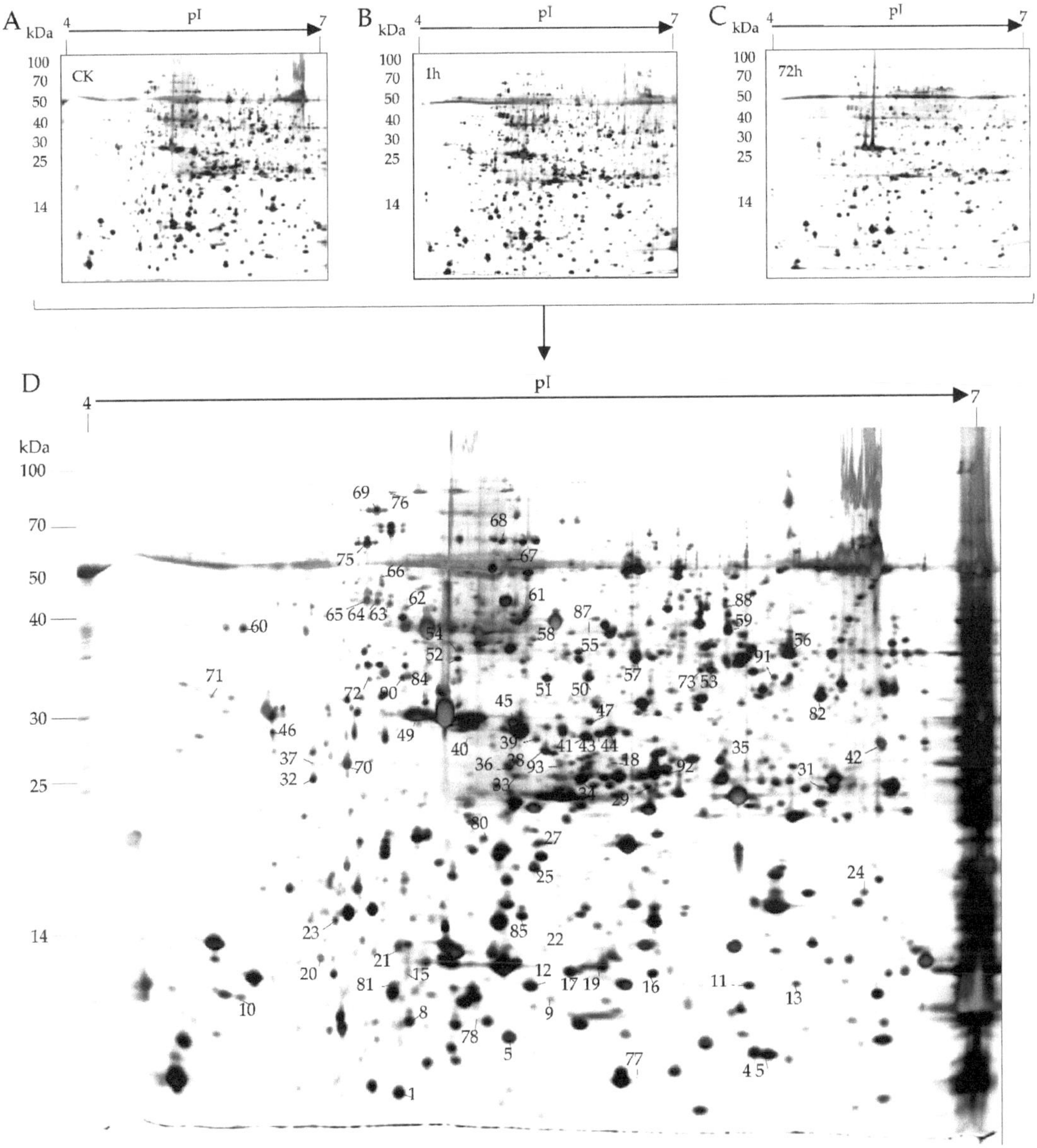

Figure 2. Representative two-dimensional electrophoresis (2-DE) gels of *A. mongolicus* leaf proteins. Total leaf proteins were extracted and separated using 2-DE. For isoelectric focusing (IEF), 400 μg of proteins was loaded onto pH 4–7 immobilized pH gradient (IPG) strips (24 cm, linear). Sodium dodecyl sulfate-polyacrylamide gel electrophoresis (SDS-PAGE) was performed using 12.5% gels. Spots were visualized via silver staining. Quantitative image analysis revealed a total of 93 spots that changed their intensities significantly ($p \leq 0.05$) by more than 1.5-fold at least at one time-point. Representative 2-DE gels of the (**A**) control group, (**B**) samples that underwent 1 h of cold stress, and (**C**) samples that underwent 72 h of cold stress. (**D**) Enlarged (**A**) with all differentially abundant proteins (DAPs) number labeled.

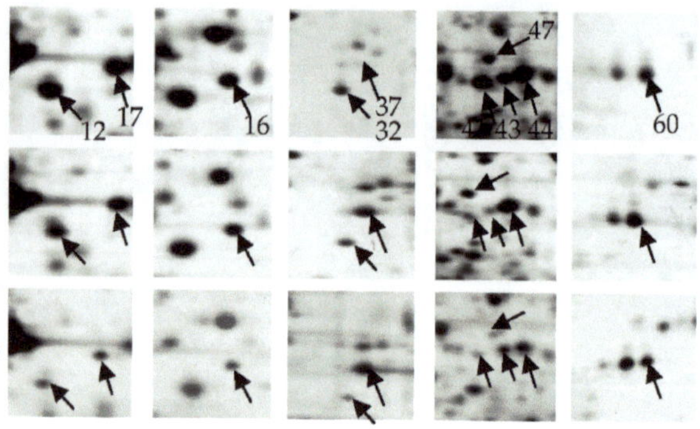

Figure 3. Close-up views of the abundance changes in 10 cold stress-responsive protein spots in the control (0 h) and cold stress-treated (1 and 72 h) groups.

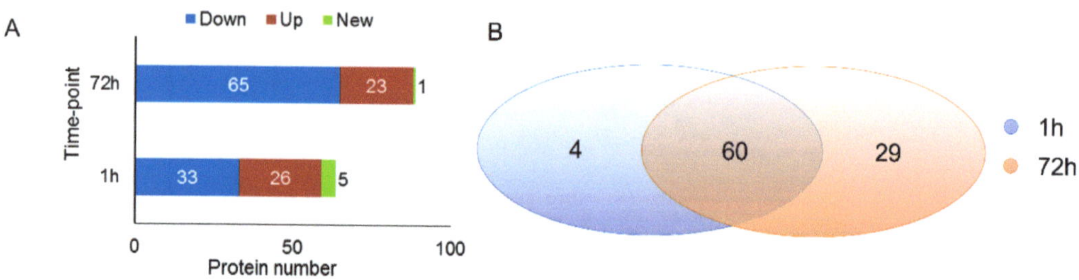

Figure 4. Change patterns and overlap of DAPs identified using 2-DE. (**A**) Numbers of DAPs with different change patterns under cold stress. Blue indicates downregulated, red indicates upregulated, and green indicates newly appeared DAPs. (**B**) Venn diagram analysis of DAPs under cold stress. Numbers of DAPs at different time-points are shown in different segments (blue for 1 h and yellow for 72 h).

3.3. Annotation, Subcellular Localization, and Functional Categorization of DAPs

Here, 93 DAP spots were excised from the gels, digested in-gel with trypsin, and analyzed using tandem mass spectrometry. In total, 79 DAP spots, representing 52 unique proteins, were successfully identified using tandem mass spectrometry (Table S2).

Some proteins were identified as more than one protein spot on a single gel. This may be due to the presence of different protein isoforms, post-translational modifications, or protein degradation. Of the 52 unique DAPs, 38 had one protein spot each, whereas the remaining 14 DAPs had two, three, four, or seven spots each. For example, ribulose-1,5-bisphosphate carboxylase/oxygenase (RuBisCO) activase was identified at protein spots 40, 49, 50, 51, 54, 58, and 73. The other identified proteins presented in multiple spots include RuBisCO large subunit (spot 27 and 67), RuBisCO small subunit (spots 9, 10, and 22), 23 kD extrinsic protein (spots 29 and 35), four spots (spots 62, 63, 64, and 65), triosephosphate isomerase (spots 18, 92, and 93), HSC70 (spots 69 and 71), cyclophilin-like peptidyl-prolyl cis-trans isomerase family protein (spots 15 and 52), chaperonin-60 alpha (spots 75, 76, 77, and 78), concanavalin A-like lectin protein kinase family protein (spots 41, 43, 45, and 46), cystatin B (spots 3 and 4,), chitinase A (spots 37 and 70), ZKT (spots 90 and 91), and ribosomal protein 40S ribosomal protein S12-2 protein (spots 16 and17).

Based on BLAST alignment to the *Arabidopsis* protein database (www.arabidopsis.org, accessed on 26 August 2022) and annotation information from the agriGO database (http://bioinfo.cau.edu.cn/agriGO/, accessed on 26 August 2022), these DAPs were classified into seven functional groups: photosynthesis, reactive oxygen species (ROS) scavenging, protein synthesis, protein processing, protein degradation, RNA processing, defense-related signaling, and others (Figure 5A; Table S2).

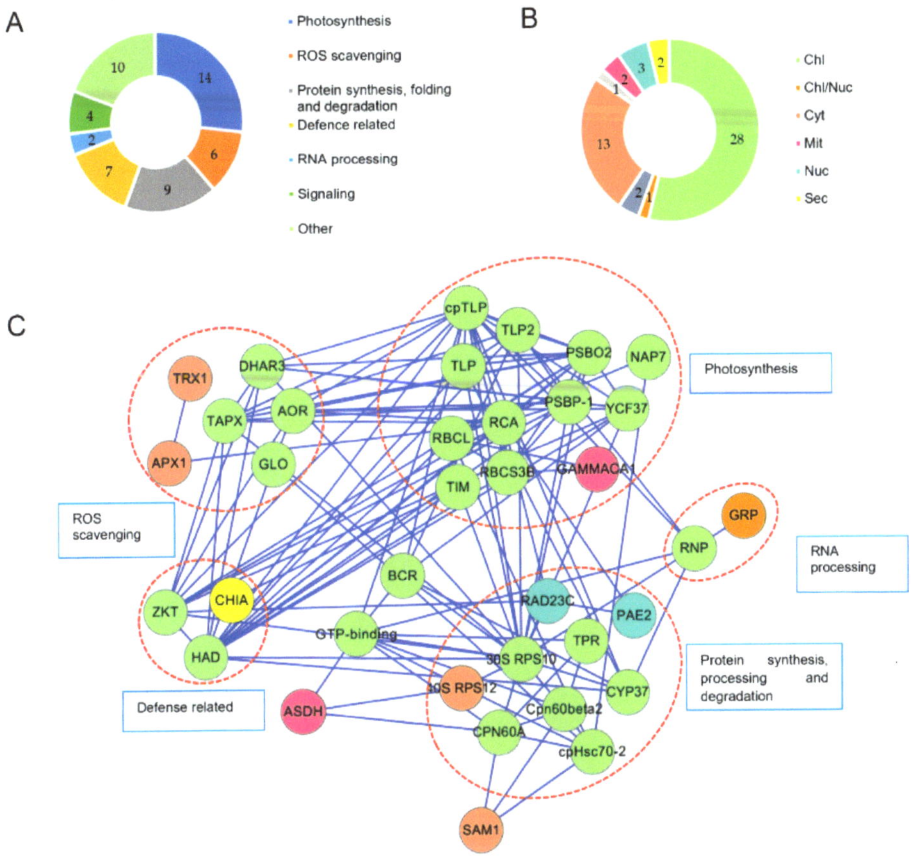

Figure 5. Functional categorization, subcellular localization, and protein–protein interaction (PPI) network of identified DAPs in *A. mongolicus* leaves under cold stress. (**A**) A total of 52 DAPs were classified into seven functional categories. The number of proteins in different functional categories is shown in the pie chart. (**B**) Subcellular localization categories of identified DAPs. Numbers of proteins with different locations are shown. Chl: chloroplast; Cyt: cytoplasm; Mit: mitochondria; Nuc: nucleus; Sec: secreted. (**C**) PPI network comprised of 36 DAPs. Nodes in different colors indicate different subcellular localization categories. Detailed information on protein names and abbreviations is given in Table S2.

Subcellular localization of 52 DAPs was predicted using five web tools (Yloc, LocTree3, ngLOC, TargetP, and Plant-mPLoc). In total, 28 DAPs were predicted to be localized in the chloroplast, 13 were predicted to be localized in the cytoplasm, 2 were predicted to be localized in the mitochondria, 3 were predicted to be localized in the nucleus, and 2 were predicted to be localized in the secretory pathway. In addition, four proteins were predicted to be localized in two organelles (Figure 5B).

PPI analysis of all DAPs identified in *A. mongolicus* revealed that the identified DAPs constituted a complex multicenter interaction network. The network comprised 36 protein nodes with 118 interactive relationships. Proteins in the PPI network were clustered into five clades: (i) photosynthesis, (ii) protein synthesis, processing, and degradation, (iii) ROS scavenging, (iv) defense-related, and (v) RNA-related proteins (Figure 5C). Notably, DAPs within the same clade and between different clades showed extensive interactions.

3.4. Regulation of Cold-Responsive Proteins at the mRNA Level

To analyze alterations in the transcriptional abundance of DAPs under cold stress, qRT-PCR analysis was performed on 12 randomly selected DAPs (Figure 6). The expression patterns of most DAP genes under cold stress were consistent with those of DAPs. Inconsistency between the expression patterns of genes and corresponding proteins has been observed in previous proteomic studies [21], which may be partly due to the post-translational modification of the protein or presence of protein isomers and transcript isomers from alternative splicing [39]. Most genes used for qRT PCR, such as comp65100_c0_seq1 and comp9843_c0_seq1, and their corresponding proteins, γ-carbonic anhydrase 1 (γ-CA1) and triosephosphate isomerase (TIM), were downregulated by cold stress. γ-CA1 and TIM are essential proteins for plant photosynthesis. Downregulation of these proteins/genes under low-temperature conditions suggests that low-temperature stress inhibits plant photosynthesis.

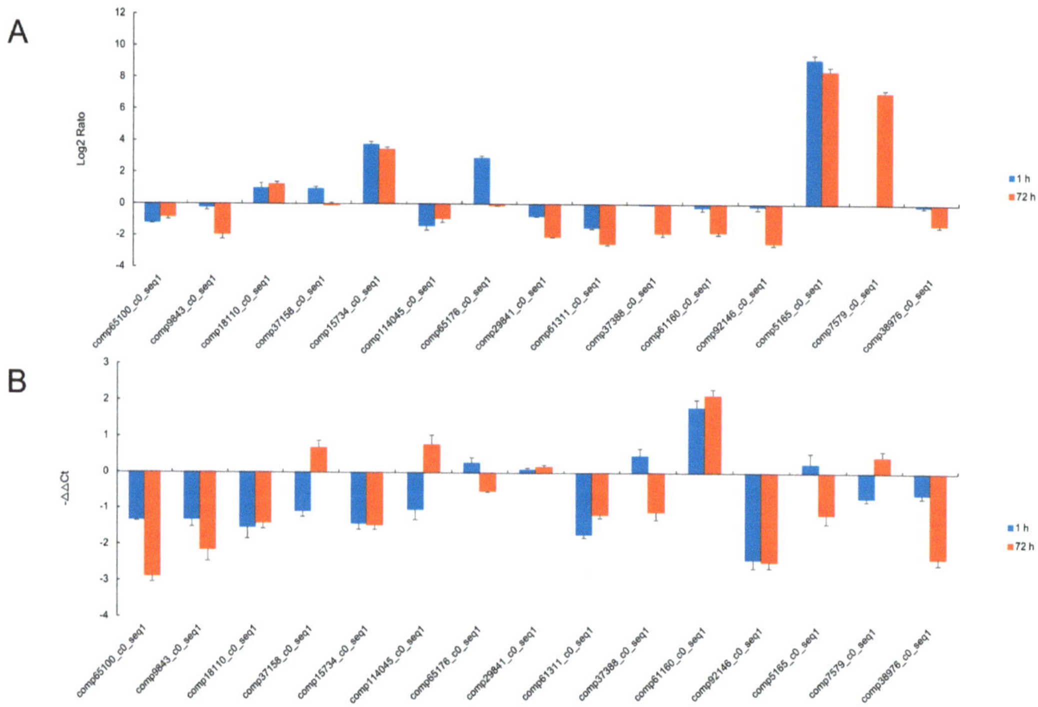

Figure 6. Alterations in the transcription levels of 12 randomly selected DAPs under cold stress determined via quantitative reverse transcription-polymerase chain reaction (qRT-PCR) analysis. (**A**) Change patterns in the protein abundance of the selected DAPs under cold stress. (**B**) Change patterns in the transcription levels of the selected DAPs under cold stress.

3.5. Sequence and Phylogenetic Analyses of AmCHIA

Our proteomic study revealed that two protein spots identified as chitinase A (AmCHIA comp15734_c0_seq1 and protein spots 37 and 70) were significantly upregulated under cold stress, indicating that AmCHIA may be involved in the cold stress response of *A. mongolicus*. AmCHIA expression levels showed the same trend as the protein levels. Therefore, AmCHIA was used for further functional analyses.

To analyze the structural features and evolutionary status of AmCHIA, the amino acid sequences of AmCHIA and its orthologs from 11 other plant species were subjected to multiple sequence alignment and phylogenetic analysis. Sequence alignment showed that AmCHIA belonged to class III of the GH18 family, harbored a conserved domain of the GH18 family (Figure 7A), and exhibited a high degree of sequence similarity with the orthologous proteins of other plant species. A signal peptide was detected at the N-terminal of AmCHIA, indicating that the AmCHIA protein was located in the apoplast. A phylogenetic tree constructed using AmCHIA and its orthologs revealed that the evolutionary relationships of CHIA were consistent with those of these plant species. The phylogenetic tree was divided into two branches (Figure 7B). AmCHIA and its orthologs from five leguminous plants, namely *Lupinus angustifolius*, *Trifolium pratense*, *Phaseolus anagularis*, and cowpea, were clustered into one branch, in which AmCHIA was clustered with its ortholog from *L. angustifolius*.

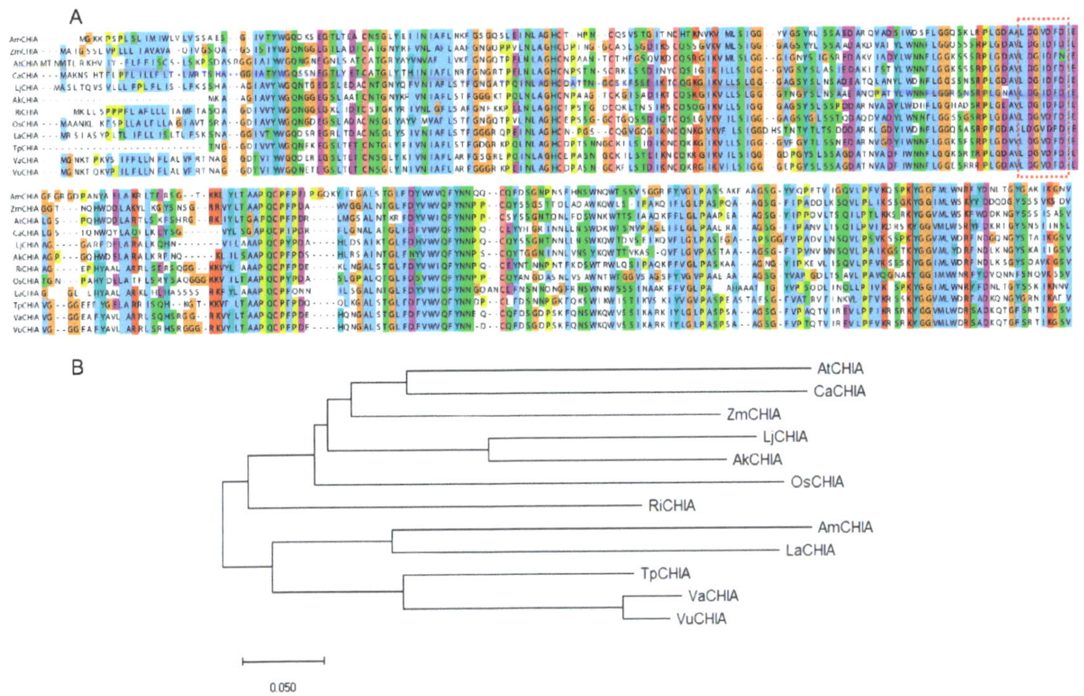

Figure 7. Sequence and phylogenetic analyses of *A. mongolicus* chitinase A (AmCHIA). (**A**) Multiple sequence alignment of AmCHIA and its orthologs from 11 other plant species. Red box indicates the GH18 signature. (**B**) Phylogenetic analysis of AmCHIA and its orthologs from 11 other plant species.

3.6. Promoter Analysis and Expression of AmCHIA under Abiotic Stresses

To understand the regulation of *AmCHIA* under environmental stress, the promoter regions of *AmCHIA* were analyzed to identify the binding motifs of known transcription factors (Table S3). The TATA box was 80 bp upstream of the start codon, and the CAAT box was 154 bp upstream of the start codon. As expected, WRKY71OS, a WRKY TF binding site associated with disease resistance, and "T/G-box", a motif involved in jasmonate response, were found at 517 and 391 upstream of the start codon, respectively, supporting the role of AmCHIA in biotic response. MYB transcription factor binding sites, such as MYB1AT and MYBPLANT, were also found in the promoter of *AmCHIA*, and these *cis*-acting elements mediate the cold, dehydration, and abscisic acid signaling pathways. Taken together, the promoter analysis indicated the involvement of *AmCHIA* in biotic and abiotic stress responses.

qRT-PCR analysis was performed to reveal the expression pattern of *AmCHIA* under abiotic and biotic stresses in leaves of *A. mongolicus* seedlings (Figure 8). The results showed that the expression of *AmCHIA* was greatly induced (more than 30-fold) by mechanical damage and cold stress at the 3 h time-point, while the expression of *AmCHIA* peaked (more than 20-fold) under drought stress at the 24 h time-point. Simultaneously, the expression of *AmCHIA* was elevated to a relatively lesser extent (less than 5-fold) at the 6 h time-point under high-salinity stress. In summary, our data indicate that the expression of *AmCHIA* was induced by various abiotic and biotic stresses, especially mechanical damage and cold stress.

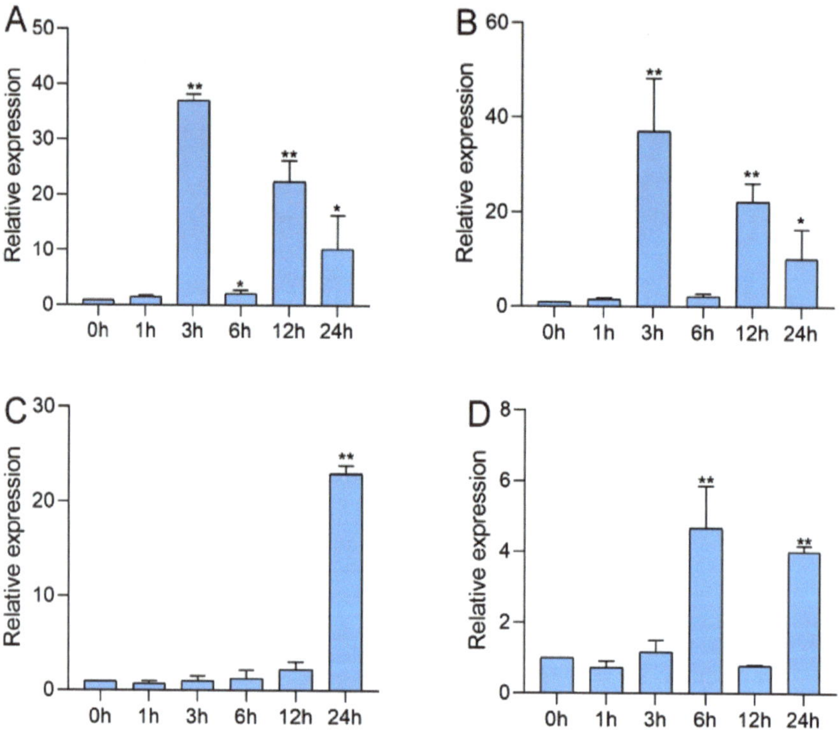

Figure 8. Expression profiles of *AmCHIA* under abiotic and biotic stresses. Expression patterns of *AmCHIA* under cold (**A**), mechanical (**B**), osmotic (**C**), and salt (**D**) stresses. Data are represented as the mean ± standard deviation (SD). $n = 3$; * $p \leq 0.05$ and ** $p \leq 0.01$.

3.7. AmCHIA Encodes a Protein with Chitinase Activity

To determine whether AmCHIA encodes a protein with chitinase activity, AmCHIA was expressed in E. coli. After the addition of IPTG (0.5 mM), a new 35.37 kD polypeptide was observed on the SDS-PAGE gels (Figure S1A), confirming the expression of His-tag-fused AmCHIA in E. coli cells. The recombinant AmCHIA protein was purified (Figure S1B), and the chitinase activity assay showed that the recombinant AmCHIA protein exhibited a specific activity of 40.46 ± 9.44 units/mg.

3.8. Expression of AmCHIA Enhances the Tolerance of E. coli Cells to Cold Stress

Plant chitinase not only has antifungal activity but also exerts a protective effect on cells under low-temperature stress. To analyze whether *AmCHIA* affects the low-temperature tolerance of cells, *AmCHIA* was expressed in *E. coli*, and the low-temperature tolerance of *E. coli* cells expressing *AmCHIA* was evaluated. Growth curve analysis indicated that compared with *E. coli* cells expressing empty vectors, the growth rate of *E. coli* expressing *AmCHIA* did not exhibit a significant change at 37 °C (Figure 9A), but it was promoted at both 4 °C or 12 °C (Figure 9B,C). This indicates that the expression of *AmCHIA* enhanced the tolerance of *E. coli* cells to cold stress.

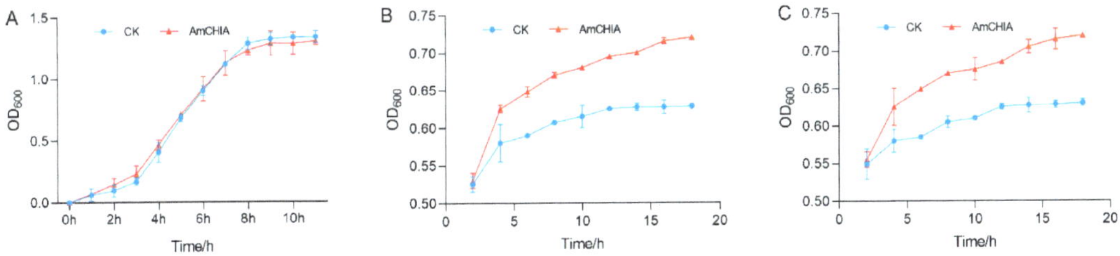

Figure 9. Growth curve analysis of *Escherichia coli* expressing AmCHIA. Growth curves of *E. coli* cells expressing *AmCHIA* and empty vectors cultured at (**A**) 37 °C, (**B**) 4 °C, and (**C**) 12 °C.

4. Discussion

Cold stress affects almost all aspects of cellular function in plants. Therefore, it is not surprising that the identified DAPs are involved in various biological functions of cells, including photosynthesis and chloroplasts [40], ROS scavenging, protein synthesis [41], folding, and degradation [42], RNA processing [43], and defense [44]. The biological significance of DAPs and their associated metabolic pathways are discussed below.

4.1. Proteins Involved in Photosynthesis

Fifty-four DAPs (approximately 60% of all DAPs) were predicted to be localized in chloroplasts, indicating that cold stress has a significant impact on the chloroplasts of *A. mongolicus* leaf cells. Among these DAPs, 14 proteins (45%) participated in photosynthesis, including photosystems, electron transport (light-dependent reactions), and carbon-fixation (light-independent reactions).

In the present study, seven spots (40, 49, 50, 51, 54, 58, and 73) were identified as photosystem II subunit O-2 (PSBO2), and two spots (29 and 35) were identified as photosystem II subunit P-1 (PSBP2). PSBP and PSBO, two extrinsic proteins of photosystem II (PS II), regulate the binding properties of the cofactors Mn^{2+}, Ca^{2+} and Cl^{-} and optimize the oxygenation reaction [45]. YCF37 contains three tetratricopeptide repeat (TPR) units and is involved in PSI assembly and stability as a molecular chaperone [46]. Non-intrinsic ABC protein 7(NAP7), which encodes an ATP-binding cassette/ATPase in plastids, is responsible for the biosynthesis and repair of Fe-S. Fe-S clusters are required for a functional cytochrome b6/f complex, ferredoxin, and photosystem I, ensuring electron flow in thylakoids [47]. In cold-stressed *A. mongolicus* leaves, the abundances of PSBP2, YCF37, and NAP7 were all downregulated, whereas the expression pattern of PSBO2 was quite different; three PSBO2 spots were upregulated and four were downregulated. In general, our results suggest that light-dependent reactions in chloroplasts are inhibited by cold stress.

RuBisCO is a key enzyme for the fixation of CO_2 in photosynthesis [48,49]. The large RuBisCO subunit contained amino acid residues related to catalysis and activation [50]. These small subunits affect the catalytic performance of RuBisCO. In the present study, two large RuBisCO subunit proteins (spots 27 and 67; spot 27 may represent a degradation fragment of the large RuBisCO subunit) increased in abundance after 1 h of cold treatment, whereas four small RuBisCO subunit proteins (spots 9, 10, 19, and 22) were downregulated under cold stress. It has been suggested that cold stress impairs RuBisCO synthesis, resulting in RuBisCO degradation. We speculate that cold stress may affect the stability of RuBisCO and promote its degradation in *A. mongolicus*. In addition, triosephosphate isomerase (TPI or TIM) is responsible for carbon fixation, which participates in the Calvin–Benson cycle [51]. Three TIM spots (spots 18, 92, and 93) were downregulated, indicating that carbon fixation was affected by cold stress.

RuBisCO activase (RCA) modulates RuBisCO activity [52]. In our study, four RCA spots (spots 62, 63, 64, and 65) were upregulated and one RCA spot (spot 33, a degradation fragment of RCA) was downregulated. Considering that the upregulation of RCA was also observed in some other low-temperature-resistant plants, including *Triticum aestivum* and *Pisum sativum* [53], but it is seldom reported in cold-sensitive crops, such as *Manihot esculenta*, *Oryza sativa* [54], and *Solanum tuberosum* [55]. It is conceivable to hypothesize that the upregulation of RCA may represent a strategy adopted by *A. mongolicus* to maintain RuBisCO activity under cold stress; hence, the upregulation of RCA should contribute to cold acclimation of *A. mongolicus*.

4.2. Proteins Involved in ROS Scavenging

At normal growth conditions, the accumulations of ROS are at a relatively low level in cells [56]. However, under stressful conditions, such as cold stress, the levels of ROS, including superoxide radicals, H_2O_2, and hydroxyl radicals, are always dramatically elevated and can damage cell components, including nucleotides, proteins, and membrane lipids. Plants have developed ROS scavenging mechanisms to maintain redox homeostasis under stressful conditions [57]. Six DAPs involved in redox homeostasis were identified in the present study. These proteins include alkenal/one oxidoreductase (AOR), chloroplast thylakoid ascorbate peroxidase (tAPX), dehydroascorbate reductase 1 (DHAR), cytosolic ascorbate peroxidase, APX1, and thioredoxin H-type 1 (TRX1).

H_2O_2 is converted to water by ascorbate peroxidase (APX) [58,59]. APX1, a cytosolic ascorbate peroxidase, was significantly upregulated in *A. mongolicus* under cold stress, indicating this ascorbate peroxidase likely plays an important role in ROS scavenging in cold-stressed *A. mongolicus* leaves.

Thylakoid tAPX scavenges hydrogen peroxide in the chloroplasts of plant cell chloroplasts. Ascorbic acid (ASC) is a major antioxidant in plants that detoxifies ROS and maintains photosynthetic function. DHAR is responsible for regenerating Asc from its oxidized state and regulating the cellular Asc redox state. DHAR affects the level of foliar ROS and photosynthetic activity during leaf development and therefore influences plant growth and leaf aging. Thioredoxin (Trx) removes H_2O_2 and regulates photosynthetic enzyme activity via irradiation with ferredoxin–thioredoxin reductase under light conditions [60]. As a chloroplastic NADPH-dependent reductase in higher plants, the AOR plays an important role in the detoxification of reactive carbonyls, which are mainly produced through stress-induced lipid peroxidation and are highly toxic to cells [61]. Despite the roles in ROS scavenging reported in other plant species, the expression of tAPX, AOR, TRX1, and DHAR2 was downregulated in leaves *A. mongolicus* under cold stress, indicating that these proteins may not be the main ROS scavengers in *A. mongolicus* leaves under cold stress. Similar results have been reported for DHAR2 and several other ascorbate peroxidases in plantain [62]. We speculated that different plants might use different ROS scavengers to cope with stress-induced ROS accumulation, and the cold-induced APX identified in our study might be essential for cold acclimation in *A. mongolicus*.

4.3. Proteins Involved in Protein Synthesis, Folding, and Degradation

Cold stress may significantly influence the protein synthesis apparatus in plants. Furthermore, cold stress increases the risk of protein misfolding, resulting in the production of nonfunctional proteins [63]. In the present study, cold stress caused the differential expression of 14 proteins in *A. mongolicus*, which were divided into three groups according to their putative functions: protein synthesis-related, protein folding-related, and protein degradation-related.

We found three DAPs involved in protein synthesis, including 40S ribosomal protein S12-2 (RPS12-2, spots 16 and 17) and 30S ribosomal protein S10 (RPS10, spot 80; Table S2). RPS12-2 is a major component of the ribosomal small subunit. Two RPS12-2 spots were downregulated under cold stress, indicating that protein synthesis was inhibited under cold stress. RPS10 belongs to the small subunit of chloroplast ribosomes, and its accumulation of RPS10 under cold treatment has been reported in *Lolium perenne* [64]. However, the biological significance of the RPS10 accumulation under cold stress requires further investigation.

Misfolded proteins may accumulate in cells under stressful conditions, and plants may employ two strategies to cope with this challenge: removing them or refolding them. Chaperonins are molecular chaperones that assist in the folding and assembly of misfolded proteins or nascent polypeptides. Commonly used chaperones include many heat shock proteins (HSPs), such as HSP and peptidyl-prolyl cis-trans isomerase. In the present study, several chaperones were upregulated under cold stress, including chaperonin-60 alpha (CPN60α) (spots 75, 76, 77, and 78), chaperonin-60β2 (Cpn60β2) (spot 68), cpHSC70s (spots 69 and 71), and cyclophilin-like peptidyl-prolyl cis-trans isomerase family protein (spots 15 and 52). The upregulation of CPN60α under cold stress was reported in the leaves of several plant species, i.e., potato, rice, and *Arabidopsis thaliana*. Hsc70 is upregulated in wheat leaves after cold treatment.

In *A. mongolicus*, most of these chaperones were upregulated only at the 1 h time-point, and we speculated that these chaperones were adopted to cope with the transitory endoplasmic reticulum stress triggered by cold stress. Notably, CPN60α was accumulated during the 72 h stress treatment. CPN60 binds to the chloroplast RuBisCO large subunit and assists in the assembly of RuBisCO [65]. Cold stress causes degradation of the RuBisCO large subunit. Thus, it is conceivable to conclude CPN60α may play a vital role in cold tolerance and acclimation in *A. mongolicus* by maintaining the structure and activity of RuBisCO.

Two DAPs involved in protein degradation were identified in this study. One of these, radiation sensitive 23 (RAD23), has been demonstrated to play an essential role in the cell cycle, morphology, and fertility of plants through the proteasomal degradation pathway [66]. Although many proteins involved in protein degradation were found to differentially accumulate under abiotic stress, our proteomics study showed that RAD23 may play an important role in the cold stress response by removing abnormal proteins and repairing DNA under stressful conditions.

4.4. Proteins Involved in the Defense Response

Several DAPs involved in defense responses were identified, including cystatin B (CYSB; spots 1, 3, and 4), chitinase A (CHIA; spots 37 and 70), MLP-like protein 31 (spot 13), and putative lactoylglutathione lyase (GLYI6) (spot 84).

Cystatins are induced in response to biotic stress-related signals and play a defensive role against many pathogens and pests. In addition, cystatins have been reported to be upregulated under environmental stresses, including cold stress in rice and *Thellungiella*, and the ectopic expression of Arabidopsis CYSB in transgenic yeast and Arabidopsis plants increased their resistance to high salinity, drought, oxidative, and cold stresses [67]. MLP-like protein 43 (MLP43) functions as a positive regulator of ABA responses and confers drought tolerance to *A. thaliana* [63]. In the present study, three CYSB spots and one MLP43 spot were downregulated under cold stress.

MLP are members of the pathogenesis-related protein 10 (PR10) family. Most MLP proteins play a role in triggering downstream signal transduction by binding ligands, such as cytokinins, brassinolides, and secondary metabolites, and participating in plant abiotic stress responses [68]. ATGLYI6 is a Ni^{2+}-dependent GLYI enzyme that condenses methyl glyoxal and glutathione into S-d-lactate glutathione, thereby reducing its cytotoxicity [69].

In this study, except CHIA, the expression of which was significantly upregulated, the expression levels of other defense-related proteins were significantly downregulated under cold stress. This finding suggests that *A. mongolicus* responds to abiotic stress (cold stress) by suppressing the expression of defense-related proteins.

4.5. *AmCHIA* Contributes to Cold Tolerance by Alleviating Cold Stress Damage

Chitin is an important natural polymer present in various organisms. Chitinase is a glycoside hydrolase that degrades chitin. The size of chitinase varies from 20 to 90 kDa, the optimal pH ≤ 7, and the isoelectric point is between 3 and 10 [70,71]. Molecular weights of different chitinases vary greatly, but most chitinase proteins have conserved domains, including the N-terminal signal peptide and catalytic domain, C-terminal chitin-binding domain, and serine/threonine-rich junction region between the N-terminus and C-terminus. Chitinases are widely distributed in plants and have several classification systems. Based on their structural characteristics, plant chitinases can be divided into classes I–VII, belonging to GH18 and GH19, the glycoside hydrolase families. In this study, AmCHIA, a cold-induced chitinase protein with chitin hydrolytic activity, was identified in *A. mongolicus*, and sequence alignment revealed that AmCHIA is a member of class III GH18 [72]. Here, the transcription of *AmCHIA* was induced under cold, osmotic, salt, and mechanical stresses. Several *cis*-acting elements were predicted in the promoter region of *AmCHIA*, supporting the involvement of *AmCHIA* in the abiotic response of *A. mongolicus*. qRT-PCR analysis revealed that *AmCHIA* was induced by mechanical damage. In the *AmCHIA* promoter region, five W-box *cis*-acting elements were identified, and these *cis*-acting elements were hypothesized to mediate the induction of *AmCHIA* after mechanical damage and insect gnawing. qRT-PCR analysis also showed that *AmCHIA* was induced by drought and low temperature. Moreover, *cis*-acting elements, such as MYB- and MYC/ICE1-binding sites, found in the *AmCHIA* promoter region were hypothesized to mediate the induction of drought and low-temperature stress by *AmCHIA* in *A. mongolicus*.

Plant chitinases include several categories of PRs that destroy the cell walls of pathogenic fungi and exoskeleton of insects, thus playing important roles in plant defense. In addition to participating in biotic stress responses, chitinases play key roles in abiotic stress responses. Some plant chitinases are involved in the cold stress responses of *Bromus inermis* and *Chimonanthus praecox* [73,74]. A class I chitinase from *Hippophae rhamnoides* protects the cells from damage caused by cold stress [75]. Here, expression of the gene encoding cold-induced AmCHIA in *E. coli* promoted the growth of *E. coli* cells under low-temperature stress, suggesting that AmCHIA contributes to cold tolerance by alleviating the damage caused by cold stress in *A. mongolicus*. However, whether AmCHIA has a molecular chaperone activity similar to that of HSPs requires further investigation.

Supplementary Materials: The following supporting information can be downloaded at: https://www.mdpi.com/article/10.3390/horticulturae9101114/s1, Table S1: The primer sequence used in qRT-PCR analysis; Table S2: Identification of differentially abundant proteins in *A. mongolicus* leaves under cold stress; Table S3: The Cis-acting elements in *AmCHIA* promoters that may be involved in environmental stress and hormone response; Figure S1: Prokaryotic expression and purification of AmCHIA.

Author Contributions: Conceptualization, Y.Z. and F.G.; Formal analysis, B.S.; Funding acquisition, Y.Z. and F.G.; Investigation, X.L., Q.L., R.W., J.B. and L.Z.; Supervision, Y.Z.; Writing—original draft, X.L., Q.L. and F.G.; Writing—review and editing, F.G. All authors have read and agreed to the published version of the manuscript.

Funding: This work was supported by the National Natural Science Foundation of China (grant numbers 31770363 and 31670335) and the Graduate Research and Practice Projects of Minzu University of China (No. BZKY2022043).

Institutional Review Board Statement: Not applicable.

Informed Consent Statement: Not applicable.

Data Availability Statement: No new data were created or analyzed in this study. Data sharing is not applicable to this article.

Conflicts of Interest: The authors declare no conflict of interest.

References

1. Aslam, M.; Fakher, B.; Ashraf, M.A.; Cheng, Y.; Wang, B.; Qin, Y. Plant low-temperature stress: Signaling and response. *Agronomy* **2022**, *12*, 702. [CrossRef]
2. Kidokoro, S.; Shinozaki, K.; Yamaguchi-Shinozaki, K. Transcriptional regulatory network of plant cold-stress responses. *Trends Plant Sci.* **2022**, *27*, 922–935. [CrossRef] [PubMed]
3. Karlova, R.; Boer, D.; Hayes, S.; Testerink, C. Root plasticity under abiotic stress. *Plant Physiol.* **2021**, *187*, 1057–1070. [CrossRef] [PubMed]
4. Chen, J.; Han, G.; Shang, C.; Li, J.; Zhang, H.; Liu, F.; Wang, J.; Liu, H.; Zhang, Y. Proteomic analyses reveal differences in cold acclimation mechanisms in freezing-tolerant and freezing-sensitive cultivars of alfalfa. *Front. Plant Sci.* **2015**, *6*, 105. [CrossRef]
5. Megha, S.; Basu, U.; Kav, N.N.V. Regulation of low temperature stress in plants by microRNAs. *Plant Cell Environ.* **2018**, *41*, 1–15. [CrossRef]
6. Zhou, Y.; Gao, F.; Liu, R.; Feng, J.; Li, H. De novo sequencing and analysis of root transcriptome using 454 pyrosequencing to discover putative genes associated with drought tolerance in *Ammopiptanthus mongolicus*. *BMC Genom.* **2012**, *13*, 266. [CrossRef]
7. Liu, M.; Chen, Y.; Lu, C.; Zhang, H.; Yin, W. Cold acclimation induced accumulation of phenolic compounds and freezing tolerance in *Ammopiptanthus mongolicus*. *For. Stud. China* **2007**, *9*, 203–207. [CrossRef]
8. Guo, H.; Li, Z.; Zhou, M.; Cheng, H. cDNA-AFLP analysis reveals heat shock proteins play important roles in mediating cold, heat, and drought tolerance in *Ammopiptanthus mongolicus*. *Funct. Integr. Genom.* **2014**, *14*, 127–133. [CrossRef]
9. Zheng, L.; Liu, Q.; Wu, R.; Songbuerbatu; Zhu, M.; Dorjee, T.; Zhou, Y.; Gao, F. The alteration of proteins and metabolites in leaf apoplast and the related gene expression associated with the adaptation of *Ammopiptanthus mongolicus* to winter freezing stress. *Int. J. Biol. Macromol.* **2023**, *240*, 124479. [CrossRef]
10. Shi, J.; Liu, M.; Chen, Y.; Wang, J.; Lu, C. Heterologous expression of the dehydrin-like protein gene *AmCIP* from *Ammopiptanthus mongolicus* enhances viability of *Escherichia coli* and tobacco under cold stress. *Plant Growth Regul.* **2016**, *79*, 71–80. [CrossRef]
11. Cao, P.; Song, J.; Zhou, C.; Weng, M.; Liu, J.; Wang, F.; Zhao, F.; Feng, D.; Wang, B. Characterization of multiple cold induced genes from *Ammopiptanthus mongolicus* and functional analyses of gene *AmEBP1*. *Plant Mol. Biol.* **2009**, *69*, 529–539. [CrossRef] [PubMed]
12. Yu, T.; Zhi, G.; Shi, J.; Chen, Y.; Shen, M.; Lu, C. The calmodulin gene *AmCaM* from *Ammopiptanthus mongolicus* confers freezing and heat tolerance in *Escherichia coli*. *Plant Growth Regul.* **2018**, *84*, 169–178. [CrossRef]
13. Juurakko, C.L.; diCenzo, G.C.; Walker, V.K. *Brachypodium* antifreeze protein gene products inhibit ice recrystallisation, attenuate ice nucleation, and reduce immune response. *Plants* **2022**, *11*, 1475. [CrossRef] [PubMed]
14. Wang, W.; Wei, L. Purification of boiling-soluble antifreeze protein from the legume *Ammopiptanthus mongolicus*. *Prep. Biochem. Biotechnol.* **2003**, *33*, 67–80. [CrossRef] [PubMed]
15. Kausar, R.; Komatsu, S. Proteomic approaches to uncover salt stress response mechanisms in Crops. *Int. J. Mol. Sci.* **2022**, *24*, 518. [CrossRef] [PubMed]
16. Gao, F.; Zhou, Y.; Zhu, W.; Li, X.; Fan, L.; Zhang, G. Proteomic analysis of cold stress-responsive proteins in *Thellungiella rosette* leaves. *Planta* **2009**, *230*, 1033–1046. [CrossRef]
17. Neilson, K.A.; Scafaro, A.P.; Chick, J.M.; George, I.S.; Van Sluyter, S.C.; Gygi, S.P.; Atwell, B.J.; Haynes, P.A. The influence of signals from chilled roots on the proteome of shoot tissues in rice seedlings. *Proteomics* **2013**, *13*, 1922–1933. [CrossRef]
18. Heidarvand, L.; Maali-Amiri, R. Physio-biochemical and proteome analysis of chickpea in early phases of cold stress. *J. Plant Physiol.* **2013**, *170*, 459–469. [CrossRef]
19. Badowiec, A.; Swigonska, S.; Weidner, S. Changes in the protein patterns in pea (*Pisum sativum* L.) roots under the influence of long- and short-term chilling stress and post-stress recovery. *Plant Physiol. Biochem.* **2013**, *71*, 315–324. [CrossRef]
20. Gao, F.; Ma, P.; Wu, Y.; Zhou, Y.; Zhang, G. Quantitative proteomic analysis of the response to cold stress in jojoba, a tropical woody crop. *Int. J. Mol. Sci.* **2019**, *20*, 243. [CrossRef]
21. Yan, S.P.; Zhang, Q.Y.; Tang, Z.C.; Su, W.A.; Sun, W.N. Comparative proteomic analysis provides new insights into chilling stress responses in rice. *Mol. Cell Proteom.* **2006**, *5*, 484–496. [CrossRef] [PubMed]
22. Tetley, R.M.; Thimann, K.V. The metabolism of oat leaves during senescence: I. respiration, carbohydrate metabolism, and the action of cytokinins. *Plant Physiol.* **1974**, *54*, 294–303. [CrossRef]
23. Khalid, M.; Rehman, H.M.; Ahmed, N.; Nawaz, S.; Saleem, F.; Ahmad, S.; Uzair, M.; Rana, I.A.; Atif, R.M.; Zaman, Q.U.; et al. Using exogenous melatonin, glutathione, proline, and glycine betaine treatments to combat abiotic stresses in crops. *Int. J. Mol. Sci.* **2022**, *23*, 12913. [CrossRef] [PubMed]
24. Gao, F.; Zhou, Y.; Huang, L.; He, D.; Zhang, G. Proteomic analysis of long-term salinity stress-responsive proteins in *Thellungiella halophila* leaves. *Chin. Sci. Bull.* **2008**, *53*, 3530–3537. [CrossRef]
25. Katayama, H.; Nagasu, T.; Oda, Y. Improvement of in-gel digestion protocol for peptide mass fingerprinting by matrix-assisted laser desorption/ionization time-of-flight mass spectrometry. *Rapid Commun. Mass Spectrom.* **2001**, *15*, 1416–1421. [CrossRef] [PubMed]
26. Bevan, M.; Bancroft, I.; Bent, E.; Love, K.; Goodman, H.; Dean, C.; Bergkamp, R.; Dirkse, W.; Van Staveren, M.; Stiekema, W.; et al. Analysis of 1.9 Mb of contiguous sequence from chromosome 4 of *Arabidopsis thaliana*. *Nature* **1998**, *391*, 485–488.
27. Zhao, Q.; Gao, J.; Suo, J.; Chen, S.; Wang, T.; Dai, S. Cytological and proteomic analyses of horsetail (*Equisetum arvense* L.) spore germination. *Front. Plant Sci.* **2015**, *6*, 441. [CrossRef]

28. Goldberg, T.; Hecht, M.; Hamp, T.; Karl, T.; Yachdav, G.; Ahmed, N.; Altermann, U.; Angerer, P.; Ansorge, S.; Balasz, K.; et al. LocTree3 prediction of localization. *Nucleic Acids Res.* **2014**, *42*, W350–W355. [CrossRef]
29. Chou, K.C.; Shen, H.B. Plant-mPLoc: A top-down strategy to augment the power for predicting plant protein subcellular localization. *PLoS ONE* **2010**, *5*, e11335. [CrossRef]
30. King, B.R.; Guda, C. ngLOC: An n-gram-based Bayesian method for estimating the subcellular proteomes of eukaryotes. *Genome Biol.* **2007**, *8*, R68. [CrossRef]
31. Emanuelsson, O.; Nielsen, H.; Brunak, S.; von Heijne, G. Predicting subcellular localization of proteins based on their N-terminal amino acid sequence. *J. Mol. Biol.* **2000**, *300*, 1005–1016. [CrossRef] [PubMed]
32. Szklarczyk, D.; Franceschini, A.; Wyder, S.; Forslund, K.; Heller, D.; Huerta-Cepas, J.; Simonovic, M.; Roth, A.; Santos, A.; Tsafou, K.P.; et al. STRING v10: Protein-protein interaction networks, integrated over the tree of life. *Nucleic Acids Res.* **2015**, *43*, D447–D452. [CrossRef] [PubMed]
33. Livak, K.J.; Schmittgen, T.D. Analysis of relative gene expression data using real-time quantitative PCR and the $2-\Delta\Delta Ct$ method. *Methods* **2001**, *25*, 402–408. [CrossRef] [PubMed]
34. Katoh, K.; Standley, D.M. MAFFT multiple sequence alignment software version 7: Improvements in performance and usability. *Mol. Biol. Evol.* **2013**, *30*, 772–780. [CrossRef]
35. Kumar, S.; Stecher, G.; Li, M.; Knyaz, C.; Tamura, K. MEGA X: Molecular evolutionary genetics analysis across computing platforms. *Mol. Biol. Evol.* **2018**, *35*, 1547–1549. [CrossRef]
36. Higo, K.; Ugawa, Y.; Iwamoto, M.; Higo, H. PLACE: A database of plant cis-acting regulatory DNA elements. *Nucleic Acids Res.* **1998**, *26*, 358–359. [CrossRef]
37. Dhanyalakshmi, K.H.; Nataraja, K.N. Universal stress protein-like gene from mulberry enhances abiotic stress tolerance in *Escherichia coli* and transgenic tobacco cells. *Plant Biol.* **2021**, *23*, 1190–1194. [CrossRef]
38. Amick, A.K.; Liu, Q.; Gautam, S.; Chupp, G.; Dela Cruz, C.S.; Sharma, L. Measurement of chitinase activity in biological samples. *J. Vis. Exp.* **2019**, *150*, e60159.
39. Rinalducci, S.; Egidi, M.G.; Karimzadeh, G.; Jazii, F.R.; Zolla, L. Proteomic analysis of a spring wheat cultivar in response to prolonged cold stress. *Electrophoresis* **2011**, *32*, 1807–1818. [CrossRef]
40. Kurepin, L.V.; Ivanov, A.G.; Zaman, M.; Pharis, R.P.; Allakhverdiev, S.I.; Hurry, V.; Hüner, N.P. Stress-related hormones and glycinebetaine interplay in protection of photosynthesis under abiotic stress conditions. *Photosynth. Res.* **2015**, *126*, 221–235. [CrossRef]
41. Choudhury, F.K.; Rivero, R.M.; Blumwald, E.; Mittler, R. Reactive oxygen species, abiotic stress and stress combination. *Plant J. Cell Mol. Biol.* **2017**, *90*, 856–867. [CrossRef] [PubMed]
42. Dresios, J.; Aschrafi, A.; Owens, G.C.; Vanderklish, P.W.; Edelman, G.M.; Mauro, V.P. Cold stress-induced protein Rbm3 binds 60S ribosomal subunits, alters microRNA levels, and enhances global protein synthesis. *Proc. Natl. Acad. Sci. USA* **2005**, *102*, 1865–1870. [CrossRef] [PubMed]
43. Nakaminami, K.; Seki, M. RNA Regulation in Plant Cold Stress Response. *Adv. Exp. Med. Biol.* **2018**, *1081*, 23–44.
44. Lantzouni, O.; Alkofer, A.; Falter-Braun, P.; Schwechheimer, C. Growth-regulating factors interact with dellas and regulate growth in cold stress. *Plant Cell* **2020**, *32*, 1018–1034. [CrossRef]
45. Bricker, T.M.; Frankel, L.K. Auxiliary functions of the PsbO, PsbP and PsbQ proteins of higher plant Photosystem II: A critical analysis. *J. Photochem. Photobiol. B Biol.* **2011**, *104*, 165–178. [CrossRef] [PubMed]
46. Wilde, A.; Lünser, K.; Ossenbühl, F.; Nickelsen, J.; Börner, T. Characterization of the cyanobacterial ycf37: Mutation decreases the photosystem I content. *Biochem. J.* **2001**, *357*, 211–216. [CrossRef] [PubMed]
47. Xu, X.M.; Møller, S.G. AtNAP7 is a plastidic SufC-like ATP-binding cassette/ATPase essential for *Arabidopsis* embryogenesis. *Proc. Natl. Acad. Sci. USA* **2004**, *101*, 9143–9148. [CrossRef]
48. Busch, F.A. Photorespiration in the context of Rubisco biochemistry, CO2 diffusion and metabolism. *Plant J. Cell Mol. Biol.* **2020**, *101*, 919–939. [CrossRef]
49. Gonçalves, A.Z.; Latansio, S.; Detmann, K.C.; Marabesi, M.A.; Neto, A.A.C.; Aidar, M.P.M.; DaMatta, F.M.; Mercier, H. What does the RuBisCO activity tell us about a C(3)-CAM plant? *Plant Physiol. Biochem.* **2020**, *147*, 172–180. [CrossRef]
50. Spreitzer, R.J.; Salvucci, M.E. Rubisco: Structure, regulatory interactions, and possibilities for a better enzyme. *Annu. Rev. Plant Biol.* **2002**, *53*, 449–475. [CrossRef]
51. Zaffagnini, M.; Michelet, L.; Sciabolini, C.; Di Giacinto, N.; Morisse, S.; Marchand, C.H.; Trost, P.; Fermani, S.; Lemaire, S.D. High-resolution crystal structure and redox properties of chloroplastic triosephosphate isomerase from *Chlamydomonas reinhardtii*. *Mol. Plant* **2014**, *7*, 101–120. [CrossRef] [PubMed]
52. Asghari, R.; Ebrahimzadeh, H. Drought stress increases the expression of wheat leaf ribulose-1, 5- bisphosphate carboxylase/oxyenase protein. *Iran. J. Sci. Technol.* **2006**, *30*, 261–267.
53. Dumont, E.; Bahrman, N.; Goulas, E.; Valot, B.; Sellier, H.; Hilbert, J.L.; Vuylsteker, C.; Lejeune-Hénaut, I.; Delbreil, B. A proteomic approach to decipher chilling response from cold acclimation in pea (*Pisum sativum* L.). *Plant Sci.* **2011**, *180*, 86–98. [CrossRef]
54. An, F.; Li, G.; Li, Q.X.; Li, K.; Carvalho, L.J.; Ou, W.; Chen, S. The comparatively proteomic analysis in response to cold stress in cassava plantlets. *Plant Mol. Biol. Rep.* **2016**, *34*, 1095–1110. [CrossRef]
55. Evers, D.; Legay, S.; Lamoureux, D.; Hausman, J.F.; Hoffmann, L.; Renaut, J. Towards a synthetic view of potato cold and salt stress response by transcriptomic and proteomic analyses. *Plant Mol. Biol.* **2012**, *78*, 503–514. [CrossRef] [PubMed]

56. Mittler, R.; Zandalinas, S.I.; Fichman, Y.; Van Breusegem, F. Reactive oxygen species signalling in plant stress responses. *Nat. Rev. Mol. Cell Biol.* **2022**, *23*, 663–679. [CrossRef] [PubMed]
57. Devireddy, A.R.; Tschaplinski, T.J.; Tuskan, G.A.; Muchero, W.; Chen, J.G. Role of reactive oxygen species and hormones in plant responses to temperature changes. *Int. J. Mol. Sci.* **2021**, *22*, 8843. [CrossRef]
58. Cunha, J.R.; Carvalho, F.E.L.; Lima-Neto, M.C.; Jardim-Messeder, D.; Cerqueira, J.V.A.; Martins, M.O.; Fontenele, A.V.; Márgis-Pinheiro, M.; Komatsu, S.; Silveira, J.A.G. Proteomic and physiological approaches reveal new insights for uncover the role of rice thylakoidal APX in response to drought stress. *J. Proteom.* **2019**, *192*, 125–136. [CrossRef]
59. Verma, D.; Upadhyay, S.K.; Singh, K. Characterization of *APX* and *APX-R* gene family in *Brassica juncea* and *B. rapa* for tolerance against abiotic stresses. *Plant Cell Rep.* **2022**, *41*, 571–592. [CrossRef]
60. Schurmann, P.; Jacquot, J.P. Plant thioredoxin systems revisited. *Annu. Rev. Plant Physiol.* **2000**, *51*, 371–400. [CrossRef]
61. Koskela, M.M.; Dahlström, K.M.; Goñi, G.; Lehtimäki, N.; Nurmi, M.; Velazquez-Campoy, A.; Hanke, G.; Bölter, B.; Salminen, T.A.; Medina, M.; et al. *Arabidopsis* FNRL protein is an NADPH-dependent chloroplast oxidoreductase resembling bacterial ferredoxin-NADP(+) reductases. *Physiol. Plant* **2018**, *162*, 177–190. [CrossRef] [PubMed]
62. Waszczak, C.; Akter, S.; Eeckhout, D.; Persiau, G.; Wahni, K.; Bodra, N.; Van Molle, I.; De Smet, B.; Vertommen, D.; Gevaert, K.; et al. Sulfenome mining in *Arabidopsis thaliana*. *Proc. Natl. Acad. Sci. USA* **2014**, *111*, 11545–11550. [CrossRef] [PubMed]
63. Kosová, K.; Vítámvás, P.; Prášil, I.T.; Renaut, J. Plant proteome changes under abiotic stress—Contribution of proteomics studies to understanding plant stress response. *J. Proteom.* **2011**, *74*, 1301–1322. [CrossRef]
64. Bocian, A.; Kosmala, A.; Rapacz, M.; Jurczyk, B.; Marczak, Ł.; Zwierzykowski, Z. Differences in leaf proteome response to cold acclimation between *Lolium perenne* plants with distinct levels of frost tolerance. *J. Plant Physiol.* **2011**, *168*, 1271–1279. [CrossRef] [PubMed]
65. Ke, X.; Zou, W.; Ren, Y.; Wang, Z.; Li, J.; Wu, X.; Zhao, J. Functional divergence of chloroplast Cpn60α subunits during *Arabidopsis* embryo development. *PLoS Genet.* **2017**, *13*, e1007036. [CrossRef]
66. Farmer, L.M.; Book, A.J.; Lee, K.H.; Lin, Y.L.; Fu, H.; Vierstra, R.D. The RAD23 family provides an essential connection between the 26S proteasome and ubiquitylated proteins in *Arabidopsis*. *Plant Cell* **2010**, *22*, 124–142. [CrossRef]
67. Zhang, X.; Liu, S.; Takano, T. Two cysteine proteinase inhibitors from *Arabidopsis thaliana*, *AtCYSa* and *AtCYSb*, increasing the salt, drought, oxidation and cold tolerance. *Plant Mol. Biol.* **2008**, *68*, 131–143. [CrossRef]
68. Wang, Y.; Yang, L.; Chen, X.; Ye, T.; Zhong, B.; Liu, R.; Wu, Y.; Chan, Z. Major latex protein-like protein 43 (*MLP43*) functions as a positive regulator during abscisic acid responses and confers drought tolerance in *Arabidopsis thaliana*. *J. Exp. Bot.* **2016**, *67*, 421–434. [CrossRef]
69. Jain, M.; Batth, R.; Kumari, S.; Mustafiz, A. *Arabidopsis thaliana* contains both Ni^{2+} and Zn^{2+} dependent glyoxalase i enzymes and ectopic expression of the latter contributes more towards abiotic stress tolerance in *E. coli*. *PLoS ONE* **2016**, *11*, e0159348. [CrossRef]
70. Menghiu, G.; Ostafe, V.; Prodanović, R.; Fischer, R.; Ostafe, R. A high-throughput screening system based on fluorescence-activated cell sorting for the directed evolution of chitinase A. *Int. J. Mol. Sci.* **2021**, *22*, 3041. [CrossRef]
71. Tabata, E.; Kashimura, A.; Wakita, S.; Sakaguchi, M.; Sugahara, Y.; Imamura, Y.; Shimizu, H.; Matoska, V.; Bauer, P.O.; Oyama, F. Acidic Chitinase-chitin complex is dissociated in a competitive manner by acetic acid: Purification of natural enzyme for supplementation purposes. *Int. J. Mol. Sci.* **2018**, *19*, 362. [CrossRef] [PubMed]
72. Filyushin, M.A.; Anisimova, O.K.; Kochieva, E.Z.; Shchennikova, A.V. Genome-wide identification and expression of chitinase class I genes in garlic (*Allium sativum* L.) Cultivars Resistant and Susceptible to *Fusarium proliferatum*. *Plants* **2021**, *10*, 720. [CrossRef] [PubMed]
73. Nakamura, T.; Ishikawa, M.; Nakatani, H.; Oda, A. Characterization of cold-responsive extracellular chitinase in bromegrass cell cultures and its relationship to antifreeze activity. *Plant Physiol.* **2008**, *147*, 391–401. [CrossRef] [PubMed]
74. Zhang, S.H.; Wei, Y.; Liu, J.L.; Yu, H.M.; Yin, J.H.; Pan, H.Y.; Baldwin, T.C. An apoplastic chitinase CpCHT1 isolated from the corolla of wintersweet exhibits both antifreeze and antifungal activities. *Biol. Plant.* **2011**, *55*, 141–148. [CrossRef]
75. Kashyap, P.; Deswal, R. A novel class I chitinase from *Hippophae rhamnoides*: Indications for participating in ICE-CBF cold stress signaling pathway. *Plant Sci.* **2017**, *259*, 62–70. [CrossRef] [PubMed]

horticulturae

Article

Strigolactone-Mediated Trehalose Enhances Salt Resistance in Tomato Seedlings

Xuefang Lu, Xiaojun Liu, Junrong Xu, Yunzhi Liu, Yuzhen Chi, Wenjin Yu and Changxia Li *

College of Agriculture, Guangxi University, Nanning 530004, China; 2217391028@st.gxu.edu.cn (X.L.); xiaojunliu405@163.com (X.L.); xujr@st.gxu.edu.cn (J.X.); 2217391027@st.gxu.edu.cn (Y.L.); 2217391004@st.gxu.edu.cn (Y.C.); yuwjin@gxu.edu.cn (W.Y.)
* Correspondence: licx@gxu.edu.cn

Abstract: Strigolactones (SLs) are newly discovered plant hormones that modulate a variety of physiological and biochemical processes and plant stress responses. In this study, SLs' synthetic analog, GR24, significantly improved the growth of tomato seedlings under salt stress, while SLs' synthesis inhibitor, TIS108, inversed the positive role of SLs, indicating that SLs could effectively enhance salt-stress resistance in tomato. To further explore the mechanism of SL-modulated trehalose (Tre) in response to salt stress, Tre metabolism was analyzed during this process. GR24 increased the endogenous Tre and starch contents and decreased the glucose (Glu) level under salt-stress conditions. Additionally, the TPS and TPP activities were enhanced by GR24 and the activity of THL was inhibited by GR24 under salt stress; thereafter, Tre biosynthesis-related genes, including *TPS1*, *TPS2*, *TPP1*, and *TPP2*, were also upregulated by GR24 under salt stress. However, the function of GR24 in Tre metabolism was inhibited by TIS108. Thus, the results indicated that GR24 improved the expression levels or activities of Tre biosynthesis-related genes or enzymes and inhibited the transcript level or activity of genes or enzymes related to Tre degradation, respectively, resulting in an increase in the endogenous Tre level and, therefore, weakening the salt toxicity of tomato seedlings.

Keywords: strigolactone; trehalose; salt; regulator pathway; genes expression

Citation: Lu, X.; Liu, X.; Xu, J.; Liu, Y.; Chi, Y.; Yu, W.; Li, C. Strigolactone-Mediated Trehalose Enhances Salt Resistance in Tomato Seedlings. *Horticulturae* **2023**, *9*, 770. https://doi.org/10.3390/horticulturae9070770

Academic Editor: Hongliang Zhu

Received: 25 May 2023
Revised: 20 June 2023
Accepted: 3 July 2023
Published: 5 July 2023

1. Introduction

Plant growth usually faces many kinds of environmental stresses. By 2050, it is expected that 50% of cultivatable land will be lost [1]. Plants are sensitive to environmental stresses, such as salt stress. There are many reasons for the increase in soil salinity, including natural occurrences, climate change, the rise in the sea level, drought evaporation, and different agricultural irrigation methods [2]. An increasing number of studies have reported that salt stress influences plant growth and development. For instance, Hasanuzzaman et al. [3] and Ahmad et al. [4] concluded that salt stress could reduce root/shoot weight, branch length, plant height, branching, leaf number, and early vegetative development. Salt stress also interferes with photosynthetic activity and respiration, reduced enzyme activity, sugar metabolism, protein and nucleic acid synthesis, impaired ion homeostasis, and osmotic and hormonal balance [5,6].

As a general rule, different crops have different ranges of adaptation to thesalt content of soil. Previous research showed that NaCl was harmful to seed germination and seedling growth in the range of 0–150 mM [7]. For instance, Hu et al. [8] determined the effect of the methylation of N^6-methyladenosine mRNA in *Arabidopsis* on its salt tolerance under 100 and 150 mM of NaCl. Therefore, the enhancement of salt tolerance in plants is a top priority in agricultural research.

Strigolactones (SLs) were first found in the root exudates of cotton [9]. Subsequently, SLs, as an important signaling molecule, were described as germination stimulants of witchweed seeds [9,10]. SLs could stimulate the spore germination, hyphal branching, and metabolic activity of beneficial microorganisms, such as arbuscular mycorrhizal fungi

(AMF) [11], in establishing a symbiotic relationship between plants and AMF, which plays an important role in adapting to stress [12]. Marro et al. [13] reported that SLs protect plants from damage to plant parasitic nematodes by enhancing mycorrhizal colonization.

Recently, SLs have been recognized as new types of plant hormones that play a regulatory role in various plants growth and development, including seedling growth [14], leaf senescence [15], photo morphogenesis [16], and the shaping of root system architecture [17]. SLs play pivotal roles in response to diverse abiotic and biotic stresses [18]. For instance, exogenous GR24, as a positive regulator, could enhance the resistance to low-light, drought, and salt in *Arabidopsis* [19–21]. In addition, the exogenous application of SLs was effective in reducing the negative influence of salt stress in sunflowers [18].

Sugars are a source of carbon units and metabolic energy, and act as signal molecules to report the carbon status in cells [22,23]. Trehalose (Tre) is a nonreducing disaccharide; its synthesis pathway requires two stages and it is highly conserved in some plants, including tomato [24,25]. Trehalose-6-phosphate synthetase (TPS) concentrates uridine diphosphate glucose (UDPG) and glucose 6-phosphate (G6P) to produce trehalose-6-phosphate (T6P). Trehalose-6-phosphate phosphatase (TPP) dephosphorylates T6P to form Tre. The upregulation of *TPS/TPP* genes increase Tre levels, thereby promoting growth and development in plant, including seed germination, root growth, and flowering [26,27].

In contrast to Tre synthesis, the trehalase THL directly decomposes Tre to produce two molecules of glucose (Glu). Furthermore, T6P hydrolase (TPHase) hydrolyzes T6P to produce G6P and Glu. The THL hydrolases include neutral trehalase (NTH) and acidic trehalase (ATH) [28,29]. The Tre pathway has close contact with abiotic stress resistance [30–32]. In plants, Tre could be produced under extreme environments, including salt-stress, drought-stress, and cold-stress environments [33]. The pre-sowing Tre treatment of seeds could promote the normal growth of radish plants under drought stress, which is associated with significant increases in chlorophyll A content, total soluble sugars, water-use efficiency, photosynthesis, free proline content and the activity of the SOD enzyme [34]. Additionally, the overexpression of the *OsTRE1* gene helps rice to adapt to salt stress [35].

The planting area of tomato (*Solanum lycopersicum* L.) is huge, and tomato is a very popular cash crop throughout the world, However, abiotic stresses seriously reduce the productivity and quality of tomato. Previous research has shown that SLs and Tre could promote the growth and development of plants under salt stress. However, we have no clear understanding of the interplay between SLs and Tre under the influence of abiotic stresses.

As a result, we carried out the current experiment and hypothesized that crosstalk between SLs and Tre is likely to play an important role in improving the growth of tomato seedlings under salt stress. This research may provide an important foundation for future studies of the interaction between SLs and Tre in plants under stress.

2. Materials and Methods

2.1. Plant Materials

In this study, the tomato "Micro-Tom" (Nanjing Ebiosci Biological Technology Co., Ltd., Nanjing, China) was selected as the plant material. The seeds were placed in a conical flask containing water, and then put in a Shaking Incubator to germinate for 3 days. The conditions for seed germination were 180 rpm at 28 °C, in darkness. The germinated seeds were planted. The four-leaf seedlings were transferred to conical flasks containing 1/2 Hogland solution to grown for 21 days. The growth conditions of the tomato seedlings were as follows: a cycle of 14 h in light at 250 μmol m^{-2} s^{-1}, photon irradiance at 25 °C and 10 h in darkness at 20 °C, with 70% relative humidity.

2.2. Treatments

Previous studies showed that salt stress stimulated by NaCl significantly inhibits the growth of tomato seedlings, especially with increased concentrations of NaCl. To study the effect of NaCl on the seedlings of tomato, NaCl concentrations were set to 150 mM NaCl [36,37]. Tomato seedlings with similar growth were selected and treated with NaCl

(150 mM), NaCl + SLs synthetic analog GR24 (15 μM), and NaCl + SLs synthesis inhibitor TIS108 (3 μM) in Hogland solution under salt stress. The tomato seedlings were immersed in different treatments for 7 days, and the Hogland solution served as the control. The concentrations of GR24 and TIS108 were selected by referring to [37] and [38]. After 7 days of treatment, the samples were frozen in liquid nitrogen, then stored quickly at −80 °C for future use.

2.3. *Measurement of Morphological Indexes*

The distance from the stem base to the growth point of a plant is defined as the plant height. Plant heights and root lengths were measured by a vernier caliper. The phenotype was captured by a Canon camera (EOS 700D).

2.4. *Determination of Tre, Starch and Glu Content*

According to the manufacturer's instructions, as provided in the Tre Content Detection Kit (Beijing Solarbio Science & Technology Co., Ltd., Beijing, China, BC0330), the content of Tre was determined. The specific steps were as follows:

(1) The foil bag was equilibrated at room temperature for 20 min; the required strip was sealed immediately and the remaining strips were placed in self-sealing bags at 4 °C.
(2) The standard wells and the sample well were set up. Fifty μL of the standard liquid of each concentration was added into the standard wells, and one standard well corresponded to one concentration.
(3) Ten μL of sample and 40 μL of sample solvent were added into the sample well, successively. No additions were made to the blank well.
(4) Apart from the blank well, 100 μL horseradish peroxidase (HRP)-labeled detection antibody was added to all standard wells and to the sample well, respectively, and the reaction well was sealed with a sealing film. The reaction well was incubated in a 37 °C water bath or incubator for 60 min.
(5) The liquid was discarded, and the board was pat dried on absorbent paper, and each microplate well was filled with washing reagent and allowed to stand for 1 min. The washing reagent was discarded, and the board was pat dried on absorbent paper, the board was washed 5 times.
(6) In each microplate well, 50 μL of substrate A and substrate B were added and incubated at 37 °C in the dark for 15 min.
(7) The 50 μL termination solution was added into each microplate well, and the OD value of each microplate well was measured at 450 nm within 15 min. The Tre content was calculated in the sample, based on the standard curve.

The content of starch was determined using iodine colorimetry [39]. An 0.2 g sample was taken and ground in 2 mL of distilled water, and the seed powder was mixed with 3.2 mL of 60% perchloric acid for approximately 10 min. After centrifugation at room temperature, 5000× *g* for 5 min, the supernatant was extracted into a 10 mL tube, and the volume was fixed with distilled water. The absorbance value was determined at 660 nm. The assay method of Glu was modified slightly, according to study of Saqib et al. [40].

The 3,5-dinitrosalicylic acid (DNS) reagent was prepared as follows: first, 3,5-dinitrosalicylic acid (6.3 g) was added in 2 M of sodium hydroxide solution (262 mL), and the mixture was stirred until completely dissolved (liquid A). Next, potassium sodium tartrate (185 g) was dissolved into 500 mL hot water (liquid B). Liquid A and liquid B were mixed, phenol and sodium sulfite (5 g each) were added and refrigerated to room temperature, and the volume was adjusted to exactly 1000 mL with distilled water. Then, 0.5 g samples were accurately weighed, ground into homogenate with 10 mL distilled water, and diluted in a 50 mL volumetric flask and mixed. After shaking well, the samples were put in a 50 °C water bath for about 10 min and refrigerated to room temperature. Then, they were centrifuged at 4 °C, 4000× *g*, for about 15 min. Two mL of supernatant was mixed with 1.5 mL DNS. Subsequently, the samples were placed into a boiling water bath for about 5 min and refrigerated to room temperature.

Then, miscible liquids were filtered into a 25 mL volumetric bottle. The absorbance value was measured at 540 nm.

2.5. Determination of TPP and TPS Activities

According to the manufacturer's instructions, as set out in the ELISA kit (Yuanmu Biotechnology, Shanghai, China), the activities of TPP and TPS were determined [37]. The extraction steps of the enzyme were performed as follows: samples (0.5 g) were ground quickly with liquid nitrogen and transferred to a test tube. PBS (9 mL, 0.01 M, pH 7.4) was added to prepare the homogenate and collected into test tubes. The homogenate was centrifuged at 4 °C, 4000× *g*, for 15 min, and the supernatant was collected for detection. Then, the enzyme activity was determined according to the following steps.

In the plate well, the 10 μL enzyme solution and 50 μL of HRP-conjugate reagent were added, and antibodies were incubated at 37 °C for about 30 min. The board was washed in the dark, then 100 μL of chromogen solution was added and incubated into the board at 37 °C for 15 min. The absorbance was monitored at 450 nm. According to the manufacturer's instructions for THL, as set out in Activities Detection Kit (Beijing Solarbio Science & Technology Co., Ltd., Beijing, China, BC2510), the activity of THL was determined.

2.6. Real-Time RT-PCR Analysis

The tomato leaves (0.5 g) were ground in liquid nitrogen in pre-cooled mortar and collected in a centrifuge tube, according to the study of Huang et al. [41]. TRIzol (1 mL) was added to the centrifuge tube and incubated at 4 °C for about 10 min. The 200 μL chloroform was added, incubated for 5 min, and centrifuged at 4 °C, 12,000× *g*, for about 15 min. The supernatant was mixed with an equal volume of isopropanol, which was incubated at −20 °C for about 1 h. Then, the supernatant was washed twice with 75% ethanol and collected into the adsorption column. Finally, RNA was dissolved with ddH_2O, which did not contain RNase. The collected total RNA was transformed into single-stranded cDNA, as recommended by the manufacturer. RNA (500 ng) that originated from various treatments was applied in the synthesis of first-strand cDNA, which happened in 10 μL reactions having 2 μL of AMV reverse transcriptase XL (AG, China) and 2.5 μM of random primer. The reaction system was as follows: 10 mL 2 × SuperReal PreMix Plus, 0.6 mL of 10 mM forward primers, 0.6 mL of 10 mM reverse primers, 2 mL of cDNA, and 6.8 mL of RNase-free ddH_2O.

According to the manufacturer's instructions, as set out in the ABI Step One Plus system (Applied Biosystems, Carlsbad, CA, USA) and SYBR® Premix Ex Taq ™ II (AG, Shanghai, China) were used to reversed transcribe the RNA. Then, gene-specific primers were applied, as shown in Table 1. The PCR cycling conditions were as follows: 95 °C for 15 min and 40 cycles of 95 °C for 10 s and 60 °C for 20 s. The actin transcript level was normalized and the relative expression level was expressed as the value of the relative expression level relative to the level in line with the control sample at a specified time. The expression level of the gene was calculated by $2^{-\Delta\Delta CT}$. Briefly, ΔCT = CT (target gene) − CT (internal reference gene). ΔΔCT = ΔCT (test group) − ΔCT (control group). Each sample was set to three biological replicates.

2.7. Statistical Analysis

All data were determined in three independent biological replicates for each experiment. Data statistical analysis was carried out via SPSS 22.0 software (SPSS Inc., Chicago, IL, USA). We adopted Duncan' s ($p < 0.05$) method for significance analysis. The boxplots were established by Origin 2021 software.

Table 1. Primer sequence for qRT-PCR.

Gene Name	Gene ID	Primer Sequence (5′–3′)
TPS1	LOC100135703	GGGCAGAAACCGAGTATGGCTAAG CGACAGAGGCGAGTTGATGGAC
TPS2	LOC101250326	GCAAGGAACGGAGAAGCTGGATG ACTCAGAAACCACCAACATGCTCTC
TPP1	LOC101259279	CCAAGCAGAGGTGTTCTTCGGATG TGTTTGTGAGCGTTCTGGTGTCC
TPP2	LOC101245612	ACTCTCAGCACCAGGTTACTCACAG ACTCTCAGCACCAGGTTACTCACAG
Actin	NC_015447	AATGAACTTCGTGTGGCTCCAGAG ATGGCAGGGGTGTTGAAGGTTTC

3. Results

3.1. Effect of GR24 and TIS108 on Tomato Seedling Growth under Salt Stress

Compared with the control, the NaCl treatment significantly reduced root length and plant height by 23.96% and 11.36%, respectively. However, NaCl + GR24 treatment significantly increased root length and plant height. The TIS108 treatment reversed the positive effect of GR24 (Figure 1a,b). Figure 1c shows a perfect match with the changes in tomato phenotypes revealed by the above results. The results suggest that the application of SLs could effectively promote the growth of tomato seedlings under salt stress.

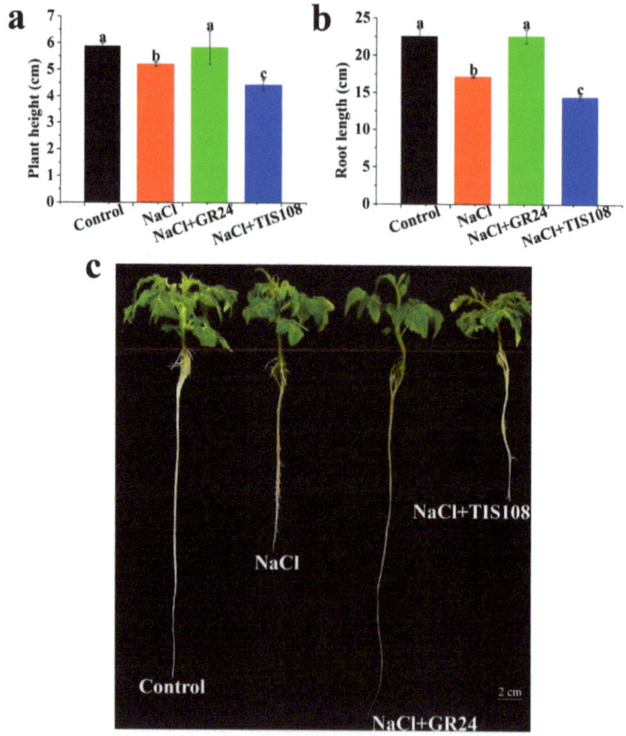

Figure 1. Effects of GR24 and TIS108 on plant height, root length, and phenotype in tomato seedlings under salt stress. (**a**) Plant height; (**b**) maximum root length; (**c**) phenotype. Every value stands for the average ± standard error (SE). At the same time period, different letters express notable differences among different treatments ($p < 0.05$). GR24: SLs′ synthetic analog; TIS108: SLs′ synthesis inhibitor.

3.2. SLs Were Involved in the Conversion among Tre, Starch, and Glu

As shown in Figure 2a, in comparison to the control, the NaCl treatment significantly reduced the Tre content, by 6.80%. However, the Tre content was raised under NaCl + GR24 treatment, even exceeding 9.55% of the control. The Tre content in the TIS108 treatment was 17.26% lower than the content in the NaCl treatment (Figure 2a). Similarly, the changes in the starch were in full agreement with the Tre levels mentioned above, under different treatments (Figure 2b). The Glu content was increased under salt stress. However, the increase was inhibited in the application of the GR24. NaCl + TIS108 treatment led to a marked increase in Glu content, by 1.75%, compared to the NaCl treatment alone (Figure 2c). These results imply that the addition of exogenous SLs promotes starch condensation to synthesize Tre and inhibits Tre being degraded to Glu, thereby protecting tomato plants from salt stress.

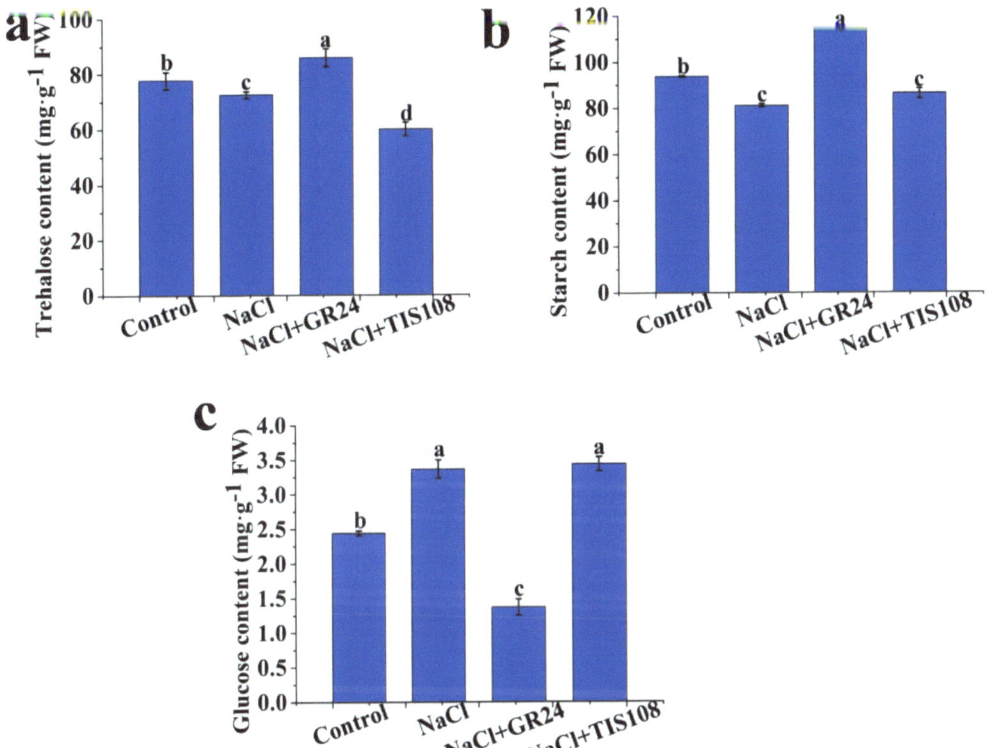

Figure 2. The GR24 and TIS108 affect the content of Tre (**a**) starch (**b**) and Glu (**c**) in tomato seedlings under salt stress. Every value stands for the average ± standard error (SE). At the same time period, different letters express notable differences among different treatments ($p < 0.05$). GR24: SLs' synthetic analog; TIS108: SLs' synthesis inhibitor.

3.3. Effect of SLs on TPS, TPP, and THL Activities under Salt Stress

As can be seen in Figure 3, the activities of TPS and TPP in the NaCl treatment were 38.17% and 42.70% lower than those in the control, but the activities of TPS and TPP increased significantly after the application of GR24. The application of TIS108 reduced the activities of TPS and TPP in NaCl-treated tomato leaves, which were about 55.64% and 87.61% of the activities with the NaCl treatment (Figure 3a,b). Salt stress resulted in a significant increase, of 83.64%, in THL activity, compared to that in the control (Figure 3c).

THL activity was reduced significantly in tomato seedlings treated with GR24 under salt stress. However, TIS108 enhanced the THL activity. This indicated that SLs play a role in controlling Tre content by upregulating its biosynthesis enzymes TPS and TPP and downregulating its degradation enzyme THL.

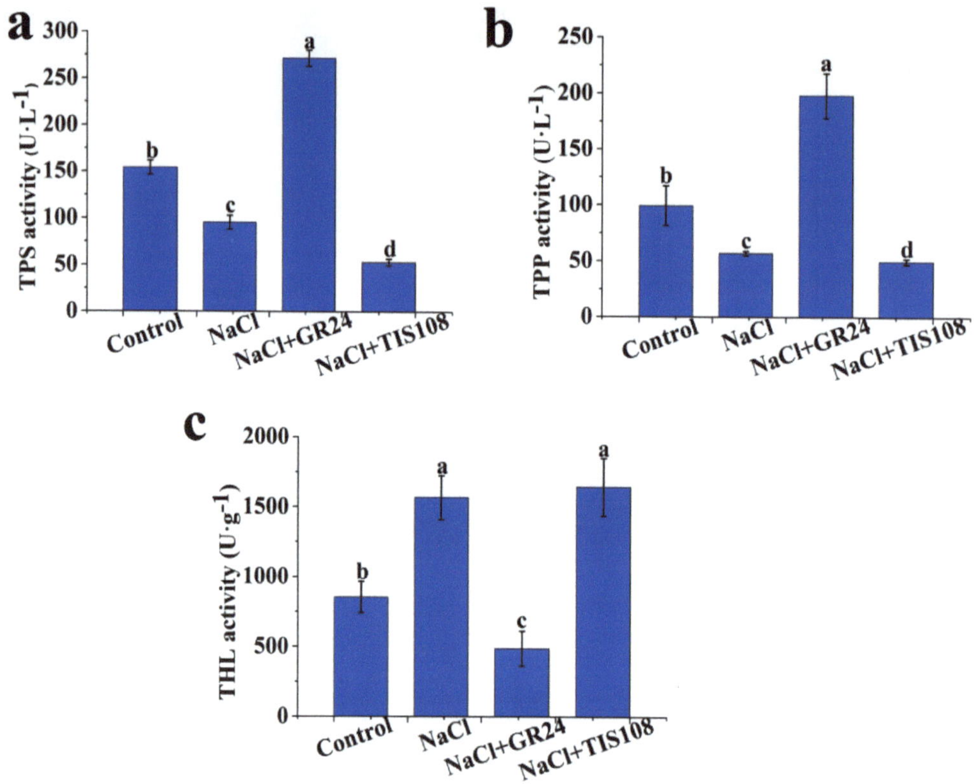

Figure 3. GR24 and TIS108 affect TPS activity (**a**), TPP activity (**b**), and THL activity (**c**) in tomato seedlings under salt stress. Every value stands for the average ± standard error (SE). At the same time period, different letters express notable differences among different treatments ($p < 0.05$). GR24: SLs' synthetic analog; TIS108: SLs' synthesis inhibitor.

3.4. SLs Regulated the Transcriptional Level of TPS1, TPS2, TPP1, and TPP2 Genes under Salt Stress

After NaCl treatment, the expression level of *TPS1* had no significant difference, compared to that of the control (Figure 4a). NaCl + GR24 treatment led to a significant upregulation in *TPS1* expression, by 604.12%, as compared to that of the NaCl treatment. Treatment with NaCl + TIS108 markedly reduced the expression level of *TPS1*. The *TPS2*, *TPP1*, and *TPP2* expressions showed very similar changes under different treatments. The expression levels of *TPS2*, *TPP1*, and *TPP2* in the NaCl treatment were 80.36%, 76.77%, and 91.39% lower than those in the control, respectively. The expression levels of *TPS2*, *TPP1*, and *TPP2* were significantly upregulated by the NaCl + GR24 treatment, compared to the NaCl treatment. However, the NaCl + TIS108 treatment significantly reduced the positive effects of *TPS2*, *TPP1*, and *TPP2* under salt stress (Figure 4b–d). The expression of these genes contributes to Tre accumulation. Thus, we concluded that SLs enhanced salt-stress resistance in tomato seedlings by upregulating *TPS1*, *TPS2*, *TPP1*, and *TPP2*, leading to Tre accumulation.

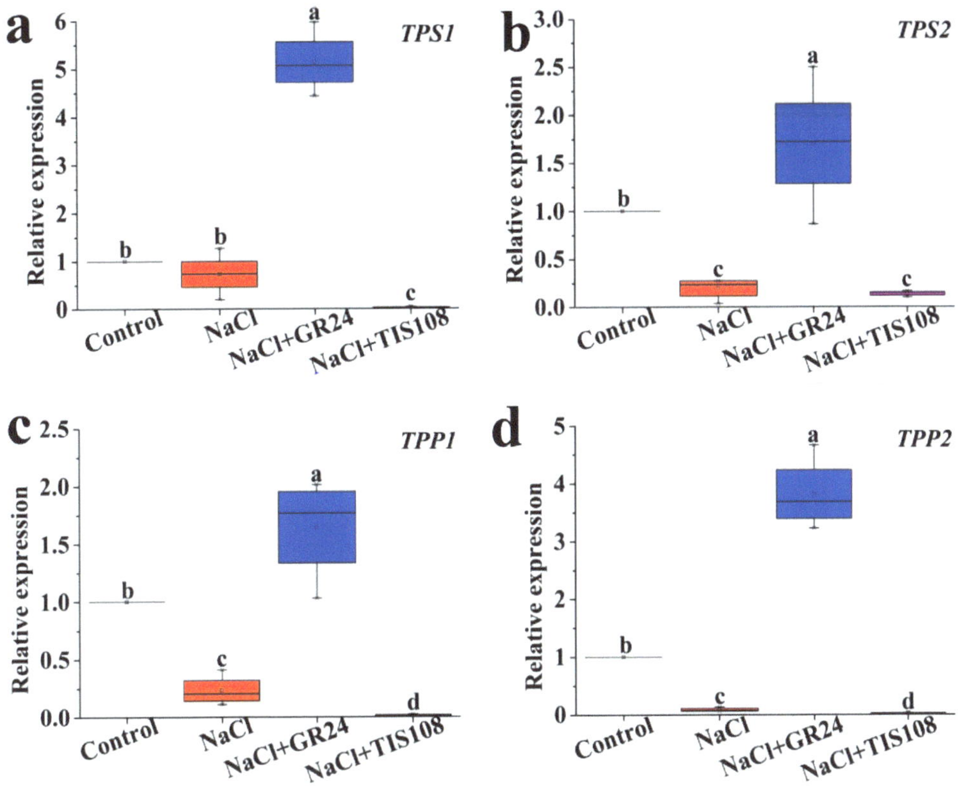

Figure 4. GR24 and TIS108 affect the expression of *TPS1* (**a**), *TPS2* (**b**), *TPP1* (**c**), and *TPP2* (**d**) genes in tomato seedlings under salt stress. Every value stands for the average ± standard error (SE). At the same time period, different letters express notable differences among different treatments ($p < 0.05$). GR24: SLs' synthetic analog; TIS108: SLs' synthesis inhibitor.

4. Discussion

Salt stress is a prevalent phenomenon that restricts plant growth. Salt stress prevents the growth of the roots and leaves of tomato seedlings, restraining photosynthesis and transpiration, resulting in a reduction in crop yield and quality [42]. Previous research indicated that SLs could efficiently relieve the oxidative damage induced by salt stress in plants [43,44]. The role of SLs is very wide, including promoting stem elongation and internode length, increasing root hairs and nodules numbers, and inhibiting adventitious roots and bud outgrowth [45]. SLs also inhibit the polar transport of auxin in stems by downregulating the *PIN1* gene, thus inhibiting branching [46]. The purpose of this study was to determine the relationship between SLs and Tre and the regulatory mechanisms of salt tolerance in tomato seedlings.

The height and root length of tomato seedlings treated with NaCl (150 mM) were significantly inhibited, compared to the control, while GR24 could relieve this inhibition (Figure 1). When the SLs' synthesis inhibitor TIS108 was added, the root length and plant height of tomato seedlings were lower than they were with NaCl treatment. It was proved that SLs could improve the salt tolerance of tomato seedlings, restoring normal growth (Figure 1). The results of [47] were consistent with our study; they reported that GR24 relieved the inhibition of growth in ornamental sunflower, led by 150 mM NaCl via raising the transpiration ratio, the stomatal opening, and the photosynthetic rate. Ma et al. [48] found that GR24 improved the photosynthesis in *Brassica napus* L., promoting the growth of shoots and roots under salt stress. Cheng et al. [46] reported that SLs stimulated the expression of ethylene (ET) biosynthetic genes, and that ET directly affected the root hair (RH) elongation by signal transduction. Furthermore, auxin efflux and polar auxin transport (PAT) affected auxin levels in RH cells, which mediated SLs' promotion of root hair elongation [49] and helped plants to absorb nutrients and water from the soil [50], thereby promoting the development of roots. Moreover, brassinosteroids (BRs), ET, and auxin, as plant growth regulators (PGR), could modulate plant development and enhance stress resistance. BRs contribute to the development of lateral roots (LRs) and adventitious roots (ARs). Moreover, BRs crosstalk with SLs through NO signaling, thereby enhancing or reducing the formation of LRs and ARs [51]. In addition, salt stress inhibits the plant height, the stem diameter, the leaf area, and the root length of tomato, while the interaction of SLs and NO reverse this inhibition by enhancing photosynthesis and antioxidant capacity [37,52]. Hira et al. [53] observed that SLs can improve the morphological, physiological, and biochemical properties of sunflower under salt stress and enhance salt tolerance. The seed dormancy of *Arabidopsis* was broken, because SLs raised the GA level. [46]. SLs increased the content of abscisic acid (ABA) in the bud to achieve the effect of inhibiting branch development [54]. Therefore, there is sufficient evidence to prove that SLs play a key role in regulating tomato adaptation to salt stress.

It is well-known that starch, Tre, and Glu are types of polysaccharide, disaccharide, and monosaccharide, respectively [55,56]. Under salt stress conditions, starch was hydrolyzed to soluble sugars, such as Glu and sucrose, which could ensure the survival of wheat in a short time [5]. Furthermore, starch could be degraded into maltose, and maltose could synthesize Tre [57]. Tre in turn degrades to Glu, which synthesizes Tre [58,59].

Many studies have proven that the accumulation of Tre is beneficial for plants in adapting to salt stress [60–62]. For example, Tre improved the salt tolerance of watermelon by maintaining the balance of cellular metabolism [60]. Yang et al. [61] also found that Tre could alleviate the damage to tomato plants caused by salt stress. At present, it has been reported that SLs interact with sugar to regulate plant growth and development. For instance, SLs play a key role in the transduction of sugar signaling molecules that participate in the development of *Arabidopsis thaliana* seedlings [62]. Takahashi et al. [63] also confirmed that SLs interfere with sugar signaling in delaying rice-leaf senescence.

Our results showed that salt stress remarkably lessened the contents of Tre and starch in tomato seedlings and raised the content of Glu, but GR24 treatment significantly reduced the Glu content and increased the Tre and starch levels, indicating that SLs might maintain salt pressure and promote Tre biosynthesis under salt stress (Figure 2). Remarkably, the degradation of Tre and starch content under salt stress may have contributed to the increase in Glu content. Therefore, our results suggest that SLs may be involved in Tre metabolism, promoting Tre biosynthesis, which regulates tomato adaptation to salt stress. The above reports imply that SLs may promote the biosynthesis of Tre by mediating the transformation among starch, Glu, and Tre, and then regulate Tre in response to salt stress. However, further research is needed in support of this hypothesis.

Tre metabolism in plants is related closely with activities of TPS, TPP, and THL. TPS and TPP play positive roles in the synthesis of Tre, while THL promotes the degradation of Tre. Under certain conditions, TPS condensed Glu to T6P, which was in turn dephosphorylated by TPP to form Tre [59]; the increase in T6P inhibited starch degradation [64]. Additionally, Kolbe et al. [65] found that the expression of *TPS* leads to a large increase in the AGPase of transgenic *Arabidopsis*. AGPase is a rate-limiting enzyme that plays a key role in starch biosynthesis, and an increase in AGPase activity can be used as a signal of starch synthesis [66]. Here, NaCl treatment significantly decreased the activities of TPS and TPP. This inhibition was reversed with the addition of GR24 (Figure 3). TIS108 treatment led to reductions in TPS and TPP activities (Figure 3). It was further confirmed that SLs increased Tre content in the tomato seedlings by increasing the activities of TPS and TPP, thereby reducing the damage to tomato seedlings caused by salt stress (Figure 3). This is consistent with other reports. The TPS/TPP pathway contributed to Tre biosynthesis, which responds to salt stress [67]. Moreover, THL was completely opposite—the activity of THL was raised under NaCl stress and NaCl + TIS108 treatment, while its activity decreased under NaCl + GR24 treatment. This indicated that SLs maintained Tre content by weakening the activity of THL. Analysis showed that increases in TPS and TPP activities and a decrease in THL activity can be used as a signal of Tre effectiveness. Therefore, we concluded that SLs regulate the response of Tre to salt stress by increasing the activities of TPS and TPP and decreasing the activities of THL.

Zhang et al. [43] found that SLs could induce the transcription of *CsMAX2* genes that help for *Arabidopsis* in adaptation to salt stress. Zhang et al. [48] also revealed that SLs could induce the transcription of genes related to tryptophan metabolism, plant hormone signaling, and photosynthesis in response to salt stress. Previous studies demonstrated that both ABA and SLs were derived from carotenoids [68,69]. Under heat stress, the application of SLs downregulated *LjNCED2*, resulting in a decrease in the ABA level [70] and promoting *Arabidopsis* seed germination [71]. Furthermore, the expression of ABA synthesis-related genes (*TaNCED1/2, TaSDR, TaZEP, TaABA3*, and *TaAAO*) led to an increase in ABA content.

The ABA signal transduction could improve salt resistance in wheat by regulating stomatal closure [72]. Exogenous ABA enhanced the freezing tolerance of grapevines by inducing an accumulation of sugars in buds and decreasing stomatal conductance [73]. For years, many transcriptomic studies have provided evidence that the expression of genes involved in Tre biosynthesis regulates plant adaptation to abiotic stress. As reported by Lei et al. [74], the *PoTPS1* gene changed Glu from UDPG to G6P, laying the foundation for the synthesis of Tre, and the content of Tre also increased with the expression of *PoTPS1*. For example, the allelic variations of *TPS* genes help with the adaptation of *Cicer arietinum* L. to salt stress [75]. In addition, the TPP genes are induced in maize seedlings under drought stress [76]. Similarly, under salt stress, *OsTPP1* is transiently expressed and accompanied by an elevation of Tre content [77]. It has been demonstrated that the expression levels of Tre biosynthesis-related genes *TPS1*, *TPS2*, *TPP1*, and *TPP2* in tomato seedlings were decreased under NaCl treatment. Surprisingly, the genes were induced by application of GR24. One key finding was that the expression levels of *TPS1*, *TPS2*, *TPP1*, and *TPP2* genes were significantly reduced in the absence of SLs (Figure 4). The molecular evidence supports the conclusion that SLs participate in Tre synthesis through the upregulation of *TPS1*, *TPS2*, *TPP1*, and *TPP2* genes, and then promote tomato-seedling growth.

This study is the first to propose and verify that SLs respond to salt stress by promoting the biosynthesis of Tre, laying the foundation for the application of SLs in agricultural practices. Future work should further explore the deep mechanisms of SLs under salt stress and excavate key genes and receptors to improve salt tolerance in tomato. SLs are very expensive, and it is difficult to apply them directly to crop production. Our results could provide a theoretical basis for the breeding work of tomato salt-tolerant cultivars.

5. Conclusions

In this article, we found that SLs plays an important role in alleviating the inhibition of tomato seedlings caused by salt stress. Additionally, we reported, for the first time, the action mechanism of SLs in enhancing salt resistance by regulating Tre metabolism. Our results indicate that GR24 enhances the accumulation of endogenous Tre and starch, promotes Glu decomposition, increases TPS and TPP activities, and decreases THL activity, and that the expression levels of *TPS1*, *TPS2*, *TPP1*, and *TPP2* genes related to Tre biosynthesis were upregulated (Figure 5).

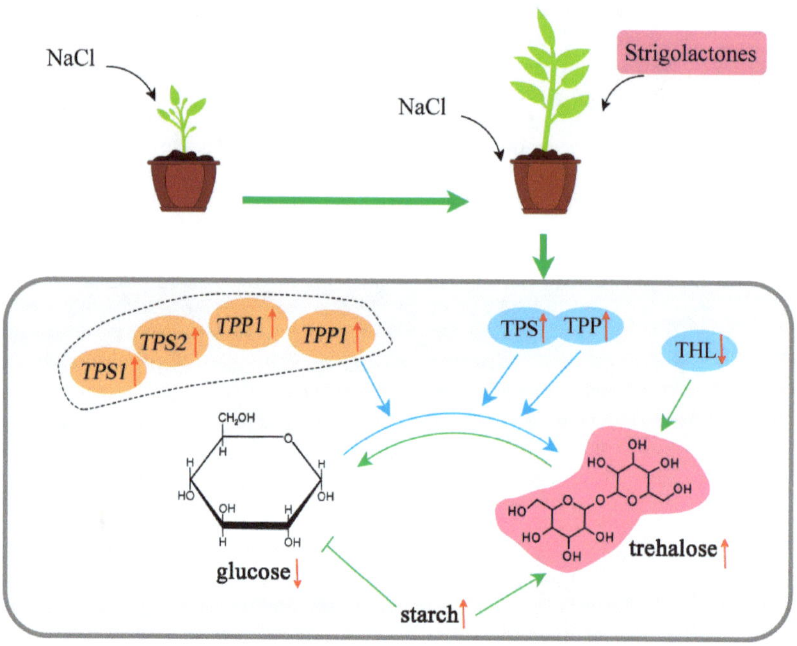

Figure 5. Role model of SLs in salt stress. Under salt stress, SLs promoted tomato seedling growth by regulating the metabolism of Tre. SLs accelerate the conversion of Glu to Tre by enhancing TPS and TPP activities. In addition, SLs inhibited the degradation of Tre by weakening THL activity. Moreover, SLs upregulated the Tre biosynthesis genes *TPS1*, *TPS2*, *TPP1*, and *TPP2*. (Green "→" indicates promotion; blue "→" indicates some active participation in the corresponding process; red "→" indicates increasing or decreasing; black "→" indicates treatment; the blocking line "⊣" indicates inhibition).

Author Contributions: X.L. (Xuefang Lu) was responsible for drawing and conceived the project and wrote the initial manuscript. X.L. (Xiaojun Liu) was also responsible for writing. J.X. and Y.L. were responsible for formatting. Y.C. was responsible for collecting data. C.L. and W.Y. were responsible for funding acquisition, obtaining resources, and writing and editing. All authors have read and agreed to the published version of the manuscript.

Funding: This research was funded by the National Nature Science Foundation of China (31660568) and by the Guangxi Science and Technology Major Project (GuikeAA22068088). C.L. received start-up funding from Guangxi University.

Data Availability Statement: Data will be made available on request.

Conflicts of Interest: The authors declare no conflict of interest.

References

1. Mahajan, S.; Tuteja, N. Cold, salinity and drought stresses: An overview. *Arch. Biochem. Biophys.* **2005**, *444*, 139–158. [CrossRef]
2. Wang, Z.; Hong, Y.; Zhu, G.; Li, Y.; Niu, Q.; Yao, J.; Hua, K.; Bai, J.; Zhu, Y.; Shi, H.; et al. Loss of salt tolerance during tomato domestication conferred by variation in a Na^+/K^+ transporter. *EMBO J.* **2020**, *39*, e103256. [CrossRef]
3. Hasanuzzaman, M.; Nahar, K.; Fujita, M. *Ecophysiology, Abiotic Stress Responses and Utilization of Halophytes*; Springer: Singapore, 2019.
4. Ahmad, P.; Azooz, M.M.; Prasad, M.N.V. *Ecophysiology and Responses of Plants under Salt Stress*; Springer: New York, NY, USA, 2013; pp. 25–87.
5. Hussain, S.; Khaliq, A.; Tanveer, M.; Matloob, A.; Hussain, H.A. Aspirin priming circumvents the salinity-induced effects on wheat emergence and seedling growth by regulating starch metabolism and antioxidant enzyme activities. *Acta Physiol. Plant.* **2018**, *40*, 68. [CrossRef]

6. Zhao, S.; Zhang, Q.; Liu, M.; Zhou, H.; Ma, C.; Wang, P. Regulation of plant responses to salt stress. *Int. J. Mol. Sci.* **2021**, *22*, 4609. [CrossRef]
7. Rofekuggaman, M.; Kubra, K.; Mahmood, S. Effect of Different Salt Concentrations (NaCl) on Seed Germination and Seedling Growth of Tomato Cv. BINA Tomato-10. *Asian Plant Res. J.* **2020**, *5*, 38–44. [CrossRef]
8. Hu, J.; Cai, J.; Park, S.J.; Lee, K.; Li, Y.; Chen, Y.; Yun, J.-Y.; Xu, T.; Kang, H. N^6-Methyladenosine mRNA Methylation Is Important for Salt Stress Tolerance in *Arabidopsis*. *Plant J.* **2021**, *106*, 1759–1775. [CrossRef]
9. Cook, C.E.; Whichard, L.P.; Turner, B.; Wall, M.E.; Egley, G.H. Germination of Witchweed (*Striga lutea Lour.*): Isolation and properties of a potent stimulant. *Science* **1966**, *154*, 1189–1190. [CrossRef]
10. Yasuda, N.; Sugimoto, Y.; Kato, M.; Inanaga, S.; Yoneyama, K. (+)-Strigol, a witchweed seed germination stimulant, from Menispermum dauricum root culture. *Phytochemistry* **2003**, *62*, 1115–1119. [CrossRef]
11. Mori, N.; Nishiuma, K.; Sugiyama, T.; Hayashi, H.; Akiyama, K. Carlactone-type strigolactones and their synthetic analogues as inducers of hyphal branching in arbuscular mycorrhizal fungi. *Phytochemistry* **2016**, *130*, 90–98. [CrossRef]
12. Pozo, M.J.; López-Ráez, J.A.; Azcón, C.; García-Garrido, J.M. Phytohormones as integrators of environmental signals in the regulation of mycorrhizal symbioses. *New Phytol.* **2015**, *205*, 1431–1436. [CrossRef]
13. Marro, N.; Caccia, M.; López-Ráez, J.A. Are Strigolactones a Key in Plant–parasitic Nematodes Interactions? An Intriguing Question. *Plant Soil* **2021**, *462*, 591–601. [CrossRef]
14. Kong, C.C.; Ren, C.G.; Li, R.Z.; Xie, Z.H.; Wang, J.P. Hydrogen peroxide and strigolactones signaling are involved in alleviation of salt stress induced by arbuscular mycorrhizal fungus in sesbania cannabina seedlings. *J. Plant Growth Regul.* **2017**, *36*, 734–742. [CrossRef]
15. Tian, M.Q.; Jiang, K.; Takahashi, I.; Li, G.D. Strigolactone-induced senescence of a bamboo leaf in the dark is alleviated by exogenous sugar. *J. Pestic. Sci.* **2018**, *43*, 173–179. [CrossRef] [PubMed]
16. Waters, M.T.; Gutjahr, C.; Bennett, T.; Nelson, D.C. Strigolactone signaling and evolution. *Annu. Rev. Plant Biol.* **2017**, *68*, 291–322. [CrossRef]
17. Tripepi, A.; Guglielminetti, L. Actions of strigolactone GR24 and DRM1 gene expression on Arabidopsis root architecture. *Russ. J. Plant Physiol.* **2017**, *64*, 845–849. [CrossRef]
18. Sarwar, Y.; Shahbaz, M. GR24 Triggered variations in morpho-physiological attributes of sunflower (*Helianthus annuus*) under salinity. *Int. J. Agric. Biol.* **2019**, *21*, 34–40.
19. Lu, T.; Yu, H.; Li, Q.; Chai, L.; Jiang, W. Improving Plant Growth and Alleviating Photosynthetic Inhibition and Oxidative Stress From Low-Light Stress With Exogenous GR24 in Tomato (*Solanum lycopersicum* L.) Seedlings. *Front. Plant Sci.* **2019**, *10*, 490. [CrossRef]
20. Zhou, X.; Tan, Z.; Zhou, Y.; Guo, S.; Sang, T.; Wang, Y.; Shu, S. Physiological mechanism of strigolactoneenhancing tolerance to low light stress in cucumber seedlings. *BMC Plant Biol.* **2022**, *22*, 30. [CrossRef]
21. Ha, C.V.; Leyva-González, M.A.; Osakabe, Y.; Tran, U.T.; Nishiyama, R.; Watanabe, Y.; Tanaka, M.; Seki, M.; Yamaguchi, S.; Dong, N.V.; et al. Positive regulatory role of strigolactone in plant responses to drought and salt stress. *Proc. Natl. Acad. Sci. USA* **2014**, *111*, 851–856. [CrossRef]
22. León, P.; Sheen, J. Sugar and hormone connections. *Trends Plant Sci.* **2003**, *8*, 110–116. [CrossRef]
23. Ruan, Y.L.; Jin, Y.; Yang, Y.J.; Li, G.J.; Boyer, J.S. Sugar input, metabolism, and signaling mediated by invertase: Roles in development, yield potential, and response to drought and heat. *Mol. Plant* **2010**, *3*, 942–955. [CrossRef] [PubMed]
24. Wingler, A. The function of trehalose biosynthesis in plants. *Phytochemistry* **2002**, *60*, 437–440. [CrossRef] [PubMed]
25. Henry, C.; Bledsoe, S.W.; Siekman, A.; Kollman, A.; Waters, B.M.; Feil, R.; Stitt, M.; Lagrimini, L.M. The trehalose pathway in maize: Conservation and gene regulation in response to the diurnal cycle and extended darkness. *J. Exp. Bot.* **2014**, *65*, 5959–5973. [CrossRef] [PubMed]
26. Eastmond, P.J.; van Dijken, A.J.H.; Spielman, M.; Kerr, A.; Tissier, A.F.; Dickinson, H.G.; Jones, J.D.G.; Smeekens, S.C.; Graham, I.A. Trehalose-6-phosphate synthase 1, which catalyses the first step in trehalose synthesis, is essential for *Arabidopsis* embryo maturation. *Plant J.* **2002**, *29*, 225–235. [CrossRef] [PubMed]
27. Ponnu, J.; Wahl, V.; Schmid, M. Trehalose-6-phosphate: Connecting plant metabolism and development. *Front. Plant Sci.* **2011**, *2*, 70. [CrossRef] [PubMed]
28. Arguelles, J.C. Physiological Roles of Trehalose in Bacteria and Yeasts: A Comparative Analysis. *Arch. Microbiol.* **2000**, *174*, 217–224. [PubMed]
29. Andersson, U.; Levander, F.; Radstrom, P. Trehalose-6-Phosphate Phosphorylase Is Part of a Novel Metabolic Pathway for Trehalose Utilization in Lactococcus Lactis. *J. Biol. Chem.* **2001**, *276*, 42707–42713. [CrossRef]
30. Garg, A.K.; Kim, J.K.; Owens, T.G.; Ranwala, A.P.; Choi, Y.D.; Kochian, L.V.; Wu, R.J. Trehalose accumulation in rice plants confers high tolerance levels to different abiotic stresses. *Proc. Natl. Acad. Sci. USA* **2002**, *99*, 15898–15903. [CrossRef]
31. Iordachescu, M.; Imai, R. Trehalose biosynthesis in response to abiotic stresses. *J. Integr. Plant Biol.* **2008**, *50*, 1223–1229. [CrossRef]
32. Fernandez, O.; Béthencourt, L.; Quero, A.; Sangwan, R.S.; Clément, C. Trehalose and plant stress responses: Friend or foe? *Trends Plant Sci.* **2010**, *15*, 409–417. [CrossRef]
33. Li, H.W.; Zang, B.S.; Deng, X.W.; Wang, X.P. Overexpression of the trehalose-6-phosphate synthase gene OsTPS1 enhances abiotic stress tolerance in rice. *Planta* **2011**, *234*, 1007–1018. [CrossRef] [PubMed]

34. Akram, N.A.; Waseem, M.; Ameen, R. Trehalose pretreatment induces drought tolerance in radish (*Raphanus sativus* L.) plants: Some key physio-biochemical traits. *Acta Physiol. Plant.* **2016**, *38*, 3. [CrossRef]
35. Islam, M.O.; Kato, H.; Shima, S.; Tezuka, D.; Matsui, H.; Imai, R. Functional identification of a rice trehalase gene involved in salt stress tolerance. *Gene* **2019**, *685*, 42–49. [CrossRef] [PubMed]
36. Redillas, M.; Park, S.H.; Lee, J.W.; Kim, Y.S.; Jin, S.J.; Jung, H.; Bang, S.W.; Hahn, T.R.; Kim, J.K. Accumulation of trehalose increases soluble sugar contents in rice plants conferring tolerance to drought and salt stress. *Plant Biotechnol. Rep.* **2012**, *6*, 89–96. [CrossRef]
37. Liu, H.; Li, C.; Yan, M.; Zhao, Z.; Huang, P.; Wei, L.; Wu, X.; Wang, C.; Liao, W. Strigolactone is involved in nitric oxide-enhanced the salt resistance in tomato seedlings. *J. Plant Res.* **2022**, *135*, 337–350. [CrossRef]
38. Ito, S.; Umehara, M.; Hanada, A.; Yamaguchi, S.; Asami, T. Effects of Strigolactone-biosynthesis Inhibitor TIS108 on Arabidopsis. *Plant Signal. Behav.* **2013**, *8*, E24193. [CrossRef]
39. Sheng, W.J.; Wei, C.X. Screening Methods for Cereal Grains with Different Starch Components: A Mini Review. *J. Cereal Sci.* **2022**, *108*, 103557. [CrossRef]
40. Saqib, A.A.N.; Whitney, P.J. Differential Behaviour of the Dinitrosalicylic Acid (DNS) Reagent towards Mono- and Di-saccharide Sugars. *Biomass Bioenergy* **2011**, *35*, 4748–4750. [CrossRef]
41. Huang, P.P.; Li, C.X.; Liu, H.W.; Zhao, Z.X.; Liao, W.B. Hydrogen Gas Improves Seed Germination in Cucumber by Regulating Sugar and Starch Metabolisms. *Horticulturae* **2021**, *7*, 456. [CrossRef]
42. Wu, H.; Fan, S.; Gong, H.; Guo, J. Roles of salicylic acid in selenium-enhanced salt tolerance in tomato plants. *Plant Soil* **2023**, *484*, 569–588. [CrossRef]
43. Zhang, R.; Dong, Y.; Li, Y.; Ren, G.; Chen, C.; Jin, X. SLs signal transduction gene *CsMAX2* of cucumber positively regulated to salt, drought and ABA stress in *Arabidopsis thaliana* L. *Gene* **2023**, *864*, 147282. [CrossRef] [PubMed]
44. Ma, C.; Bian, C.; Liu, W.; Sun, Z.; Xi, X.; Guo, D.; Liu, X.; Tian, Y.; Wang, C.; Zheng, X. Strigolactone alleviates the salinity-alkalinity stress of Malus hupehensis seedlings. *Front. Plant Sci.* **2022**, *13*, 901782. [CrossRef] [PubMed]
45. Fan, X.Q.; Li, H.; Guo, Y.S.; Sun, H.C.; Wang, S.Y.; Qi, Q.; Jiang, X.N.; Wang, Y.; Xu, X.F.; Qiu, C.P.; et al. Integrated Multi-omics Analysis Uncovers Roles of Mdm-miR164b-MdORE1 in Strigolactone-mediated Inhibition of Adventitious Root Formation in Apple. *Plant Cell Environ.* **2022**, *45*, 3582–3603. [CrossRef] [PubMed]
46. Cheng, X.; Ruyter-Spira, C.; Bouwmeester, H. The interaction between strigolactones and other plant hormones in the regulation of plant development. *Front. Plant Sci.* **2013**, *4*, 199. [CrossRef]
47. Ahsan, M.; Zulfiqar, H.; Farooq, M.A.; Ali, S.; Tufail, A.; Kanwal, S.; Shaheen, M.R.; Sajid, M.; Gul, H.; Jamal, A.; et al. Strigolactone (GR24) Application Positively Regulates Photosynthetic Attributes, Stress-Related Metabolites and Antioxidant Enzymatic Activities of Ornamental Sunflower (*Helianthus Annuus* Cv. Vincent's Choice) under Salinity Stress. *Agriculture* **2023**, *13*, 50. [CrossRef]
48. Ma, N.; Hu, C.; Wan, L.; Hu, Q.; Xiong, J.; Zhang, C. Strigolactones Improve Plant Growth, Photosynthesis, and Alleviate Oxidative Stress under Salinity in Rapeseed (*Brassica napus* L.) by Regulating Gene Expression. *Front. Plant Sci.* **2017**, *8*, 1671. [CrossRef]
49. Koltai, H.; Dor, E.; Hershenhorn, J.; Joel, D.M.; Weininger, S.; Lekalla, S.; Shealtiel, H.; Bhattacharya, C.; Eliahu, E.; Resnick, N.; et al. Strigolactones' effect on root growth and roothair elongation may be mediated by auxin-efflux carriers. *J. Plant Growth Regul.* **2010**, *29*, 129–136. [CrossRef]
50. Villaécija-Aguilar, J.A.; Hamon-Josse, M.; Carbonnel, S.; Kretschmar, A.; Schmidt, C.; Dawid, C.; Bennett, T.; Gutjahr, C. *SMAX1/SMXL2* regulate root and root hair development downstream of *KAI2*-mediated signalling in *Arabidopsis*. *PLoS Genet.* **2019**, *15*, e1008327. [CrossRef]
51. Altamura, M.M.; Piacentini, D.; Rovere, F.D.; Fattorini, L.; Falasca, G.; Betti, C. New Paradigms in Brassinosteroids, Strigolactones, Sphingolipids, and Nitric Oxide Interaction in the Control of Lateral and Adventitious Root Formation. *Plants* **2023**, *12*, 413. [CrossRef]
52. Zhang, X.; Zhang, L.; Ma, C.; Su, M.; Wang, J.; Zheng, S.; Zheng, T. Exogenous strigolactones alleviate the photosynthetic inhibition and oxidative damage of cucumber seedlings under salt stress. *Sci. Hortic.* **2022**, *297*, 110962. [CrossRef]
53. Hira, Z.; Muhammad, S.; Muhammad, A.; Muhammad, N.; Hammad, N.; Muhammad, A.; Shah, F. Strigolactone (GR24) Induced Salinity Tolerance in Sunflower (*Helianthus annuus* L.) by Ameliorating Morpho-Physiological and Biochemical Attributes Under In Vitro Conditions. *J. Plant Growth Regul.* **2021**, *40*, 2079–2091.
54. Wang, L.; Wang, B.; Yu, H.; Guo, H.; Lin, T.; Kou, L.; Wang, A.; Shao, N.; Ma, H.; Xiong, G.; et al. Transcriptional regulation of strigolactone signalling in *Arabidopsis*. *Nature* **2020**, *583*, 277–281. [CrossRef] [PubMed]
55. Ho Do, M.; Seo, Y.S.; Park, H.Y. Polysaccharides: Bowel health and gut microbiota. *Crit. Rev. Food Sci. Nutr.* **2021**, *61*, 1212–1224. [CrossRef]
56. Ballell, L.; Young, R.J.; Field, R.A. Synthesis and evaluation of mimetics of UDP and UDP-alpha-D-galactose, dTDP and dTDP-alpha-D-glucose with monosaccharides replacing the key pyrophosphate unit. *Org. Biomol. Chem.* **2005**, *3*, 1109–1115. [CrossRef] [PubMed]
57. Wang, J.H.; Tsai, M.Y.; Lee, G.C.; Shaw, J.F. Construction of a recombinant thermostable beta-amylase-trehalose synthase bifunctional enzyme for facilitating the conversion of starch to trehalose. *J. Agric. Food Chem.* **2007**, *55*, 1256–1263. [CrossRef] [PubMed]

58. Delorge, I.; Janiak, M.; Carpentier, S.; Van Dijck, P. Fine tuning of trehalose biosynthesis and hydrolysis as novel tools for the generation of abiotic stress tolerant plants. *Front. Plant Sci.* **2014**, *5*, 147. [CrossRef]
59. Henry, C.; Bledsoe, S.W.; Griffiths, C.A.; Kollman, A.; Paul, M.J.; Sakr, S.; Lagrimini, L.M. Differential Role for Trehalose Metabolism in Salt-Stressed Maize. *Plant Physiol.* **2015**, *169*, 1072–1089. [CrossRef]
60. Yuan, G.; Sun, D.; An, G.; Li, W.; Si, W.; Liu, J.; Zhu, Y. Transcriptomic and Metabolomic Analysis of the Effects of Exogenous Trehalose on Salt Tolerance in Watermelon (*Citrullus lanatus*). *Cells* **2022**, *11*, 2338. [CrossRef]
61. Yang, Y.; Xie, J.; Li, J.; Zhang, J.; Zhang, X.; Yao, Y.; Wang, C.; Niu, T.; Bakpa, E.P. Trehalose alleviates salt tolerance by improving photosynthetic performance and maintaining mineral ion homeostasis in tomato plants. *Front. Plant Sci.* **2022**, *13*, 974507. [CrossRef]
62. Li, G.D.; Pan, L.N.; Jiang, K.; Takahashi, I.; Nakamura, H.; Xu, Y.W.; Asami, T.; Shen, R.F. Strigolactones Are Involved in Sugar Signaling to Modulate Early Seedling Development in Arabidopsis. *Plant Biotechnol.* **2016**, *33*, 87–97. [CrossRef]
63. Takahashi, I.; Jiang, K.; Asami, T. Counteractive effects of sugar and strigolactone on leaf senescence of rice in darkness. *Agronomy* **2021**, *11*, 1044. [CrossRef]
64. Martins, M.C.; Hejazi, M.; Fettke, J.; Steup, M.; Feil, R.; Krause, U.; Arrivault, S.; Vosloh, D.; Figueroa, C.M.; Ivakov, A.; et al. Feedback inhibition of starch degradation in Arabidopsis leaves mediated by trehalose 6-phosphate. *Plant Physiol.* **2013**, *163*, 1142–1163. [CrossRef] [PubMed]
65. Kolbe, A.; Tiessen, A.; Schluepmann, H.; Paul, M.; Ulrich, S.; Geigenberger, P. Trehalose 6-phosphate Regulates Starch Synthesis via Posttranslational Redox Activation of ADP-glucose Pyrophosphorylase. *Proc. Natl. Acad. Sci. USA* **2005**, *102*, 11118–11123. [CrossRef] [PubMed]
66. Saripalli, G.; Gupta, P.K. AGPase: Its role in crop productivity with emphasis on heat tolerance in cereals. *Theor. Appl. Genet.* **2015**, *128*, 1893–1916. [CrossRef] [PubMed]
67. Stracke, C.; Meyer, B.H.; Hagemann, A.; Jo, E.; Lee, A.; Albers, S.V.; Cha, J.; Bräsen, C.; Siebers, B. Salt Stress Response *of Sulfolobus acidocaldarius* Involves Complex Trehalose Metabolism Utilizing a Novel Trehalose-6-Phosphate Synthase (TPS)/Trehalose-6-Phosphate Phosphatase (TPP) Pathway. *Appl. Environ. Microbiol.* **2020**, *86*, e01565-20. [CrossRef]
68. Haider, I.; Aneo-Jimenez, B.; Bruno, M.; Bimbo, A.; Floková, K.; Abuauf, H.; Ntui, V.O.; Guo, X.; Charnikhova, T.; Al-Babili, S.; et al. The Interaction of Strigolactones with Abscisic Acid during the Drought Response in Rice. *J. Exp. Bot.* **2018**, *69*, 2403–2414. [CrossRef]
69. Wu, S.; Ma, X.Q.; Zhou, A.Q.; Valenzuela, A.; Zhou, K.; Li, Y.R. Establishment of Strigolactone-producing Bacterium-yeast Consortium. *Sci. Adv.* **2021**, *7*, eabh4048. [CrossRef]
70. Liu, J.; He, H.; Vitali, M.; Visentin, I.; Charnikhova, T.; Haider, I.; Cardinale, F. Osmotic stress represses strigolactone biosynthesis in lotus japonicus roots: Exploring the interaction between strigolactones and ABA under abiotic stress: An international journal of plant biology. *Planta* **2015**, *241*, 1435–1451. [CrossRef]
71. Toh, S.; Kamiya, Y.; Kawakami, N.; Nambara, E.; McCourt, P.; Tsuchiya, Y. Thermoinhibition uncovers a role for strigolactones in *Arabidopsis* seed germination. *Plant Cell Physiol.* **2012**, *53*, 107–117. [CrossRef]
72. Zhang, L.; Zhao, L.J.; Wang, L.T.; Liu, X.Y.; Yu, Z.; Liu, J.; Wu, W.Z.; Ding, L.; Xia, C.; Zhang, L.C.; et al. *TabZIP60* Is Involved in the Regulation of ABA Synthesis-mediated Salt Tolerance through Interacting with *TaCDPK30* in Wheat (*Triticum aestivum* L.). *Planta* **2023**, *257*, 107. [CrossRef]
73. Wang, H.R.; Blakeslee, J.J.; Jones, M.L.; Chapin, L.J.; Dami, I.E. Exogenous Abscisic Acid Enhances Physiological, Metabolic, and Transcriptional Cold Acclimation Responses in Greenhouse-Grown Grapevines. *Plant Sci.* **2020**, *293*, 110437. [CrossRef] [PubMed]
74. Lei, M.; Wu, X.; Zhang, J.; Wang, H.; Huang, C. Gene cloning, expression, and characterization of trehalose-6-phosphate synthase from Pleurotus ostreatus. *J. Basic Microbiol.* **2017**, *57*, 580–589. [CrossRef] [PubMed]
75. Kumar, T.; Tiwari, N.; Bharadwaj, C.; Roorkiwal, M.; Reddy, S.P.P.; Patil, B.S.; Kumar, S.; Hamwieh, A.; Vinutha, T.; Bindra, S.; et al. A comprehensive analysis of Trehalose-6-phosphate synthase (TPS) gene for salinity tolerance in chickpea (*Cicer arietinum* L.). *Sci. Rep.* **2022**, *12*, 16315. [CrossRef]
76. Acosta-Pérez, P.; Camacho-Zamora, B.D.; Espinoza-Sánchez, E.A.; Gutiérrez-Soto, G.; Zavala-García, F.; Abraham-Juárez, M.J.; Sinagawa-García, S.R. Characterization of Trehalose-6-phosphate Synthase and Trehalose-6-phosphate Phosphatase Genes and Analysis of its Differential Expression in Maize (*Zea mays*) Seedlings under Drought Stress. *Plants* **2020**, *9*, 315. [CrossRef] [PubMed]
77. Pramanik, M.H.R.; Imai, R. Functional Identification of a Trehalose 6-phosphate Phosphatase Gene That Is Involved in Transient Induction of Trehalose Biosynthesis during Chilling Stress in Rice. *Plant Mol. Biol.* **2005**, *58*, 751–762. [CrossRef] [PubMed]

MDPI AG
Grosspeteranlage 5
4052 Basel
Switzerland
Tel.: +41 61 683 77 34

Horticulturae Editorial Office
E-mail: horticulturae@mdpi.com
www.mdpi.com/journal/horticulturae